Quantum Optics and the Spectroscopy of Solids

Fundamental Theories of Physics

An International Book Series on The Fundamental Theories of Physics: Their Clarification, Development and Application

Volume 83

Quantum Optics and the Spectroscopy of Solids

Concepts and Advances

edited by

T. Hakioğlu

and

A.S. Shumovsky

Physics Department,
Bilkent University,
Ankara, Turkey

Springer-Science+Business Media, B.V.

A C.I.P. Catalogue record for this book is available from the Library of Congress.

ISBN 978-90-481-4797-7 978-94-015-8796-9 (eBook)
DOI 10.1007/978-94-015-8796-9

Printed on acid-free paper

Dedicated to the memory of Prof. Asım Orhan Barut.

Prof. A.O. Barut (1926-1994)

(photograph: Courtesy of Dr. İrfan Açıkgöz)

Table of Contents

Preface

Prior to the invention of the laser in the 60s, the development in optics was primarily dependent on the development in fundamental quantum mechanics. During the 50s, progress was primarily made in the direction of optical coherence by von Laue, Schrödinger, and Zernike and then by Glauber who has constructed the final form of the quantum theory of coherence. An exceptional case was that of photon bunching as a result of intensity correlation function measurements. This was a major leap in the field, primarily taken by Hanbury-Brown and Twiss in the 50s.

The path paved by this first correlation function measurement has brought new insights into the field of optics in conjunction with such concepts as field coherence and correlations. The HBT experiment received considerable attention from a wide spectrum of physics researchers. With the invention of lasers in the 60s, we came across the realization of a non-classical state, and the examination of its coherence and correlation properties was promising for the possible realization and existence of other non-classical states. Starting from the 60s, advances in laser research stimulated major developments in fundamental quantum mechanics in conceptual as well as experimental directions, and the discipline was often referred to as quantum optics. The idea of squeezed states and the phenomenon of squeezing in the phase space of the field quadrature operators were initiated by the works of Robinson and Takahashi in mid 60s and later by Stoler, Yuen, Caves et al. during the 70s and early 80s. During the mid 70s Mandel and coworkers successfully observed the first photon antibunching effect which was predicted earlier by Glauber in the quantum theory of coherence. The first experimental observation of squeezing was reported by Slusher, Hollberg, Yurke *et al.* in 1985. This was just the beginning of a rich gold-mine in which more non-classical states could be conceptualized and engineered. Number-phase squeezing was introduced by Kigatawa and Yamamoto in 1986 and the race for maximum squeezing began. In the beginning of the 90s, laboratories in Europe and the USA perfected the idea of phase-space squeezing by employing sophisticated techniques, such as quantum non-demolition measurements, which was introduced in the early 70s by Braginsky. About the same time, to study the realistic decoherence effects of pure quantum states interacting with a classical reservoir, Schrödinger cat and kitten states were conceptualized. The developments in non-classical states and their experimental realizations also stimulated major recent progress in disciplines such as atom optics, atomic interferometry, and laser cooling in optical molasses.

Another direction of studying pure quantum effects was paved by the two-level atomic dipoles interacting with a single mode of radiation field. The first example of two-level system was used by Einstein in 1917 in his classical work on spontaneous radiation. Dicke in 1954 studied the cooperative effects in a cluster of two-level atoms interacting with a continuous spectrum of radiation. The quantum properties of light in a cavity interacting with a single two-level system was studied by Jaynes and Cummings in the early 60s. Despite its simplicity, the so-called single-atom maser turned out to be an important tool both in generating and in understanding the quantum behavior of atom-field interactions in charged ion and neutral atom traps using highly excited Rydberg atoms. Photon antibunching, super and subradiation, sub Poissonian statistics, individual as well as collective collapses, and revivals in the time dependence of the photon number are major effects one can study with this simplest atom-field interaction model. The experimental work on single two-level atoms in high quality cavities was initiated by Walther during the mid 80s.

When these atoms are laser-cooled down to virtually zero temperatures, several non-linear dispersive bistabilities in the cavity mode intensity have been recently observed.

Progress in the early 90s on homodyne measurements and detection on the phase-space distribution functions comprise yet another fundamental development in the examination of non-classical states. The basic idea is to extend the conventional tomography to measure the statistics and distribution of phase space operators in the quantum domain by means of rotated quadrature operators. This promising field, although still in its infancy, is undergoing a rapid development as new techniques are introduced to increase the dimensionality of the phase space from two (one degree of freedom) to four (two degrees of freedom), and to higher dimensions.

It is an impossible task to collect all major developments in quantum optics and spectroscopy in a publication of this size. In these proceedings, contributions from eleven authors are presented in a comprehensive textbook style, ranging from orthodox quantum optics to applications in solid state physics and atomic spectroscopy.

In the first chapter, by O. Keller, *Aspects of Local Field Electrodynamics in Condensed Matter* are presented. He starts from microscopic Maxwell-Lorentz equations plus a linear equation describing the material-dependent induced current densities non-local in time and spatial extent. The standard theory of local-field electrodynamics is then discussed and the discourse is extended to the transverse and longitudinal fields, emphasizing the fundamental difference between them. The coupled antenna theory is discussed and the local field effects in mesoscopic media are presented. Applications to one and zero dimensional mesoscopic systems and the photon drag phenomenon in mesoscopic rings are finally examined.

In the second chapter, by I. Kulik, *Persistent Currents and Persistent Charges in Nanostructures* are discussed. The main part of this chapter is devoted to the quantum effects arising from the interaction of the electromagnetic field with mesoscopic metallic particles and rings. Magnetic (time-independent) and electric (time-dependent) Aharanov-Bohm and quantum high frequency effects are then investigated in coupled systems, such as optical fiber and mesoscopic conducting loops.

In the third chapter, by A.S. Shumovsky, *Quantum Optics and Solid State Spectroscopy* are presented. There, the possibility of strongly correlated bosonic excitations, such as squeezed states of photons, is considered and their measurement in optical spectroscopy is elaborated.

The fourth chapter, by H. Walther, discusses *Laser Experiments with Single Atoms in Cavities and Traps.* It starts with a review of the one-atom maser in a high quality cavity including the derivation of the model as well as the cavity photon distribution. A discussion on the quantum jumps of the micromaser field is then followed by atomic interferometry of micromasers. Experiments with ionic traps and ion storage rings are then discussed. Finally, experiments with single laser-cooled ions are presented.

The fifth chapter, by G. Compagno, R. Passante, and F. Persico, is entitled *Dressed States in Atoms and in Excitons.* The authors discuss the concept of dressed atoms in analogy with polaronic excitons in solid state physics. They discuss the topic in the context of a generalized two-level system and examine the population of dressed states. The observational time interval and dynamical time scales in the measurement of such effects are then compared. The authors suggest that the formation of the virtual cloud can be experimentally observed, and the fundamental implications of this result to test partially explored QED processes such as positronium creation and annihilation are emphasized.

The sixth chapter, by A. İmamoğlu, H. Schmidt, R.J. Ram, K. Campman, and A. Gossard, is on *Electron Coherence in Quantum Well Intersubband Transitions*, and stresses the importance of absorption as a fundamental limitation to the novel properties of non-linear devices. They primarily analyze the electron coherence and interference effects in intersubband transitions. The authors propose a double quantum-well structure in order to eliminate the absorption by Fano-type destructive quantum interference and concentrate on two new types of optical devices: semiconductor lasers without inversion and resonant nonlinear devices that generate into a transparent medium.

The seventh chapter, by T. Hakioğlu, is entitled *Interaction of Two-Level Atomic System with a Single-Mode Radiation.* The author discusses the celebrated Dicke model in the equivalent field mode limit and illustrates domains of qualitatively different physical behavior as the number of atoms and the number of excitations vary arbitrarily. The commensurability/incommensurability of the eigenvalues is discussed in a wide range, starting from strong to weak field limits. Interplay between various timescales and their effect on the collective collapses and revivals are examined. The accuracy of the rotating wave approximation is limited by the total number of excitations. Superficial instabilities in the spectrum can be observed if one arbitrarily increases the number of excitations remaining in the rotating wave approximation. It is shown that instabilities are not chronic but can only be cured in the fully unitary model. This requires the proper handling of the counterrotating terms.

In the eighth chapter, G.M. D'Ariano discusses *Quantum Estimation Theory and Optical Detection.* This fundamental problem establishes the necessary marriage between conventional quantum mechanics and the quantum theory of measurement. The author starts with a discussion of commuting and non-commuting phase operator measures and elaborates on Naimark's theorem in two examples. After discussing various detection schemes (direct, balanced homodyne, heterodyne), he discusses the theory of joint measurement of two non-commuting observables. The chapter is continued with a discussion of the quantum estimation theory, and concludes with the study of quantum non-destructive measurements from a generalized point of view.

The ninth chapter, by G.M. D'Ariano, is entitled *Measuring Quantum States.* The author starts with the discussion of the central limit theorem and illustrates the limitations in its use in the calculation of moments of random variables. In the section devoted to quantum homodyne tomography, the author introduces the idea of tomography and extends the discussion to its application in quantum optics. Practical difficulties arising in the numerical extraction of the Wigner function from the marginal probabilities are then clarified. An exact method for the measurement of the density matrix is introduced for certain representations. The author then discusses various measurable representations of the kernel of the Wigner function and calculates their bounds for quantum efficiency. The formalism is then extended to finite resolution, followed by an investigation of the principal problems arising in the recovery of the full Wigner function for a system with a single degree of freedom. A brief discussion of the quantum and classical Radon transform is followed by a conclusion and remarks on experimental applications.

The tenth chapter, by P. Tombesi, is entitled *Optical Tomography and Macroscopic Coherence.* This recently introduced technique is a strong potential tool in understanding the state of a quantum system in the framework of phase-space-analysis. The author first discusses generalized (s-parametrized) Wigner functions and marginal probabilities. Quadrature phase measurements from a homodyne scheme are discussed using a general non-ideal quantum efficiency. The tomographic reconstruction of the Wigner function from

the repeated measurements of the marginal probability distribution and the practical problems arising therefrom are illustrated. The formalism is then generalized to displaced and squeezed quadratures respectively. The non-ideal detector and its influence on the measurement scheme are discussed next. The chapter is concluded with a discussion of detecting optical Schrödinger cats and interference fringes. We believe that the three chapters by D'Ariano and Tombesi will give the reader a good account of the principles of tomography and quantum phase space analysis.
The eleventh chapter, by A. Miranowicz, is on *Harmonic Oscillator States in Finite Dimensional Hilbert Space.* Hilbert spaces in a finite dimension started receiving attention in physics after the introduction of a unitary phase formalism by Pegg and Barnett. The author analyzes finite dimensional coherent states in two different approaches, first by simply truncating the number state expansion of coherent states and later by the action of a generalized finite dimensional Hilbert space displacement operator. The number-phase Wigner function is also analyzed using these finite bases.
Chapter twelve is by A. Serpengüzel, S. Arnold, G. Grieffel, and J.A. Lock, and is entitled *Optical Spectroscopy of Microcavities.* The authors discuss light coupling in microspheres and other microcavities with morphology dependent resonances and they examine efficient coupling mechanisms.

The first idea of organizing a summer school on Quantum Optics and Spectroscopy of Solids (QOSS/95) arose from our discussion with Prof. Barut, when the three of us met in Ankara in September 1994. After his death on the 6th of December 1994, we were honored to turn the organization into a memorial meeting dedicated to his unique personality, and scientific achievements. We should pay our tribute to him for giving the idea and motivating us to start this large organization.
This volume is largely composed of lectures given by the invited lecturers in the *Summer School on Quantum Optics and Spectroscopy of Solids* (QOSS/95), held at Bilkent University, Physics Department, on 2-10 July 1995, with over 100 participants and 13 lecturers. The funding was provided by the Scientific and Technical Research Council of Turkey (TÜBİTAK), the International Center for Theoretical Physics (ICTP), and Bilkent University. We are grateful to all of them for their generous support without which the organization would not have become a reality.
We at last but not least thank all of our lecturers and participants for their participation in this significant event.

Bilkent/Ankara
September 1996

T. Hakioğlu
A.S. Shumovsky

Biography of Prof. Asım Orhan Barut

After a rich, dynamic life searching for the fundamentals of reality, Prof. Asım Orhan Barut died on the 6th of December 1994. His early death was a shock to the Physics community.

Born in 1926, he lived his early childhood in the small eastern provincial town of Malatya/Turkey. In 1944, in the midst of war in Europe, the prospect of hardship and a tormented life could not stop him from his devotion to learning and scientific knowledge, which gave him an impulse to study experimental Physics in the Swiss Federal Institute of Technology in Zurich. There, he had the chance to attend courses by Planchaarel, Hopf, Stiefel, Eckmann, Pfluger, Saxer, Gonsath, Pauli, Scherr, Wentzel, Busch, Ackeret and Ziegler. He received his diploma in 1948 and in the following five years studied experimental physics. His first scientific publication was entitled *Die Laufzeit, Elektronenbahnen, Kathodenfeldstarke und Potential der Raumladungsdiode für jede Anfangsgeschwindigkeit, Anfangsrichtung und Strom*, which appeared in *Angew. Math. Phys.* **2**, 35 (1951).

During his experimental research years he had a gradually changing interest from experimental to theoretical physics. In 1953 and 1954 he studied pure Mathematics and Physics in the University of Chicago, where he visited as a Rockefeller fellow. There, he attended one of the last lectures delivered by Enrico Fermi on quantum mechanics. This period had an immense influence on his later scientific interests on the fundamentals of quantum mechanics. He always remembered Bergen Davis' comment: *I don't believe you young fellows understand it any better than I do; but you stick together and say the same thing.* He believed that understanding quantum mechanics is equivalent to calculating Planck's constant or the hyperfine coupling. He met W. Heisenberg in 1965. Together with Fermi these two people made a permanent mark on his mode of science and scientific thinking.

He was an assistant professor at Reed College at Oregon in 1954 and 1955, at Montreal University in 1955 and 1956, and associate professor at Syracuse University in New York state in 1956 and 1959. He visited CERN at Geneva in 1959-1960. In 1961 and 1962, he was at the Lawrence Radiation Laboratory and Physics Department of the University of California/Berkeley. There he largely benefited from the traditional scientific contacts between the faculty and young researchers. For him, the time spent there was one of the most fruitful intervals in his academic life. Witnessing the benefits of such contacts, Barut always believed in the importance of conveying scientific knowledge to young generations in the coherent smooth progress of science.

Later, he joined the Physics Faculty in The University of Colorado at Boulder. He held an office on the tenth floor in the Gamow tower overlooking the mountains.

His early theoretical interest was on quantum theory and statistical mechanics. The time was ripe for new ideas on symmetry. In the second half of the fifties, he pointed out in his work *On the Symmetry of Elementary Particles, Nuovo Cimento* **110**, 1146 (1958) the importance of the octet symmetry, which appeared three years earlier than the celebrated work of Gell-Mann and Neumann (1961).

In the sixties, he wrote *Strong Reflection Principle for Each Fermion* on the (V-A) theory of weak interactions. It was not accepted for publication by Physical Review Letters and this idea was developed later by other groups. In his guidance of young researchers he used to say that *if your work is out of the mainstream, you may have a harder time to get your ideas published. On the other hand, working with ideas of the past (unless unnoticed) may not get you enough interest in your work.*

Barut published more than 550 papers in various scientific journals in a wide spectrum of subjects. His contributions to the problems of theoretical physics can be summarized as follows:

Theory of Dynamical Groups: Symmetry groups were developed by van der Waerden, Wigner, and von Neumann who were interested in the symmetry properties of atoms. Barut extended the application of the group representation to include the solution of certain dynamical problems and solved the radial part of the hydrogen atom by using dynamical group representations. Later, this development led to the discovery of new coherent states enabling us to define the space-time symmetries with the dynamical ones. Most of these developments are summarized in his book *Dynamical Groups and Generalized Symmetries*, University of Canterbury Press (1972), New Zealand.

Infinite Component Wave Equations: Dirac developed the four component spinor to describe the wave function of the single relativistic electron. This equation is not valid for composite systems such as proton, hydrogen atom, or systems with more than one electron. Barut showed that relativistic composite systems of fermions are described by an infinite component wave function.

Group Theory of the Periodic Table: Barut developed a new way of looking at the periodic table of the elements, incorporating Madelung's rule and its O(4,2) group symmetry and he obtained the operators that related the neutral elements and their ions.

Radiative Processes and Radiation Reaction in Classical Electrodynamics: He derived the Lorentz-Dirac equation by the analytical continuation method, using only the retarded Green's function of the electromagnetic field and generalized the Lorentz-Dirac equation to the spinning classical particle system.

The S-Matrix Theory of the Electomagnetic and Weak Interactions: Heisenberg's S-matrix theory was applied only to short range strong interactions. Barut developed the S-matrix theory for the long range electromagnetic and weak interactions, and within the framework of this theory obtained the electron's anomalous magnetic moment.

Self Field Quantum Electrodynamics: He obtained the nonlinear integro-differential equations for the relativistic bound states of hydrogen, muonium, positronium, and many fermion systems. These equations give the bound state spectrum of this kind of binary atoms with all the corrections up to the order α^5. They also give a finite value for the vacuum polarization of the bound states and which opens a new scheme in the mass renormalization procedure. The calculation of the Lamb shift and all radiative effects opened up new directions for exploring the foundations of relativistic quantum electrodynamics.

Classical Theory of the Zitterbewegung: Barut developed a classical model of the spinning particles that have internal degrees of freedom and the quantization of this system gives the Dirac equation and higher spin wave equations.

Extended Objects: He introduced a new magnetic-charge model of the proton, called "dyonium" and obtained the superselection rule for magnetic monopoles. He also introduced the concept of "dynamical electromagnetic string" as well as the field theory of the strings with the point masses located at their end points.

Theory of the Single Events and Wavelets: Barut initiated a new deterministic "quantum theory of single events", distinct from the standard probabilistic quantum mechanics, along with an accompanying resolution of quantum paradoxes.

Barut's important contributions to mathematical physics can be summarized as follows:

Relativistic Scattering Amplitudes: He constructed the invariant scattering amplitudes using the representations of the Poincaré group. It is well known that this work initiated a whole new field in high energy physics. He also developed the theory of the analytical continuation of the angular momentum to complex values, and applications of this method in relativistic field theory.

Relativistic Wave Equations: The construction of the irreducible $(2j+1)$ component spinors leads to the development of the relativistic wave equations. Later he also obtained another set of relativistic wave equations by quantizing the classical model of zitterbewegung.

Lagrangians with Higher Order Derivatives: He generalized the least action principle to the Lagrangians that have higher derivatives, and later this theory was applied to the lepton families.

Properties of Noncompact Groups: Barut was interested in the representations of the noncompact groups and their applications to physical problems as conformal or dynamical symmetries, and derived the algebraic scattering theory.

Nonlinear dynamical systems: He studied the symmetries of the completely integrable N-body problems in three dimensions.

Barut's contributions to high energy physics are the following:

Symmetries of the Elementary Particles: Barut developed the octet symmetry diagrams for the mesons and baryons in 1958. Later he proposed a simple model of the matter, which is composed from the known stable particles, electron, and neutrino, without any hypothetical constituents and discussed the symmetries of this model.

Dynamical Models of the Elementary Particles: In this model, all unstable particles are described as the bound states of electron and neutrino by the short range spin dependent electromagnetic forces.

Most of his books are classics and remain in demand. They have appeared in several editions and are translated into several languages.

Electrodynamics and Classical Theory of Fields and Particles

The Theory of the Scattering Matrix

Dynamical Groups and Generalized Symmetries

Theory of the Group Representations and Applications, (coauthored by R. Raczka)

Geometry and Physics.

He received the Alexander von Humboldt Award in 1975. In 1982, the Republic of Turkey bestowed upon Barut its Medal of Science. He was awarded Honorary D.Sc. degrees by Karadeniz University at Trabzon, Turkey, also in 1982, and İnönü University at Malatya in Turkey, 1989. Since 1964, he was a fellow of the American Physical Society.

He visited the Trieste International Centre for Theoretical Physics and other research centers frequently, and organized a lot of international conferences all over the world. During these visits and conferences, he collaborated with a large number of physicists from different nations.

We, his students, collaborators, colleagues and friends will always remember his remarkable scholarship, vitality, humanity, clear vision, uncanny ability to coherently mold several facts into a single, harmonious whole, and his organizational ability in science and teaching.

Nuri Ünal, *Akdeniz University, Physics Department, P.K.510, 07200, Antalya, Turkey*

ASPECTS OF LOCAL-FIELD ELECTRODYNAMICS IN CONDENSED MATTER

OLE KELLER
Institute of Physics, Aalborg University,
Pontoppidanstræde 103, DK-9220 Aalborg Øst, Denmark

Abstract. Taking as a starting point the microscopic Maxwell-Lorentz equations plus a linear material equation relating induced current densities and prevailing fields in a spatially and temporally nonlocal manner, the standard theory of local-field electrodynamics in condensed-matter systems is reviewed and related to textbook formulations. A recently established theory, which takes into account the fundamental difference between the electrodynamics of transverse and longitudinal fields, is discussed, and by means of a many-body constitutive equation describing the response to the transverse part of the prescribed external field, a loop equation for the transverse local field is derived and analyzed. A so-called coupled-antenna theory, adequate for investigations of local-field effects in mesoscopic media, is presented. Finally, the fundamental theory is applied to studies of (i) the linear electrodynamics of quantum wells, (ii) the possibilities of making so-called quantum dots of light in which electromagnetic fields are confined to foci substantially smaller in spatial extension than those possible within the scope of classical diffraction theory, and (iii) the photon-drag phenomenon in mesoscopic rings.

1. Introduction

In the memory of Prof. Asım Orhan Barut, the Physics Department of Bilkent University arranged a summer school on "Advances in Quantum Optics and Spectroscopy of Solids" in Ankara in the period 2–10 July, 1995. As one of the participants, I gave four lectures on the theory of local-field electrodynamics in condensed matter. In my first lecture, general principles and conceptual problems were addressed. Main emphasis was devoted to a discussion of the electrodynamics in transverse and longitudinal fields, and on the basis of an electromagnetic vacuum propagator formalism the important integral relation between the local field and the induced current density was introduced and discussed. Next, within the framework of nonrelativistic quantum mechanics, linear and nonlocal response theory was used to derive a material relation between the induced current density and the sum of the transverse local field and the longitudinal external field. By combining the two aforementioned relations a loop equation for the transverse part of the electromagnetic field was established, and it was demonstrated that the fundamental tensor-product structure of the linear conductivity response function formally allows one to establish an explicit solution of the loop equation. It was stressed that the explicit solution recently has turned out to be extremely useful for the studies of local-field effects in mesoscopic media. Finally,

T. Hakioğlu and A. S. Shumovsky (eds.), Quantum Optics and the Spectroscopy of Solids, 1–44.

the transverse current density domain concept was introduced, and it was shown to be of fundamental importance for separating system and source-particle dynamics. In my second lecture, I discussed a bright new subject for which local-field effects play an important role viz. the formation of so-called quantum dots of light by means of the phase conjugation of middle- and near-field components of the electromagnetic field radiated from the tip of an optical near-field microscope. Special attention was drawn to the electromagnetic field-confinement problem, the size dependence of the optical polarizability of mesoscopic particles, and the theory for phase conjugation of evanescent waves. The phrase "quantum dot of light" is used for foci smaller in spatial extension than allowed by classical diffraction theory. In my third lecture, it was outlined how the basic local-field theory can be applied in studies of the linear and nonlinear electrodynamics of quantum wells. In particular, I emphasized that new insight into the jump conditions of the electromagnetic field across interfaces carrying field-induced interface current and charge densities may be obtained from local-field theory. Also the new possibilities of using local-field formalisms to describe the electrodynamics of (thin) films from the quantum limit, where electronic confinement effects dominate, right up to the macroscopic region, where classical field interference effects dominate, were mentioned. In my fourth lecture, I discussed the idea of creating photon-drag currents by nonlinear local-field interactions in mesoscopic metal and semiconductor rings, and a newly established theory for the phenomenon that might be useful for investigating the spectroscopy of non-simply connected mesoscopic matter.

The present article has been prepared in the wake of my lectures at the Barut memorial meeting. I have not here aimed at giving a balanced review of local-field electrodynamics as such, nor of the applications presented in lectures two to four. Instead I have attempted to supplement (review) articles I have written recently on the subject with new material of importance for obtaining (i) a better overview of the field and (ii) a closer relation to textbook expositions. Fundamental aspects of theory as well as a number of specific studies in linear and nonlinear electrodynamics are covered in my two review articles "Local-field studies in the nonlinear optics of mesoscopic systems" in O. Keller (ed.), Studies in Classical and Quantum Nonlinear Optics (Nova Science Publishers, New York, 1995) p. 269, and "Local fields in the electrodynamics of mesoscopic media", Phys. Rep. (in press). A recent account of the local-field electrodynamics in quantum wells including also a new discussion of jump conditions for the electromagnetic field across a quantum well sheet is given in my articles "Sheet-model description of the linear optical response of quantum wells", J. Opt. Soc. Amer. 12, 987 (1995), and "Optical response of a quantum-well sheet: internal electrodynamics", J. Opt. Soc. Amer. 12, 997 (1995). A description of the fundamental theory behind the formation of light dots can be found in my paper entitled "Quantum dots of light", which is to appear in the proceedings from the 2nd Mediterranean Workshop and Topical Meeting on "Novel Optical Materials and Applications" held in Cetraro in Italy in the period May 28 – June 2, 1995. The proceedings are to appear as a special issue (Vol. 5, no. 1 and 2) of J. Nonlinear Optical Physics and Materials (World Scientific, 1996). Readers interested in the photon-drag phenomenon in mesoscopic rings may consult my Phys. Rep. article for an heuristic approach, or my upcoming review article from the proceedings of the Third International Aalborg Summer School on Nonlinear Optics (World Scientific, 1996).

In chapter 2 of the present paper I discuss the conductivity response formalism in the time-space domain. Starting from the microscopic Maxwell-Lorentz equations the free-space electromagnetic propagator, the selfconsistent and external conductivity response, as well as the relation between the selfconsistent and external conductivity tensors are studied. To relate the present formalism to textbook expositions, I introduce dielectric (permittivity) and magnetic permeability functions in such a manner that the basic ambiguity inherent in the definitions of the dielectric and magnetic permeability tensors appears clearly. Although this ambiguity has been known to exist for a long time it is almost always overlooked in the literature, sometimes with fatal conceptual consequences. To stress the freedom of choice one has in electrodynamic response theory, two illustrative examples are investigated. In chapter 3, the integral relation between the local field and the induced current density is established and discussed stressing from the outset the difference between the transverse and longitudinal electrodynamics in the so-called transverse current-density domain. In chapter 4, the many-body constitutive relation describing the linear current density response to the sum of the transverse local field and the longitudinal external field, and the loop equation for the transverse part of the local field are introduced and discussed. In chapter 5, the so-called coupled-antenna theory is presented and its main ingredients, i.e. the tensor-product structure of the combined para- and diamagnetic conductivity response functions, and the subsequent transformation of the loop (integral) equation for the transverse part of the local field into a set of linear and inhomogeneous algebraic equations among the amplitude strengths with which each of the microscopic antennas radiate are analyzed. In the coupled-antenna approach each electronic transition in the (mesoscopic) medium under consideration acts as an antenna. The paper is finished in chapter 6 with a brief discussion of three examples taken from the physics of mesoscopic systems. Hence, I start by a brief analysis of the linear electrodynamics of quantum wells, I go on with a conceptual study of quantum dots of light, and I finish with a few considerations on the possible photon drag in mesoscopic rings.

In the context of this memorial meeting for Prof. Barut, I would like to emphasize that the programme layed forward in this paper for studying local-field phenomena in condensed matter in a sense is closely related to the works of Prof. Barut on the electrodynamics of atoms and atomic particles. In these works a formulation of quantum electrodynamics that is not second quantized, but rather based on self-fields is taken as a starting point, see e.g. Refs. 1–6. In the self-field picture of atomic physics, one views radiative corrections as arising from radiation reaction effects due to the interaction of a particle with its own self-field. In the modern approach to the understanding of local-field phenomena in condensed matter self-field interactions also play a fundamental role as we shall realize in the following.

2. Conductivity response formalism in the time-space domain

2.1. THE MICROSCOPIC MAXWELL-LORENTZ EQUATIONS

As a starting point for the description of the linear and nonlocal response of a system of material particles to an impressed electromagnetic field we take the microscopic Maxwell-Lorentz equations, i.e.

$$\vec{\nabla} \times \vec{E}(\vec{r},t) = -\frac{\partial \vec{B}(\vec{r},t)}{\partial t}, \tag{1}$$

$$\vec{\nabla} \times \vec{B}(\vec{r},t) = \mu_0 \vec{J}(\vec{r},t) + \frac{1}{c_0^2}\frac{\partial \vec{E}(\vec{r},t)}{\partial t}, \tag{2}$$

$$\vec{\nabla} \cdot \vec{E}(\vec{r},t) = \frac{1}{\varepsilon_0}\rho(\vec{r},t), \tag{3}$$

$$\vec{\nabla} \cdot \vec{B}(\vec{r}, t) = 0, \tag{4}$$

where $\vec{E}(\vec{r}, t)$ and $\vec{B}(\vec{r}, t)$ are the electric and magnetic fields prevailing at the space point $\vec{r}$ at time t. In the microscopic formulation of electrodynamics the material properties are described completely in terms of the total current density, $\vec{J}(\vec{r}, t)$, and charge density, $\rho(\vec{r}, t)$. These quantities are not independent but related via the equation of continuity for the electric charge, viz.,

$$\vec{\nabla} \cdot \vec{J}(\vec{r}, t) + \frac{\partial \rho(\vec{r}, t)}{\partial t} = 0. \tag{5}$$

The equation of continuity can be derived from the microscopic Maxwell-Lorentz equations. Thus, by taking the divergence of Eq. (2) one obtains

$$\mu_0 \vec{\nabla} \cdot \vec{J}(\vec{r}, t) + \frac{1}{c_0^2} \frac{\partial}{\partial t} \left(\vec{\nabla} \cdot \vec{E}(\vec{r}, t) \right) = 0, \tag{6}$$

since $\vec{\nabla} \cdot \vec{\nabla} \times \vec{B}(\vec{r}, t) = 0$. By inserting the expression for $\vec{\nabla} \cdot \vec{E}$ given in Eq. (3) into Eq. (6), and by utilizing that the phase velocity of light in vacuum is given by

$$c_0 = (\varepsilon_0 \mu_0)^{-1/2}, \tag{7}$$

where ε_0 and μ_0 are the vacuum permittivity and permeability, respectively, one readily obtains Eq. (5). The form of the equation of continuity in Eq. (5) guarantees that the principle of electric charge conservation is built into the microscopic Maxwell-Lorentz equations.

2.2. EXTERNAL SOURCES

Let us consider the situation where a medium consisting of a system of electrically charged particles is excited by a so-called external (e) electromagnetic field produced by a source distribution of external charges and currents of densities $\rho^e(\vec{r}, t)$ and $\vec{J}^e(\vec{r}, t)$, respectively. The external electric ($\vec{E}^e(\vec{r}, t)$) and magnetic ($\vec{B}^e(\vec{r}, t)$) fields originating from these charges and their motions satisfy the following set of microscopic Maxwell-Lorentz equations:

$$\vec{\nabla} \times \vec{E}^e(\vec{r}, t) = -\frac{\partial \vec{B}^e(\vec{r}, t)}{\partial t}, \tag{8}$$

$$\vec{\nabla} \times \vec{B}^e(\vec{r}, t) = \mu_0 \vec{J}^e(\vec{r}, t) + \frac{1}{c_0^2} \frac{\partial \vec{E}^e(\vec{r}, t)}{\partial t}, \tag{9}$$

$$\vec{\nabla} \cdot \vec{E}^e(\vec{r}, t) = \frac{1}{\varepsilon_0} \rho^e(\vec{r}, t), \tag{10}$$

$$\vec{\nabla} \cdot \vec{B}^e(\vec{r}, t) = 0. \tag{11}$$

In many cases, the charges of the medium and the source are well separated in space. In these cases, where each distribution of charges constitutes a closed system, the equation of continuity

$$\vec{\nabla} \cdot \vec{J}^e(\vec{r}, t) + \frac{\partial \rho^e(\vec{r}, t)}{\partial t} = 0 \tag{12}$$

holds for the source charges (and a corresponding one for the charges of the medium, of course).

When the external electromagnetic field interacts with the charged particles of our medium, induced (i) charge and current densities $\rho^i(\vec{r},t)$ and $\vec{J}^i(\vec{r},t)$ are produced inside the medium. In turn, these give rise to induced electric, $\vec{E}^i(\vec{r},t)$, and magnetic, $\vec{B}^i(\vec{r},t)$, fields. Thus, in terms of the external and induced fields the selfconsistent (prevailing) fields in the medium are given by

$$\vec{E}(\vec{r},t) = \vec{E}^e(\vec{r},t) + \vec{E}^i(\vec{r},t), \tag{13}$$

and

$$\vec{B}(\vec{r},t) = \vec{B}^e(\vec{r},t) + \vec{B}^i(\vec{r},t). \tag{14}$$

The selfconsistent fields inside the medium (M) satisfy the microscopic Maxwell-Lorentz equations in (1)–(4) with

$$\vec{J}(\vec{r},t) = \vec{J}^i(\vec{r},t), \qquad \vec{r} \in M, \tag{15}$$

$$\rho(\vec{r},t) = \rho^i(\vec{r},t), \qquad \vec{r} \in M, \tag{16}$$

provided that no charge and current densities exist in the medium in absence of the external field.

The induced motion of the particles of our medium will give rise not only to an electromagnetic field inside the medium but also outside. However, if the induced field in the source region is sufficiently small the motion of the external charges will not be affected. In this case $\rho^e(\vec{r},t)$ and $\vec{J}^e(\vec{r},t)$ can be considered as *prescribed* quantities and the external fields $\vec{E}^e(\vec{r},t)$ and $\vec{B}^e(\vec{r},t)$ can be obtained directly from Eqs. (8)–(11).

Although we have considered the medium and source regions to be spatially separated (the precise meaning of this phrase will be discussed in section 3.1) so far, this is not necessarily always so. Thus, in the more general case one should take

$$\vec{J}(\vec{r},t) = \vec{J}^e(\vec{r},t) + \vec{J}^i(\vec{r},t), \tag{17}$$

and

$$\rho(\vec{r},t) = \rho^e(\vec{r},t) + \rho^i(\vec{r},t) \tag{18}$$

in the overlap region. As soon as such an overlap exists one should be extremely cautious when formulating a response formalism. Firstly, it may not be meaningful any longer to distinguish the external charges and the charges of the medium. Secondly, an exchange of particles may occur between the two systems so that the equation of continuity for the external charges (Eq. (12)), say, is no longer valid (conservation of the total charge implies that Eq. (5) is always fulfilled, of course). Thirdly, even if the external charges are distinguishable from the charges of our medium, the quantities $\rho^e(\vec{r},t)$ and $\vec{J}^e(\vec{r},t)$ will in general be so strongly affected by the induced fields that they no longer can be considered as prescribed quantities.

2.3. FREE-SPACE ELECTROMAGNETIC PROPAGATOR – AN HEURISTIC APPROACH

Let us return to the case where the external charge and current densities can be considered as prescribed quantities. To determine the associated external electromagnetic field one takes the curl ($\vec{\nabla}\times$) of Eq. (8) to find $\vec{\nabla}\times\vec{E}^e = -\frac{\partial}{\partial t}(\vec{\nabla}\times\vec{B}^e)$. By inserting in the resulting equation the expression for $\vec{\nabla}\times\vec{B}^e$ given in Eq. (9) one obtains the following wave equation for the external electric field:

$$\vec{\nabla}\times(\vec{\nabla}\times\vec{E}^e(\vec{r},t))+\frac{1}{c_0^2}\frac{\partial^2\vec{E}^e(\vec{r},t)}{\partial t^2}=-\mu_0\frac{\partial\vec{J}^e(\vec{r},t)}{\partial t}, \tag{19}$$

or in dyadic notation

$$\left[\vec{\nabla}\vec{\nabla}-\vec{\vec{U}}\left(\nabla^2-\frac{1}{c_0^2}\frac{\partial}{\partial t^2}\right)\right]\cdot\vec{E}^e(\vec{r},t)=-\mu_0\frac{\partial\vec{J}^e(\vec{r},t)}{\partial t}, \tag{20}$$

where $\vec{\vec{U}}$ is the unit tensor of dimension 3×3. Since the partial differential equation in (20) necessarily dictates a linear relation between the external field and the current density, let us write the acceptable solution for $\vec{E}^e(\vec{r},t)$ in the quite general form

$$\vec{E}^e(\vec{r},t)=\mu_0\int_{-\infty}^{\infty}\vec{\vec{D}}_0(\vec{r},\vec{r}\,',t,t')\cdot\frac{\partial\vec{J}^e(\vec{r}\,',t')}{\partial t'}d^3r'dt'. \tag{21}$$

By means of the Dirac delta functions $\delta(\vec{r}-\vec{r}\,')$ and $\delta(t-t')$ the right hand side of Eq. (20) can be cast in a form similar to that of Eq. (21). Thus,

$$\frac{\partial\vec{J}^e(\vec{r},t)}{\partial t}=\int_{-\infty}^{\infty}\vec{\vec{U}}\cdot\frac{\partial\vec{J}^e(\vec{r}\,',t')}{\partial t'}\delta(\vec{r}-\vec{r}\,')\delta(t-t')d^3r'dt'. \tag{22}$$

Substituting Eqs. (21) and (22) into Eq. (20) and interchanging the order of appearance of the dyadic differential operator $\vec{\nabla}\vec{\nabla}-\vec{\vec{U}}(\nabla^2-c_0^{-2}\partial^2/\partial t^2)$ and the volume and time integrations one obtains

$$\int_{-\infty}^{\infty}\left\{\left[\vec{\nabla}\vec{\nabla}-\vec{\vec{U}}\left(\nabla^2-\frac{1}{c_0^2}\frac{\partial^2}{\partial t^2}\right)\right]\cdot\vec{\vec{D}}_0(\vec{r},\vec{r}\,',t,t')\right.$$
$$\left.+\vec{\vec{U}}\delta(\vec{r}-\vec{r}\,')\delta(t-t')\right\}\cdot\frac{\partial\vec{J}^e(\vec{r}\,',t')}{\partial t'}d^3r'dt'=\vec{0}. \tag{23}$$

Since this equation has to hold for an arbitrary time-space distribution of the external current density, the dyadic free-space Green's function $\vec{\vec{D}}_0(\vec{r},\vec{r}\,',t,t')$ must obey the following differential equation:

$$\left[\vec{\nabla}\vec{\nabla}-\vec{\vec{U}}\left(\nabla^2-\frac{1}{c_0^2}\frac{\partial^2}{\partial t^2}\right)\right]\cdot\vec{\vec{D}}_0(\vec{r},\vec{r}\,',t,t')=-\vec{\vec{U}}\delta(\vec{r}-\vec{r}\,')\delta(t-t'). \tag{24}$$

It is important to note that the interchange of the differential operator and the volume integral sign has serious implications when the point of observation, $\vec{r}$, is located inside the source region (see section 3.1). Here, we shall be interested in the external field at points, $\vec{r}$, outside the source region, only. The form of Eq. (24) indicates that the free-space Green's function depends on the differences $\vec{r}-\vec{r}\,'$ and $t-t'$ only, i.e.

$$\vec{\vec{D}}_0(\vec{r},\vec{r}\,',t,t')=\vec{\vec{D}}_0(\vec{r}-\vec{r}\,',t-t'). \tag{25}$$

From a physical point of view the underlying assumption is that the particle-free vacuum should exhibit translational invariance in space and time. Thus, for a space-time translational invariant vacuum the electromagnetic field, prevailing at given time-space point and caused by the radiation from a current-density source located at another time-space point, can necessarily depend on the relative space-time coordinates only.

To solve Eq. (24) let us start by introducing Fourier integral representations for the time dependences of $\overleftrightarrow{D}_0(\vec{r}-\vec{r}\,',t-t')$ and $\delta(t-t')$, i.e.

$$\overleftrightarrow{D}_0(\vec{r}-\vec{r}\,',t-t') = \frac{1}{2\pi}\int_{-\infty}^{\infty} \overleftrightarrow{D}_0(\vec{r}-\vec{r}\,';\omega)e^{-i\omega(t-t')}d\omega \tag{26}$$

and

$$\delta(t-t') = \frac{1}{2\pi}\int_{-\infty}^{\infty} e^{-i\omega(t-t')}d\omega. \tag{27}$$

The simplicity of the integral expression in Eq. (26) can be misleading since it turns out that the integral cannot be performed along the real axis of the complex ω-plane. It can be shown that the possible contour choices for the integration are restricted by the postulate of causality. Leaving the question of the contour choice open for the time being one obtains by combining Eqs. (24), (26), and (27) the following differential equation for the Fourier transform $\overleftrightarrow{D}_0(\vec{r}-\vec{r}\,';\omega)$:

$$\left\{\vec{\nabla}\vec{\nabla} - \overleftrightarrow{U}\left[\nabla^2 + \left(\frac{\omega}{c_0}\right)^2\right]\right\}\cdot \overleftrightarrow{D}_0(\vec{r}-\vec{r}\,';\omega) = -\overleftrightarrow{U}\delta(\vec{r}-\vec{r}\,'). \tag{28}$$

It is well known that the dyadic free-space Green's function, $\overleftrightarrow{D}_0(\vec{r}-\vec{r}\,';\omega)$, of Eq. (28) can be written in the form

$$\overleftrightarrow{D}_0(\vec{r}-\vec{r}\,';\omega) = -\left[\overleftrightarrow{U} + \left(\frac{c_0}{\omega}\right)^2\vec{\nabla}\vec{\nabla}\right]d_0(|\vec{r}-\vec{r}\,'|;\omega), \tag{29}$$

where the so-called scalar free-space Green's function $d_0(|\vec{r}-\vec{r}\,'|;\omega)$ satisfies the inhomogeneous differential equation

$$\left[\nabla^2 + \left(\frac{\omega}{c_0}\right)^2\right]d_0(|\vec{r}-\vec{r}\,'|;\omega) = -\delta(|\vec{r}-\vec{r}\,'|). \tag{30}$$

By means of the substitution $\vec{r}-\vec{r}\,' = \vec{R}$, Eq. (30) takes the form

$$\left[\nabla^2_{\vec{R}} + \left(\frac{\omega}{c_0}\right)^2\right]d_0(|\vec{R}|;\omega) = -\delta(|\vec{R}|), \tag{31}$$

where $\nabla^2_{\vec{R}}$ is the Laplace operator in the $\vec{R}$-coordinates. In spherical coordinates Eq. (31) becomes

$$\frac{1}{R}\frac{d^2}{dR^2}(Rd_0(R;\omega)) + \left(\frac{\omega}{c_0}\right)^2 d_0(R;\omega) = -\delta(R). \tag{32}$$

For $R \neq 0$, the right hand side of Eq. (32) is zero, and the complete solution for $d_0(R;\omega)$ is given by

$$d_0(R;\omega) = C^o\frac{\exp(i\frac{\omega}{c_0}R)}{R} + C^i\frac{\exp(-i\frac{\omega}{c_0}R)}{R}, \qquad R \neq 0 \tag{33}$$

where C^o and C^i are yet unknown constants. To understand the physical interpretation attached to the two terms in Eq. (33), this equation is inserted into Eq. (29), and this in turn into Eq. (26). Hence, one obtains

$$\vec{\vec{D}}_0(\vec{r}-\vec{r}',t-t') = -\frac{1}{2\pi}\int_{-\infty}^{\infty}\left[\vec{\vec{U}}+\left(\frac{c_0}{\omega}\right)^2\vec{\nabla}\vec{\nabla}\right]$$
$$\times\left(C^o\frac{\exp\left\{i\omega\left[\frac{|\vec{r}-\vec{r}'|}{c_0}-(t-t')\right]\right\}}{|\vec{r}-\vec{r}'|}+C^i\frac{\exp\left\{-i\omega\left[\frac{|\vec{r}-\vec{r}'|}{c_0}-(t-t')\right]\right\}}{|\vec{r}-\vec{r}'|}\right)d\omega, \quad (34)$$

for $|\vec{r}-\vec{r}'| \neq 0$. It appears from Eq. (34) that the free-space dyadic Green's function is composed of a linear superposition of different(complex) frequency components. The spectrum appearing in the superposition is not unique but depends on the path of integration in the complex ω-plane. The wave pattern belonging to a given ω consists of two wavelets propagating respectively inwards (i) towards the source point and outwards (o) away from the source point. The amplitudes of these components are proportional to C^i and C^o, respectively. Thus, being interested in the radiation from the external source distribution we keep only the outgoing (OUT) part, $\vec{\vec{D}}_0^{\mathrm{OUT}}(\vec{r}-\vec{r}',t-t')$ of the Green's function, i.e. C^i is set equal to zero in Eq. (34). The constant C^o is determined by integration of Eq. (31) over a sphere of (infinitesimal) radius δ centered at $R=0$. Thus, by utilizing Gauss' theorem the first term on the left hand side of Eq. (31) becomes, with $\vec{e}_{\vec{R}}=\vec{R}/R$

$$\int_V \vec{\nabla}_{\vec{R}}\cdot\vec{\nabla}_{\vec{R}} d_0(R;\omega)dV =$$
$$\int_S \vec{e}_{\vec{R}}\cdot\vec{\nabla}_{\vec{R}} d_0(R;\omega)dS = \int_S \vec{e}_{\vec{R}}\cdot\vec{e}_{\vec{R}}\frac{\partial d_0(R;\omega)}{\partial R}dS = \left[4\pi R^2\frac{\partial d_0(R;\omega)}{\partial R}\right]_{R=\delta}, \quad (35)$$

denoting the volume and surface of the sphere by V and S, respectively. Using the result in Eq. (35) and the fact that $\int_V \delta(R)dV = 1$, one obtains upon integration of Eq. (31)

$$4\pi\delta^2\left[\frac{\partial d_0(R;\omega)}{\partial R}\right]_{R=\delta}+\left(\frac{\omega}{c_0}\right)^2\int_0^\delta 4\pi R^2 d_0(R;\omega)dR = -1. \quad (36)$$

By inserting in this equation the outgoing part

$$d_0^{\mathrm{OUT}}(R;\omega) = C^o\frac{\exp(i\frac{\omega}{c_0}R)}{R} \quad (37)$$

of $d_0(R;\omega)$ it is readily realized that the second term on the left hand side of Eq. (36) vanishes in the limit $\delta\to 0$, and that the first term gives $-4\pi C^o$, so that finally $C^o = (4\pi)^{-1}$. Having calculated C^o, we have obtained the final expression for the outgoing free-space dyadic Green's function, viz.

$$\vec{\vec{D}}_0^{\mathrm{OUT}}(\vec{r}-\vec{r}',t-t') =$$
$$-\frac{1}{8\pi^2}\int_{-\infty}^{\infty}\left[\vec{\vec{U}}+\left(\frac{c_0}{\omega}\right)^2\vec{\nabla}\vec{\nabla}\right]\frac{\exp\left\{i\omega\left[\frac{|\vec{r}-\vec{r}'|}{c_0}-(t-t')\right]\right\}}{|\vec{r}-\vec{r}'|}d\omega. \quad (38)$$

The knowledge of the explicit expression for the outgoing free-space Green's function in principle allows us to determine the external field stemming from a prescribed external current density distribution, cf. Eq. (21). Thus, one has

$$\vec{E}^e(\vec{r},t) = \mu_0 \int_{-\infty}^{\infty} \overleftrightarrow{D}_0^{\text{OUT}}(\vec{r}-\vec{r}\,',t-t') \cdot \frac{\partial \vec{J}^e(\vec{r}\,',t')}{\partial t'} d^3r'dt' \tag{39}$$

for observation points outside the source region.

2.4. THE SELFCONSISTENT CONDUCTIVITY RESPONSE

In order to determine the electric and magnetic fields in a system of material particles placed in an external electromagnetic field produced by an assembly of moving external charges we have seen that it is necessary in the general case to solve the microscopic Maxwell-Lorentz equations

$$\vec{\nabla} \times \vec{E}(\vec{r},t) = -\frac{\partial \vec{B}(\vec{r},t)}{\partial t}, \tag{40}$$

$$\vec{\nabla} \times \vec{B}(\vec{r},t) = \mu_0 \left(\vec{J}^e(\vec{r},t) + \vec{J}^i(\vec{r},t)\right) + \frac{1}{c_0^2}\frac{\partial \vec{E}(\vec{r},t)}{\partial t}, \tag{41}$$

$$\vec{\nabla} \cdot \vec{E}(\vec{r},t) = \frac{1}{\varepsilon_0}\left(\rho^e(\vec{r},t)) + \rho^i(\vec{r},t)\right) \tag{42}$$

$$\vec{\nabla} \cdot \vec{B}(\vec{r},t) = 0. \tag{43}$$

However, since the set of equations in (40)–(43) is not closed it does not permit one to express the selfconsistent electric, $\vec{E}(\vec{r},t)$, and magnetic, $\vec{B}(\vec{r},t)$, fields in terms of $\rho^e(\vec{r},t)$ and $\vec{J}^e(\vec{r},t)$. To close the system of equations it is necessary to introduce a so-called material (constitutive) relation. In the standard description one invokes a constitutive relation between the induced current density, $\vec{J}^i(\vec{r},t)$ and either the selfconsistent electric field, $\vec{E}(\vec{r},t)$, or the external field, $\vec{E}^e(\vec{r},t)$. In the present section we relate $\vec{J}^i(\vec{r},t)$ to $\vec{E}(\vec{r}\,',t)$ via a material relation of the form

$$\vec{J}^i(\vec{r},t) = \int_{-\infty}^{\infty} \overleftrightarrow{\sigma}(\vec{r},\vec{r}\,',t,t') \cdot \vec{E}(\vec{r}\,',t')d^3r'dt'. \tag{44}$$

By adopting a constitutive equation of this kind we have limited ourselves to a study of the linear electrodynamic properties of the system, essentially. The second-rank tensor $\overleftrightarrow{\sigma}(\vec{r},\vec{r}\,',t,t')$, which relates $\vec{J}^i(\vec{r},t)$ linearly to $\vec{E}(\vec{r}\,',t')$ we shall name "the selfconsistent conductivity response tensor" or, from time to time simply "the conductivity tensor". The form of Eq. (44) indicates that we have allowed the relation between $\vec{J}^i$ and $\vec{E}$ to be nonlocal in time and space. A response is nonlocal in time and space if the induced current density prevailing at the space point $\vec{r}$ at time t depends not only on the selfconsistent field at the same time-space point but also on the field prevailing at neighbouring time-space points $(\vec{r}\,',t')$. From a physical point of view the selfconsistent conductivity response tensor can be considered as a weight function determining the strength of the contribution to the induced current density at $(\vec{r},t)$ from the selfconsistent electric field at $(\vec{r}\,',t')$.

Let us assume that by some method we have been able to calculate the linear conductivity tensor $\overleftrightarrow{\sigma}(\vec{r},\vec{r}\,',t,t')$. Even then it is not possible to determine the induced current density $\vec{J}^i(\vec{r},t)$ from Eq. (44) alone since also the selfconsistent field $\vec{E}(\vec{r},t)$ is not known beforehand. Stated in a slightly different manner one may say that the constitutive relation gives a constraint between two yet unknown functions $\vec{J}^i(\vec{r},t)$ and $\vec{E}(\vec{r},t)$.

To obtain the selfconsistent electric field one takes the curl of Eq. (40) and inserts into the resulting equation the expression for $\vec{\nabla}\times\vec{B}$ given in Eq. (41). Hence, one gets

$$\vec{\nabla}\times(\vec{\nabla}\times\vec{E}(\vec{r},t))+\frac{1}{c_0^2}\frac{\partial^2\vec{E}(\vec{r},t)}{\partial t^2}=-\mu_0\frac{\partial\vec{J}^i(\vec{r},t)}{\partial t}-\mu_0\frac{\partial\vec{J}^e(\vec{r},t)}{\partial t}. \tag{45}$$

By inserting into this equation the constitutive relation in Eq. (44) one obtains in dyadic notation

$$\begin{aligned}&\left[\vec{\nabla}\vec{\nabla}-\overleftrightarrow{U}\left(\nabla^2-\frac{1}{c_0^2}\frac{\partial^2}{\partial t^2}\right)\right]\cdot\vec{E}(\vec{r},t)\\ &\qquad+\mu_0\frac{\partial}{\partial t}\int_{-\infty}^{\infty}\overleftrightarrow{\sigma}(\vec{r},\vec{r}',t,t')\cdot\vec{E}(\vec{r},t)d^3r'dt'=-\mu_0\frac{\partial\vec{J}^e(\vec{r},t)}{\partial t}.\end{aligned} \tag{46}$$

Equation (46) is a linear, inhomogeneous integro-differential equation for the selfconsistent electric field. For a prescribed external current density distribution, this equation in principle can be solved using a Green's function technique analogous to that described in section 2.3. As the outcome of such a Green's function approach one obtains, cf. Eq. (21)

$$\vec{E}(\vec{r},t)=\mu_0\int_{-\infty}^{\infty}\overleftrightarrow{G}(\vec{r},\vec{r}',t,t')\cdot\frac{\partial\vec{J}^e(\vec{r}',t')}{\partial t'}d^3r'dt'. \tag{47}$$

The dyadic Green's function $\overleftrightarrow{G}(\vec{r},\vec{r}',t,t')$ appearing in Eq. (47) is not a free-space Green's function since it incorporates in its propagation characteristics for instance the electrodynamic screening properties of the medium ($\overleftrightarrow{G}(\vec{r},\vec{r}',t,t')$ depends on ($\overleftrightarrow{\sigma}(\vec{r},\vec{r}',t,t')$)). For a more elaborated and exemplified discussion of screened electromagnetic propagators (Green's functions) the reader is referred to Refs. 7 and 8.

The expression for $\vec{E}(\vec{r},t)$ given in Eq. (47) represents a particular solution to the inhomogeneous integro-differential equation in (46). To this solution one may add an arbitrary solution to the homogeneous part of the equation and still have a solution to Eq. (46). In cases where there exists a field inside our system in the absence of $\partial\vec{J}^e(\vec{r},t)/\partial t$ it is necessary to add an appropriate solution to the homogeneous part of the integro-differential equation in Eq. (46) to the right hand side of Eq. (47).

Once the selfconsistent electric field, $\vec{E}(\vec{r},t)$, has been obtained the induced current density can be calculated from Eq. (44). Furthermore, the induced charge density, $\rho^i(\vec{r},t)$, can be determined from Eq. (42), i.e.

$$\rho^i(\vec{r},t)=\varepsilon_0\vec{\nabla}\cdot\vec{E}(\vec{r},t)-\rho^e(\vec{r},t), \tag{48}$$

assuming the external charge density to be a prescribed quantity. Finally, one may obtain the prevailing magnetic field in the standard manner from

$$\left(\nabla^2-\frac{1}{c_0^2}\frac{\partial^2}{\partial t^2}\right)\vec{B}(\vec{r},t)=-\mu_0\vec{\nabla}\times\left(\vec{J}^e(\vec{r},t)+\vec{J}^i(\vec{r},t)\right). \tag{49}$$

2.5. THE EXTERNAL CONDUCTIVITY RESPONSE

In the previous section the field-induced current density was related to the selfconsistent electric field by means of the linear and nonlocal constitutive equation in (44). Experimentally, it is usually the response to a prescribed external field which is being studied. From this point of view it seems natural to introduce a linear and nonlocal relation between the field-induced current density and the external field, viz.

$$\vec{J}^i(\vec{r},t) = \int_{-\infty}^{\infty} \overleftrightarrow{\sigma}^e(\vec{r},\vec{r}\,',t,t') \cdot \vec{E}^e(\vec{r}\,',t')d^3r'dt'. \tag{50}$$

The quantity $\overleftrightarrow{\sigma}^e$ is called the external (e) conductivity response tensor. It is a true response function in the sense that there exists necessarily a causal relationship between $\vec{J}^i$ and $\vec{E}^e$. Such a relation does not exist between $\vec{J}^i$ and $\vec{E}$ since the selfconsistent (prevailing) electric field is created in part by the induced current density stemming from the external field *and* the induced field (calculated selfconsistently).

Let us imagine that the two conductivity tensors $\overleftrightarrow{\sigma}(\vec{r},\vec{r}\,',t,t')$ and $\overleftrightarrow{\sigma}^e(\vec{r},\vec{r}\,',t,t')$ have been determined. If so, it will be advantageous to use the constitutive relation in Eq. (50) to calculate the induced current density. The reason for this stems from the fact that the right hand side of Eq. (50) is known (one only needs to carry out integrations of a product of given functions $\overleftrightarrow{\sigma}^e(\vec{r},\vec{r}\,',t,t')$ and $\vec{E}^e(\vec{r}\,',t')$ over time (t') and space ($\vec{r}\,'$)). The constitutive equation in Eq. (44), on the other hand, still contains an unknown local field $\vec{E}(\vec{r}\,',t')$. As discussed in section 2.4, the constitutive relation in (44) has to be combined with the wave equation in (45) before the field-induced current density finally can be obtained. A further advantage of Eq. (50) is associated with the fact that also the boundary conditions for the electromagnetic field and the particle field are incorporated in $\overleftrightarrow{\sigma}^e(\vec{r},\vec{r}\,',t,t')$. Although it looks as if the constitutive relation in Eq. (50) is much more adequate for a practical calculation of $\vec{J}^i(\vec{r},t)$ than is Eq. (44), one should remember that this is the case *provided* the response function $\overleftrightarrow{\sigma}^e$ is known. To obtain an explicit expression for the external conductivity tensor is in general a more difficult task than to determine the selfconsistent conductivity tensor. Thus, $\overleftrightarrow{\sigma}^e$ will be a more complicated "object" than $\overleftrightarrow{\sigma}$, because it incorporates, via the boundary conditions for the field, the geometry of the material system under study, and the field-propagation characteristics associated with local-field effects in the medium. In many situations the external field is slowly varying in space across the domain in which $\overleftrightarrow{\sigma}^e(\vec{r},\vec{r}\,',t,t')$ is effectively different from zero. If so, the constitutive relation in Eq. (50) may be simplified by a Taylor series expansion (containing a few terms) of $\vec{E}^e(\vec{r}\,',t')$ around $\vec{r}$. Even if the external field is slowly varying in space the local field, $\vec{E}(\vec{r},t)$, normally varies so rapidly in space that it is meaningless to make use of a few-term Taylor expansion of $\vec{E}(\vec{r},t')$ around $\vec{r}$, cf. the discussion in section 5.2.

Once the induced current density has been obtained from Eq. (50), the selfconsistent field can be calculated via Eq. (45) inserting for $\vec{J}^i(\vec{r},t)$ the expression in Eq. (50). The resulting partial and inhomogeneous differential equation, which in dyadic notation reads

$$\left[\vec{\nabla}\vec{\nabla} - \overleftrightarrow{U}\left(\nabla^2 - \frac{1}{c_0^2}\frac{\partial^2}{\partial t^2}\right)\right] \cdot \vec{E}(\vec{r},t) =$$
$$-\mu_0\frac{\partial \vec{J}^e(\vec{r},t)}{\partial t} - \mu_0\frac{\partial}{\partial t}\int_{-\infty}^{\infty} \overleftrightarrow{\sigma}^e(\vec{r},\vec{r}\,',t,t') \cdot \vec{E}^e(\vec{r}\,',t')d^3r'dt', \tag{51}$$

is adequately solved by a Green's function technique (see chapter 3 and Ref. 9). For field points outside the medium under study and the domain of the external sources the formalism described in section 2.3 readily shows that

$$\vec{E}(\vec{r},t) = \vec{E}^e(\vec{r},t) + \mu_0 \int_{-\infty}^{\infty} \overleftrightarrow{D}_0^{\mathrm{OUT}}(\vec{r}-\vec{r}\,',t-t') \cdot \left\{ \frac{\partial}{\partial t'} \int_{-\infty}^{\infty} \overleftrightarrow{\sigma}^e(\vec{r}\,',\vec{r}\,'',t',t'') \cdot \vec{E}^e(\vec{r}\,'',t'') d^3r'' dt'' \right\} d^3r' dt', \tag{52}$$

with $\vec{E}^e$ given by Eq. (39).

2.6. THE RELATION BETWEEN THE SELFCONSISTENT AND EXTERNAL CONDUCTIVITY RESPONSE FUNCTIONS

In the preceding two sections we have studied the linear electrodynamic response of a system of material particles on the basis of a constitutive equation involving either the selfconsistent field or the external field. In the present section we shall establish an integral-equation relation among the associated response tensors $\overleftrightarrow{\sigma}(\vec{r},\vec{r}\,',t,t')$ and $\overleftrightarrow{\sigma}^e(\vec{r},\vec{r}\,',t,t')$. To do this we set out to obtain a relation between the selfconsistent and external electric fields. Thus, by inserting into Eq. (51) the expression for $-\mu_0 \partial \vec{J}^e(\vec{r},t)/\partial t$ given i Eq. (20) one finds

$$\left[\vec{\nabla}\vec{\nabla} - \overleftrightarrow{U}\left(\nabla^2 - \frac{1}{c_0^2}\frac{\partial^2}{\partial t^2}\right)\right] \cdot \left[\vec{E}(\vec{r},t) - \vec{E}^e(\vec{r},t)\right] = -\mu_0 \frac{\partial}{\partial t} \int_{-\infty}^{\infty} \overleftrightarrow{\sigma}^e(\vec{r},\vec{r}\,',t,t') \cdot \vec{E}^e(\vec{r}\,',t') d^3r' dt'. \tag{53}$$

By using the appropriate (see section 3.1 and Ref. 9) free-space electromagnetic Green's function, $\overleftrightarrow{G}_0(\vec{r},\vec{r}\,',t,t')$, it is realized that $\vec{E}(\vec{r},t)$ and $\vec{E}^e(\vec{r},t)$ are related by the following integral equation:

$$\vec{E}(\vec{r},t) = \vec{E}^e(\vec{r},t) + \mu_0 \int_{-\infty}^{\infty} \overleftrightarrow{G}_0(\vec{r},\vec{r}\,',t-t') \cdot \frac{\partial \overleftrightarrow{\sigma}^e(\vec{r}\,',\vec{r}\,'',t',t'')}{\partial t'} \cdot \vec{E}^e(\vec{r}\,'',t'') d^3r'' d^3r' dt'' dt'. \tag{54}$$

In passing, I stress that $\overleftrightarrow{G}_0(\vec{r},\vec{r}\,',t,t')$ is different from $\overleftrightarrow{D}_0(\vec{r},\vec{r}\,',t,t')$. The difference originates in the fact that the field point $\vec{r}$ may be located *inside* the charge distribution of our medium, so that an extra so-called transverse self-field term has to be added to the vacuum propagator $\overleftrightarrow{D}_0$. For a fully account of this aspect the reader is referred to Ref. 9. The field-induced current density written in the two alternative forms (Eqs. (44) and (50)) gives

$$\int_{-\infty}^{\infty} \overleftrightarrow{\sigma}(\vec{r},\vec{r}\,',t,t') \cdot \vec{E}(\vec{r}\,',t') d^3r' dt' = \int_{-\infty}^{\infty} \overleftrightarrow{\sigma}^e(\vec{r},\vec{r}\,'t,t') \cdot \vec{E}^e(\vec{r}\,',t') d^3r' dt'. \tag{55}$$

By insertion of the expression given in Eq. (54) for $\vec{E}(\vec{r},t)$ into Eq. (55) one obtains

$$\int_{-\infty}^{\infty} \left[\overleftrightarrow{\sigma}(\vec{r},\vec{r}\,',t,t') - \overleftrightarrow{\sigma}^e(\vec{r},\vec{r}\,',t,t')\right] \cdot \vec{E}^e(\vec{r}\,',t') d^3r' dt' = -\mu_0 \int \overleftrightarrow{\sigma}(\vec{r},\vec{r}\,',t,t') \cdot \overleftrightarrow{G}_0(\vec{r}\,'\vec{r}\,'',t',t'') \cdot \frac{\partial \overleftrightarrow{\sigma}^e(\vec{r}\,'',\vec{r}\,''',t'',t''')}{\partial t''} \cdot \vec{E}^e(\vec{r}\,''',t''') \times d^3r''' d^3r'' d^3r' dt''' dt'' dt'. \tag{56}$$

By interchanging the dummy variables $\vec{r}''' \leftrightarrow \vec{r}'$ and $t''' \leftrightarrow t'$, and successively the order of integrations on the right hand side of Eq. (56) we get

$$\int_{-\infty}^{\infty} [\vec{\sigma}(\vec{r},\vec{r}',t,t') - \vec{\sigma}^e(\vec{r},\vec{r}',t,t')] \cdot \vec{E}^e(\vec{r}',t')d^3r'dt' =$$
$$-\mu_0 \int_{-\infty}^{\infty} \vec{\sigma}(\vec{r},\vec{r}''',t,t''') \cdot \vec{G}_0(\vec{r}''',\vec{r}'',t''',t'') \cdot \frac{\partial \vec{\sigma}^e(\vec{r}'',\vec{r}',t'',t')}{\partial t''} \cdot \vec{E}^e(\vec{r}',t')$$
$$\times d^3r'''d^3r''d^3r'dt'''dt''dt'. \tag{57}$$

Since this equation must hold for an arbitrary external field distribution, $\vec{E}^e(\vec{r}',t')$, it is realized that the selfconsistent and external conductivity tensors are related by the integral equation

$$\vec{\sigma}(\vec{r},\vec{r}',t,t') = \vec{\sigma}^e(\vec{r},\vec{r}',t,t')$$
$$-\mu_0 \int_{-\infty}^{\infty} \vec{\sigma}(\vec{r},\vec{r}''',t,t''') \cdot \vec{G}_0(\vec{r}''',\vec{r}'',t''',t'') \cdot \frac{\partial \vec{\sigma}^e(\vec{r}'',\vec{r}',t'',t')}{\partial t''} d^3r'''d^3r''dt'''dt''. \tag{58}$$

As one would have anticipated, it has turned out that the interconnection between the response tensors $\vec{\sigma}$ and $\vec{\sigma}^e$ is a complicated one. However, for systems exhibiting translational invariance in time and space it can be shown that the relation between the external and selfconsistent conductivity tensors becomes a simple algebraic one in the wave vector-frequency domain (the Fourier domain of space-time).

2.7. DIELECTRIC-PERMITTIVITY AND MAGNETIC-PERMEABILITY FUNCTIONS

In studies of the electrodynamic response of condensed matter, use is often made of the polarization and magnetization vectors, denoted by $\vec{P}$ and $\vec{M}$, respectively, see e.g. Ref. 10. These vectors are introduced in an indirect way by a split of the field induced current density $\vec{J}^i(\vec{r},t)$ into two parts, i.e.

$$\vec{J}^i(\vec{r},t) = \frac{\partial \vec{P}(\vec{r},t)}{\partial t} + \vec{\nabla} \times \vec{M}(\vec{r},t), \tag{59}$$

where $\partial\vec{P}/\partial t$ and $\vec{\nabla}\times\vec{M}$ are the so-called polarization and magnetization current densities. At once, one should notice that the division in Eq. (59) does not define $\vec{P}$ and $\vec{M}$ uniquely. Thus, if $\vec{N}(\vec{r},t)$ is an arbitrary vector, differentiable in time and space, it is readily realized that a replacement of $\vec{P}$ and $\vec{M}$ in Eq. (59) by the new vectors $\vec{P}'$ and $\vec{M}'$ given by

$$\vec{P}'(\vec{r},t) = \vec{P}(\vec{r},t) - \vec{\nabla} \times \vec{N}(\vec{r},t), \tag{60}$$

and

$$\vec{M}'(\vec{r},t) = \vec{M}(\vec{r},t) + \frac{\partial \vec{N}(\vec{r},t)}{\partial t} \tag{61}$$

does not alter the induced current density. Since, from a microscopic point of view, basically only $\vec{J}^i(\vec{r},t)$ is a physically meaningful quantity, the "weak" definitions of the polarization and magnetization vectors involve a certain arbitrariness. Part of this arbitrariness can be removed noting that $\vec{\nabla}\cdot\vec{P}' = \vec{\nabla}\cdot\vec{P}$ since $\vec{\nabla}\cdot\vec{\nabla}\times\vec{N} = 0$. Thus, since the divergence of the polarization vector is an invariant quantity we fix it by the *definition*

$$\rho^i(\vec{r},t) \equiv -\vec{\nabla} \cdot \vec{P}(\vec{r},t), \tag{62}$$

where $\rho^i(\vec{r},t)$ is the field-induced charge density of the medium under consideration, cf. section 2.2.

Let us analyse the consequences of introducing the polarization and magnetization vectors via the relations in Eqs. (59) and (62) a bit further. By taking the divergence on both sides of Eq. (59), using in turn Eq. (62) and the fact that $\vec{\nabla}\cdot\vec{\nabla}\times\vec{M}=0$, one obtains

$$\vec{\nabla}\cdot\vec{J}^i(\vec{r},t)+\frac{\partial\rho^i(\vec{r},t)}{\partial t}=0. \tag{63}$$

This is just the equation of continuity for the induced electric charge. Such an equation necessarily must exist in a system where the external charges can be distinguished from the induced ones, cf. the discussion in section 2.2. Further insight is gained by dividing each of the vector fields into their divergence-free (subscript T) and irrotational (subscript L) parts, i.e. in generic form

$$\vec{W}(\vec{r},t)=\vec{W}_{\mathrm{T}}(\vec{r},t)+\vec{W}_{\mathrm{L}}(\vec{r},t), \tag{64}$$

where

$$\vec{\nabla}\cdot\vec{W}_{\mathrm{T}}(\vec{r},t)=0, \tag{65}$$

and

$$\vec{\nabla}\times\vec{W}_{\mathrm{L}}(\vec{r},t)=\vec{0}. \tag{66}$$

Using the Helmholtz theorem, which states that a vector field $\vec{E}$ can be completely specified in terms of its divergence $\vec{\nabla}\cdot\vec{W}=\vec{\nabla}\cdot\vec{W}_{\mathrm{L}}$ and curl $\vec{\nabla}\times\vec{W}=\vec{\nabla}\times\vec{W}_{\mathrm{T}}$, it is realized that the arbitrariness of the polarization vector is attached to the divergence-free part, $\vec{P}_{\mathrm{T}}$, of $\vec{P}$ only, since the irrotational part, $\vec{P}_{\mathrm{L}}$, is fixed by

$$\rho^i(\vec{r},t)=-\vec{\nabla}\cdot\vec{P}_{\mathrm{L}}(\vec{r},t). \tag{67}$$

A division of Eq. (59) into its L and T parts gives

$$\vec{J}^i_{\mathrm{L}}(\vec{r},t) = \frac{\partial\vec{P}_{\mathrm{L}}(\vec{r},t)}{\partial t}, \tag{68}$$

$$\vec{J}^i_{\mathrm{T}}(\vec{r},t) = \frac{\partial\vec{P}_{\mathrm{T}}(\vec{r},t)}{\partial t}+\vec{\nabla}\times\vec{M}_{\mathrm{T}}(\vec{r},t), \tag{69}$$

since $\vec{\nabla}\times\vec{M}_{\mathrm{L}}=\vec{0}$. The relation in Eq. (69) emphasizes the fact that only the transverse part of $\vec{M}$ appears in the description. The arbitrariness in our definition of the polarization and magnetization vectors is completely attached to $\vec{P}_{\mathrm{T}}$ and $\vec{M}_{\mathrm{T}}$, the only restriction being the requirement that the divergence-free current density in Eq. (69) has to be invariant.

Let us assume we have made a choice for $\vec{P}_{\mathrm{T}}$ and $\vec{M}_{\mathrm{T}}$. For this choice we *define* a new field $\vec{D}$ called the electric displacement (or simply the $\vec{D}$-field) via

$$\vec{D}(\vec{r},t)=\varepsilon_0\vec{E}(\vec{r},t)+\vec{P}(\vec{r},t). \tag{70}$$

Also a new magnetic field $\vec{H}$ (the $\vec{H}$-field) *defined* by the relation

$$\vec{H}(\vec{r},t)=\frac{1}{\mu_0}\vec{B}(\vec{r},t)-\vec{M}(\vec{r},t). \tag{71}$$

is introduced. With the aid of these "new" fields the induced current and charge densities can be written as follows:

$$\vec{J}^i(\vec{r},t) = \frac{\partial \vec{D}(\vec{r},t)}{\partial t} - \vec{\nabla} \times \vec{H}(\vec{r},t) + \frac{1}{\mu_0}\vec{\nabla} \times \vec{B}(\vec{r},t) - \varepsilon_0 \frac{\partial \vec{E}(\vec{r},t)}{\partial t}, \tag{72}$$

and

$$\rho^i(\vec{r},t) = -\vec{\nabla} \cdot \vec{D}(\vec{r},t) + \varepsilon_0 \vec{\nabla} \cdot \vec{E}(\vec{r},t). \tag{73}$$

By inserting the above-given expressions for $\vec{J}^i$ and ρ^i into the two material-dependent Maxwell-Lorentz equations ((41) and (42)), the system of Maxwell equations takes the form

$$\vec{\nabla} \times \vec{E}(\vec{r},t) = -\frac{\partial \vec{B}(\vec{r},t)}{\partial t}, \tag{74}$$

$$\vec{\nabla} \times \vec{H}(\vec{r},t) = \vec{J}^e(\vec{r},t) + \frac{\partial \vec{D}(\vec{r},t)}{\partial t}, \tag{75}$$

$$\vec{\nabla} \cdot \vec{D}(\vec{r},t) = \rho^e(\vec{r},t), \tag{76}$$

$$\vec{\nabla} \cdot \vec{B}(\vec{r},t) = 0. \tag{77}$$

It is important to emphasize here that although the set of Maxwell equations in (74)–(77) appear to be identical to those used in every textbook on the macroscopic electrodynamic properties of condensed matter, in fact, they are radically different for three reasons. Firstly, we do not define $\vec{P}$ and $\vec{M}$ as the electric and magnetic dipole moments per unit volume. In a multipole description of the electrodynamic response of a system of charges $\vec{P}$ and $\vec{M}$ contain also higher-order moments than the dipole moments. Keeping only the electric and magnetic dipoles, the description of the electrodynamics of condensed matter becomes incomplete. Secondly, since no spatial averaging is performed in Eqs. (74)–(77) these equations are still microscopic ones. Thirdly, the possibility for different choices of $\vec{P}_{\mathrm{T}}$ and $\vec{M}_{\mathrm{T}}$ is still left open in the present formulation. In many (mainly European) textbooks the "external" charge and current density distributions are identified as those belonging to the "free" carriers (conduction electrons, mobile charges, etc.) of the system under study. To the extent that the distinction between so-called free carries and the rest of the charges is a reasonable one, which unfortunately is often not the case in optics, the textbook case mentioned above can of course be covered as a special case of Eqs. (74)–(77).

In order to close the system of Maxwell equations in (74)–(77) one introduces the following two linear and nonlocal material relations:

$$\vec{D}(\vec{r},t) = \varepsilon_0 \int_{-\infty}^{\infty} \overleftrightarrow{\varepsilon}(\vec{r},\vec{r}',t,t') \cdot \vec{E}(\vec{r},t') d^3r' dt', \tag{78}$$

and

$$\vec{B}(\vec{r},t) = \mu_0 \int_{-\infty}^{\infty} \overleftrightarrow{\mu}(\vec{r},\vec{r}',t,t') \cdot \vec{H}(\vec{r}',t') d^3r' dt'. \tag{79}$$

The second-rank tensor quantities $\overleftrightarrow{\varepsilon}(\vec{r},\vec{r}',t,t')$ and $\overleftrightarrow{\mu}(\vec{r},\vec{r}',t,t')$ are called the relative dielectric (permittivity) function and the relative magnetic permeability function, respectively. For brevity, we shall in the following as is often done leave out the word relative when referring to $\overleftrightarrow{\varepsilon}$ and $\overleftrightarrow{\mu}$. According to the preceding discussion concerning the arbitrariness in the definition of $\vec{P}_{\mathrm{T}}$ and $\vec{M}_{\mathrm{T}}$ it follows that the dielectric and magnetic permeability functions in general have no unique physical meaning. This conceptually important statement is often overlooked in textbooks. The basic ambiguity inherent in the definitions of the dielectric and magnetic permeability functions does not in general make every possible choice of $\overleftrightarrow{\varepsilon}$ and $\overleftrightarrow{\mu}$ equally practical in a given calculation, of course.

In passing one should note that since no physical significance can be attached in any manner to the irrotational part of $\vec{M}$, also $\vec{H}_L$ is an unphysical quantity. This conclusion is in agreement with the fact that only $\vec{\nabla} \times \vec{H} = \vec{\nabla} \times \vec{H}_T$ appears in the Maxwell equation (75). Contrarily, a definite physical meaning is given to the irrotational part of the $\vec{D}$-field since $\vec{\nabla} \cdot \vec{D}_L = \vec{\nabla} \cdot \vec{D} = \rho^e$, cf. Eq. (76).

2.8. ELECTRODYNAMIC RESPONSE THEORY WITH THE CHOICE $\overleftrightarrow{\mu}(\vec{r}, \vec{r}', t, t') = \delta(\vec{r} - \vec{r}')\delta(t - t')\overleftrightarrow{U}$

In the previous section the polarization and magnetization vectors $\vec{P}$ and $\vec{M}$ were introduced to describe the linear and nonlocal electrodynamic response of a system of charged particles. On the basis of the equations (59) and (62) relating $\vec{P}$ and $\vec{M}$ to the induced current and charge densities it was argued that different choices, leading to identical physical results, were possible for $\vec{P}$ and $\vec{M}$. In the following a choice frequently encountered in electrodynamic studies of condensed media is discussed. Thus, let us put the divergence-free part, $\vec{M}_T$, of the magnetization vector $\vec{M}$ to zero (the irrotational part of $\vec{M}$ is irrelevant, cf. Eq. (59)). This implies

$$\vec{J}^i(\vec{r}, t) = \frac{\partial \vec{P}(\vec{r}, t)}{\partial t}, \tag{80}$$

and consequently one finds by differentiation of Eq. (70) with respect to time

$$\frac{\partial \vec{D}(\vec{r}, t)}{\partial t} = \varepsilon_0 \frac{\partial \vec{E}(\vec{r}, t)}{\partial t} + \vec{J}^i(\vec{r}, t). \tag{81}$$

With the choice in Eq. (80) the Maxwell equations still retain the form given in Eqs. (74)–(77).

Because the irrotational part of $\vec{M}$ does not enter the theory at all one may as well let it be zero from the outset. Doing this Eq. (71) becomes

$$\vec{B}(\vec{r}, t) = \mu_0 \vec{H}(\vec{r}, t). \tag{82}$$

By comparing this relation to Eq. (79), it is realized that

$$\overleftrightarrow{\mu}(\vec{r}, \vec{r}', t, t') = \delta(\vec{r} - \vec{r}')\delta(t - t')\overleftrightarrow{U} \tag{83}$$

for the present choice, $\vec{M} = \vec{0}$. Even though the magnetic permeability is "unity", it should be emphasized that the approach adopted in this section does not imply at all that the induced magnetic properties of the medium are neglected. These properties, as well as the electric properties, simply are included in a new dielectric function $\overleftrightarrow{\varepsilon}(\vec{r}, \vec{r}', t, t')$ defined in terms of the material relation

$$\vec{D}(\vec{r}, t) = \varepsilon_0 \int_{-\infty}^{\infty} \overleftrightarrow{\varepsilon}(\vec{r}, \vec{r}', t, t') \cdot \vec{E}(\vec{r}', t') d^3r' dt'. \tag{84}$$

By inserting Eq. (84) into Eq. (81) it appears that the induced current density can be written as follows:

$$\vec{J}^i(\vec{r},t) = \varepsilon_0 \int_{-\infty}^{\infty} \left[\frac{\partial \overleftrightarrow{\varepsilon}(\vec{r},\vec{r}',t,t')}{\partial t} - \delta(\vec{r}-\vec{r}')\frac{\partial \delta(t-t')}{\partial t} \right] \cdot \vec{E}(\vec{r}',t') d^3r' dt'. \tag{85}$$

By comparing this equation to Eq. (44) it is seen that there exists a simple relationship between the selfconsistent conductivity response tensor and the dielectric tensor $\overleftrightarrow{\varepsilon}(\vec{r},\vec{r}',t,t')$, viz.

$$\overleftrightarrow{\sigma}(\vec{r},\vec{r}',t,t') = \varepsilon_0 \left(\frac{\partial \overleftrightarrow{\varepsilon}(\vec{r},\vec{r}',t,t')}{\partial t} - \delta(\vec{r}-\vec{r}')\frac{\partial \delta(t-t')}{\partial t} \overleftrightarrow{U} \right). \tag{86}$$

For systems exhibiting translational invariance in time, a time-Fourier transformed and spatially local version of the relation in Eq. (86) is often used in textbooks in an uncritical manner to express the "conductivity" in terms of the "dielectric function". It appears from the preceding analysis that for such a relation to hold microscopically precise (specific) definitions of the conductivity and dielectric tensors are needed.

2.9. ELECTROMAGNETIC RESPONSE FORMALISM BASED ON THE CHOICE $\vec{D}(\vec{r},t) = \varepsilon_0 \vec{E}^e(\vec{r},t)$ and $\vec{H}(\vec{r},t) = \mu_0^{-1} \vec{B}^e(\vec{r},t)$

In section 2.5, the external conductivity response function $\overleftrightarrow{\sigma}^e(\vec{r},\vec{r}',t,t')$ was introduced and it was pointed out that $\overleftrightarrow{\sigma}^e$ is a true response function in the sense that it relates the external field (stimulus), $\vec{E}^e$, in a causal manner to the induced current density (response), $\vec{J}^i$. In this section we shall define the dielectric function $\overleftrightarrow{\varepsilon}(\vec{r},\vec{r}',t,t')$ and the magnetic permeability $\overleftrightarrow{\mu}(\vec{r},\vec{r}',t,t')$ in such a way that the so-called inverse dielectric function, denoted by $\overleftrightarrow{\varepsilon}^{-1}(\vec{r},\vec{r}',t,t')$ and to be defined below, and $\overleftrightarrow{\mu}(\vec{r},\vec{r}',t,t')$ are causal response functions.

We begin by considering the $\vec{D}$-field. From Eq. (76) one has

$$\vec{\nabla} \cdot \vec{D}_{\rm L}(\vec{r},t) = \rho^e(\vec{r},t), \tag{87}$$

since $\vec{\nabla} \cdot \vec{D}_{\rm T} = 0$. If one compares Eq. (87) to the relation

$$\varepsilon_0 \vec{\nabla} \cdot \vec{E}^e_{\rm L}(\vec{r},t) = \rho^e(\vec{r},t), \tag{88}$$

obtained from Eq. (10) using that $\vec{\nabla} \cdot \vec{E}^e_{\rm T} = 0$, it is realized that the irrotational part of the $\vec{D}$-field is given by

$$\vec{D}_{\rm L}(\vec{r},t) = \varepsilon_0 \vec{E}^e_{\rm L}(\vec{r},t). \tag{89}$$

In turn, it appears from this equation that if the irrotational part of the external field is prescribed, the irrotational part of $\vec{D}$ is completely fixed. Now, in the view of Eq. (89) let us investigate whether it is possible to make the choice

$$\vec{D}_{\rm T}(\vec{r},t) = \varepsilon_0 \vec{E}^e_{\rm T}(\vec{r},t) \tag{90}$$

for the divergence-free part of the electric displacement field, and in case, let us analyse the consequences of this choice. First of all, however, let me emphasize that combining Eqs. (89) and (90) one has

$$\vec{D}(\vec{r},t) = \varepsilon_0 \vec{E}^e(\vec{r},t), \tag{91}$$

a relation which apart from a trivial constant (ε_0) identifies the $\vec{D}$-field with the external field in the present case.

With the assumption that the $\vec{D}$-field can be given by Eq. (91) it follows that the polarization vector is completely specified, viz.

$$\vec{P}(\vec{r},t) = \varepsilon_0 \left(\vec{E}^e(\vec{r},t) - \vec{E}(\vec{r},t)\right). \tag{92}$$

We knew beforehand of course that the irrotational part of $\vec{P}$ was fixed by the induced charge density ρ^i, cf. Eq. (67). Thus, from Eq. (92) one obtains immediately

$$\vec{\nabla} \cdot \vec{P}_L(\vec{r},t) = \varepsilon_0 \left(\vec{\nabla} \cdot \vec{E}^e(\vec{r},t) - \vec{\nabla} \cdot \vec{E}(\vec{r},t)\right). \tag{93}$$

Inserting on the right hand side of this equation the expression for $\vec{\nabla} \cdot \vec{E}^e$ and $\vec{\nabla} \cdot \vec{E}$ taken from Eqs. (10) and (42) one regains Eq. (67). It was realized on the basis of the discussion in section 2.7 that the arbitrariness in the definition of the polarization and magnetization vectors is attached to $\vec{P}_T$ and $\vec{M}_T$ only. Since the choice for $\vec{D}_T$ given in Eq. (90) fixes $\vec{P}_T$, as already emphasised in Eq. (92), also $\vec{M}_T$ is now fixed. In implicit form $\vec{M}_T$ is given via

$$\vec{\nabla} \times \vec{M}_T(\vec{r},t) = \vec{J}_T^i(\vec{r},t) - \varepsilon_0 \frac{\partial \vec{E}_T^e(\vec{r},t)}{\partial t} + \varepsilon_0 \frac{\partial \vec{E}_T(\vec{r},t)}{\partial t}, \tag{94}$$

as one realizes by combining Eq. (69) and the divergence-free part of Eq. (92). The unique expression for $\vec{\nabla} \times \vec{M}_T$, in turn, allows us to establish the enforced formula for the divergence-free part of the $\vec{H}$-field. Hence, from the definition in Eq. (71) one obtains

$$\vec{\nabla} \times \vec{M}_T(\vec{r},t) = \frac{1}{\mu_0} \vec{\nabla} \times \vec{B}_T(\vec{r},t) - \vec{\nabla} \times \vec{H}_T(\vec{r},t). \tag{95}$$

Next, by combination of Eqs. (94) and (95) and use of the expression $\vec{J}_T = \vec{J}_T^e + \vec{J}_T^i$ (see Eq. (17)) one finds

$$\vec{\nabla} \times \vec{H}_T(\vec{r},t) - \vec{J}_T^e(\vec{r},t) - \varepsilon_0 \frac{\partial \vec{E}_T^e(\vec{r},t)}{\partial t} = \frac{1}{\mu_0} \vec{\nabla} \times \vec{B}_T(\vec{r},t) - \vec{J}_T(\vec{r},t) - \varepsilon_0 \frac{\partial \vec{E}_T(\vec{r},t)}{\partial t}. \tag{96}$$

On the basis of the divergence-free part of the Maxwell-Lorentz equation in (2) it is realized that the right hand side of Eq. (96) vanishes. This implies that

$$\vec{\nabla} \times \vec{H}_T(\vec{r},t) = \vec{J}_T^e(\vec{r},t) + \varepsilon_0 \frac{\partial \vec{E}_T^e(\vec{r},t)}{\partial t}. \tag{97}$$

If this equation is compared with the divergence-free part of the Maxwell-Lorentz equation in (9) for the external fields, it is realized that

$$\vec{H}_T(\vec{r},t) = \frac{1}{\mu_0} \vec{B}_T^e(\vec{r},t). \tag{98}$$

Since, as we already know, the irrotational part of the $\vec{H}$-field does not enter the theory one can of course chose $\vec{H}_L$ freely. Thus, with the choice

$$\vec{H}_L(\vec{r},t) = \frac{1}{\mu_0} \vec{B}_L^e(\vec{r},t) \tag{99}$$

one ends up with the relation

$$\vec{H}(\vec{r},t) = \frac{1}{\mu_0}\vec{B}^e(\vec{r},t). \tag{100}$$

In conclusion, it has been demonstrated that (i) the choice $\vec{D} = \varepsilon_0\vec{E}^e_T$ (Eq. (90)) *is* possible, and (ii) the auxiliary fields $\vec{D}$ and $\vec{H}$, apart from trivial constants, can be identified with the external electric and magnetic fields (Eqs. (91) and (100)).

Before proceeding our study of the electromagnetic response formalism based on the choice $\vec{D} = \varepsilon_0\vec{E}^e$, let us indicate an alternative and shorter route (but maybe less persuasive) for arriving at Eq. (98) from Eq. (90). Hence, by inserting Eq. (91) into Eq. (75) one obtains

$$\vec{\nabla} \times \vec{H}(\vec{r},t) = \vec{J}^e(\vec{r},t) + \varepsilon_0\frac{\partial \vec{E}^e(\vec{r},t)}{\partial t}. \tag{101}$$

By comparing this equation with the Maxwell-Lorentz equation (9) for the external fields it is readily realized that

$$\vec{\nabla} \times \vec{H}(\vec{r},t) = \frac{1}{\mu_0}\vec{\nabla} \times \vec{B}^e(\vec{r},t). \tag{102}$$

This equation immediately leads to the relation in Eq. (98), and hence with the choice in Eq. (99) to Eq. (100), q.e.d.

On the basis of the choices in Eqs. (91) and (100) the material relations in Eqs. (78) and (79) take the form

$$\vec{E}^e(\vec{r},t) = \int_{-\infty}^{\infty} \overleftrightarrow{\varepsilon}^e(\vec{r},\vec{r}',t,t')\vec{E}(\vec{r}',t')d^3r'dt', \tag{103}$$

and

$$\vec{B}(\vec{r},t) = \int_{-\infty}^{\infty} \overleftrightarrow{\mu}^e(\vec{r},\vec{r}',t,t')\vec{B}^e(\vec{r}',t')d^3r'dt'. \tag{104}$$

In the two equations above we have added a superscript e to the new response functions to indicate that these relate the selfconsistent (local) electric and magnetic fields to the corresponding external ones. As for the relation between the induced current density and the external electric field (Eq. (50)) there must necessarily exist a causal relationship between the electric fields in Eq. (103), and between the magnetic fields in Eq. (104). The causal response function for the magnetic part is $\overleftrightarrow{\mu}^e(\vec{r},\vec{r}',t,t')$. For the electric part, the relevant causal response tensor in principle can be obtained by solving the integral equation (103) with respect to the local field. Thus, in a formal sense

$$\vec{E}(\vec{r},t) = \int_{-\infty}^{\infty} \overleftrightarrow{\varepsilon}_e^{-1}(\vec{r},\vec{r}',t,t') \cdot \vec{E}^e(\vec{r}',t')d^3r'dt', \tag{105}$$

where $\overleftrightarrow{\varepsilon}_e^{-1}$ is the "name" of the causal response function. An integral relation between $\overleftrightarrow{\varepsilon}^e$ and $\overleftrightarrow{\varepsilon}_e^{-1}$ can be obtained by (i) inserting Eq. (105) into Eq. (103), i.e.

$$\vec{E}^e(\vec{r},t) = \int_{-\infty}^{\infty} \overleftrightarrow{\varepsilon}^e(\vec{r},\vec{r}',t,t') \cdot \overleftrightarrow{\varepsilon}_e^{-1}(\vec{r}',\vec{r}'',t',t'') \cdot \vec{E}^e(\vec{r}'',t'')d^3r'd^3r''dt'dt'', \tag{106}$$

and (ii) making use of the identity

$$\vec{E}^e(\vec{r},t) = \int_{-\infty}^{\infty} \delta(\vec{r}-\vec{r}'')\delta(t-t'')\overleftrightarrow{U} \cdot \vec{E}^e(\vec{r}'',t'')d^3r''dt''. \tag{107}$$

Thus, since Eqs. (106) and (107) must hold for arbitrary external field distributions one obtains

$$\int_{-\infty}^{\infty} \overleftrightarrow{\varepsilon}^{e}(\vec{r},\vec{r}'',t,t'') \cdot \overleftrightarrow{\varepsilon}_{e}^{-1}(\vec{r}'',\vec{r}',t'',t')d^3r''dt'' = \delta(\vec{r}-\vec{r}')\delta(t-t')\overleftrightarrow{U}. \quad (108)$$

Because $\overleftrightarrow{\varepsilon}_{e}^{-1}$ and $\overleftrightarrow{\mu}^{e}$ play the role as causal response function it is of interest to establish the connection between these tensors and the external conductivity $\overleftrightarrow{\sigma}^{e}$, which itself is a causal response function. Here, we shall limit ourselves to a presentation of the relation between $\overleftrightarrow{\varepsilon}_{e}^{-1}$ and $\overleftrightarrow{\sigma}^{e}$. This relation may adequately be derived from Eqs. (54) and (105). Thus, by rewriting Eq. (54) as follows:

$$\vec{E}(\vec{r},t) = \int_{-\infty}^{\infty}\Big[\overleftrightarrow{U}\delta(\vec{r},-\vec{r}'')\delta(t-t'') + \mu_0 \cdot \int_{-\infty}^{\infty} \overleftrightarrow{G}_0(\vec{r},\vec{r}',t,t') \cdot \frac{\partial\overleftrightarrow{\sigma}^{e}(\vec{r}',\vec{r}'',t't'')}{\partial t'}d^3r'dt'\Big] \cdot \vec{E}^{e}(\vec{r}'',t'')d^3r''dt'', \quad (109)$$

one finds immediately upon comparison with Eq. (105) the result

$$\overleftrightarrow{\varepsilon}_{e}^{-1}(\vec{r},\vec{r}',t,t') = \overleftrightarrow{U}\delta(\vec{r}-\vec{r}')\delta(t-t') + \mu_0\int_{-\infty}^{\infty}\overleftrightarrow{G}_0(\vec{r},\vec{r}'',t,t'') \cdot \frac{\partial\overleftrightarrow{\sigma}^{e}(\vec{r}'',\vec{r}',t'',t')}{\partial t''}d^3r''dt''. \quad (110)$$

In order to compare with the relationship between $\overleftrightarrow{\varepsilon}$ and $\overleftrightarrow{\sigma}$ given in Eq. (86), we shall derive also an equation connecting $\overleftrightarrow{\varepsilon}^{e}$ and $\overleftrightarrow{\sigma}$. To this end one combines Eqs. (20) and (46) to eliminate the term containing the external current density, i.e.

$$\left[\vec{\nabla}\vec{\nabla} - \overleftrightarrow{U}\left(\nabla^2 - \frac{1}{c_0^2}\frac{\partial^2}{\partial t^2}\right)\right] \cdot (\vec{E}(\vec{r},t) - \vec{E}^{e}(\vec{r},t)) = -\mu_0\frac{\partial}{\partial t}\int_{-\infty}^{\infty} \overleftrightarrow{\sigma}(\vec{r},\vec{r}',t,t') \cdot \vec{E}(\vec{r}',t')d^3r'dt'. \quad (111)$$

By means of the free-space propagator $\overleftrightarrow{G}_0(\vec{r},\vec{r}',t,t')$, Eq. (111) can be converted into the integral form

$$\vec{E}^{e}(\vec{r},t) = \vec{E}(\vec{r},t) - \mu_0\int_{-\infty}^{\infty}\overleftrightarrow{G}_0(\vec{r},\vec{r}'',t,t'') \cdot \frac{\partial\overleftrightarrow{\sigma}(\vec{r}'',\vec{r}',t'',t')}{\partial t''} \cdot \vec{E}(\vec{r}',t')d^3r''d^3r'dt''dt'. \quad (112)$$

By a comparison of this equation to Eq. (103) it is realized that the relation we are seeking has the form

$$\overleftrightarrow{\varepsilon}^{e}(\vec{r},\vec{r}',t,t') = \overleftrightarrow{U}\delta(\vec{r}-\vec{r}')\delta(t-t') - \mu_0\int_{-\infty}^{\infty}\overleftrightarrow{G}_0(\vec{r},\vec{r}'',t,t'') \cdot \frac{\partial\overleftrightarrow{\sigma}(\vec{r}'',\vec{r}',t'',t')}{\partial t''}d^3r''dt''. \quad (113)$$

The relationship between the external dielectric function $\overleftrightarrow{\varepsilon}^{e}$ and the selfconsistent conductivity response function $\overleftrightarrow{\sigma}$ is considerably more complicated than the one between the dielectric function $\overleftrightarrow{\varepsilon}$, introduced in Eq. (84), and $\overleftrightarrow{\sigma}$ (compare Eqs. (86) and (113)). This is not surprising after all since the relation between $\overleftrightarrow{\varepsilon}^{e}$ and $\overleftrightarrow{\sigma}$ necessarily must involve (via $\overleftrightarrow{G}_0$) the propagation characteristics of the induced field. Readers interested in further aspects of the conductivity response formalism may consult the excellent description given in Ref. 11. The results obtained in sections 2.4–2.9 are not new but the analysis deviate in part from previous ones.

3. Integral relation between the local field and induced current density

3.1. TRANSVERSE AND LONGITUDINAL ELECTRODYNAMICS

Let us now consider the electrodynamics inside our medium more closely. Starting from Eq. (45), with $\vec{J}^e(\vec{r},t) = 0$ in accordance with Eq. (15), one obtains after a Fourier transformation in time the following wave equation for the local field in the frequency-space domain:

$$\vec{\nabla} \times \left(\vec{\nabla} \times \vec{E}(\vec{r};\omega)\right) - \left(\frac{\omega}{c_0}\right)^2 \vec{E}(\vec{r};\omega) = i\mu_0\omega\vec{J}^i(\vec{r},t). \tag{114}$$

To proceed from here one divides the field and current density into their divergence-free (T) and rotational-free (L) parts, i.e.

$$\vec{E}(\vec{r};\omega) = \vec{E}_T(\vec{r};\omega) + \vec{E}_L(\vec{r};\omega), \tag{115}$$

and

$$\vec{J}^i(\vec{r};\omega) = \vec{J}^i_T(\vec{r};\omega) + \vec{J}^i_L(\vec{r};\omega), \tag{116}$$

where $\vec{\nabla}\cdot\vec{E}_T = \vec{\nabla}\cdot\vec{J}^i_T = 0$ and $\vec{\nabla}\times\vec{E}_L = \vec{\nabla}\times\vec{J}^i_L = \vec{0}$. By inserting Eqs. (115) and (116) into Eq. (114) this splits into two, viz.

$$(\nabla^2 + q_0^2)\vec{E}_T(\vec{r};\omega) = -i\mu_0\omega\vec{J}^i_T(\vec{r};\omega), \tag{117}$$

and

$$q_0^2\vec{E}_L(\vec{r};\omega) = -i\mu_0\omega\vec{J}^i_L(\vec{r};\omega), \tag{118}$$

where $q_0 = \omega/c_0$ is the vacuum wave number of light.

It is adequate at this point to turn Eq. (119) into an integral relation between the transverse parts of the local field and induced current density. Before doing so let me however emphasize an important aspect concerning the domain of integration. Hence, due to the fact that the relation between the transverse part of the current density and the current density is a nonlocal one, viz.

$$\vec{J}^i_T(\vec{r};\omega) = \vec{\nabla}\times\left[\vec{\nabla}\times\int_V \frac{\vec{J}^i(\vec{r}\,';\omega)d^3r'}{4\pi|\vec{r}-\vec{r}\,'|}\right], \tag{119}$$

the spatial domain, V_T, occupied by the transverse current density does not coincide completely with the domain, V, in which the current density itself is (effectively) different from zero. The relation between $\vec{J}^i_L$ and $\vec{J}^i$ also is nonlocal, namely

$$\vec{J}^i_L(\vec{r};\omega) = -\vec{\nabla}\left[\vec{\nabla}\cdot\int_V \frac{\vec{J}^i(\vec{r}\,';\omega)d^3r'}{4\pi|\vec{r}-\vec{r}\,'|}\right], \tag{120}$$

and since $\vec{J}^i_T(\vec{r};\omega) + \vec{J}^i_L(\vec{r};\omega) = \vec{J}^i(\vec{r};\omega)$ it appears that in those regions of space outside V where $\vec{J}^i_T$ is different from zero also $\vec{J}^i_L$ is nonvanishing. In turn this means that there is no need of distinguishing between the domains V_L occupied by the longitudinal and transverse current densities, respectively. In a formal sense the equations in (119) and (120) can be written as follows:

$$\vec{J}^i_T(\vec{r};\omega) = \int_V \overleftrightarrow{\delta}_T(\vec{r}-\vec{r}\,')\cdot\vec{J}^i(\vec{r}\,';\omega)d^3r', \tag{121}$$

and

$$\vec{J}^{i}_{\rm L}(\vec{r};\omega) = \int_V \overleftrightarrow{\delta}_{\rm L}(\vec{r}-\vec{r}\,')\cdot\vec{J}^{i}(\vec{r}\,';\omega)d^3r', \tag{122}$$

where $\overleftrightarrow{\delta}_{\rm T}(\vec{r}-\vec{r}\,')$ and $\overleftrightarrow{\delta}_{\rm L}(\vec{r}-\vec{r}\,')$ are the so-called transverse and longitudinal delta functions (tensors).

As described in some detail in Ref. 9, it is possible to turn the differential equation in (117) into an integral relation between $\vec{E}_{\rm T}$ and $\vec{J}^{i}_{\rm T}$, viz.

$$\vec{E}_{\rm T}(\vec{r};\omega) = \vec{\Sigma}_{\rm T}(\vec{r};\omega) + \frac{1}{3i\varepsilon_0\omega}\vec{J}^{i}_{\rm T}(\vec{r};\omega) - i\mu_0\omega\int_{\varepsilon\to 0}^{V_{\rm T}} \overleftrightarrow{D}_0(\vec{r}-\vec{r}\,';\omega)\cdot\vec{J}^{i}_{\rm T}(\vec{r}\,';\omega)d^3r', \tag{123}$$

for observation points inside $V_{\rm T}$ ($\vec{r}\in V_{\rm T}$), where

$$\vec{\Sigma}_{\rm T}(\vec{r};\omega) = \int_{\Sigma_{\rm T}}\left[\vec{E}_{\rm T}(\vec{r}\,';\omega)\cdot\frac{\partial\overleftrightarrow{D}_0(\vec{r}-\vec{r}\,';\omega)}{\partial n'} - \overleftrightarrow{D}_0(\vec{r}-\vec{r}\,';\omega)\cdot\frac{\partial\vec{E}_{\rm T}(\vec{r}\,';\omega)}{\partial n'}\right]dS'. \tag{124}$$

The integration in Eq. (123) extends over the transverse current-density domain but an infinitesimal small spherical exclusion volume $\varepsilon(\vec{r})$ centered on the point $\vec{r}$ inside $V_{\rm T}$ at which we want to calculate $\vec{E}_{\rm T}(\vec{r},\omega)$ is cut out. In Eq. (124), $\Sigma_{\rm T}$ denotes the closed surface effectively bounding $V_{\rm T}$, and $\partial/\partial n' \equiv \vec{n}\,'\cdot\vec{\nabla}'$, where $\vec{n}\,'(\vec{r}\,')$ is the outwards directed unit normal from $V_{\rm T}$. On the basis of a microscopic Ewald-Oseen extinction theorem it is possible to identify $\vec{\Sigma}_{\rm T}(\vec{r};\omega)$ with the necessarily transverse source field originating in the external sources (see Ref. 9), i.e.

$$\vec{\Sigma}_{\rm T}(\vec{r};\omega) = \vec{E}^{e}_{\rm T}(\vec{r};\omega), \quad \vec{r}\in V_{\rm T} \tag{125}$$

cf. Eq. (39). In passing I note that in Eqs. (123) and (124) it is actually the *outgoing* free-space dyadic propagator $\overleftrightarrow{D}_0^{\rm OUT}(\vec{r}-\vec{r}\,';\omega)$ which enters, see Eq. (38). *In the following I omit for notational simplicity the superscript* 'OUT' *on the free-space Green's function.*

Formally, it is possible to rewrite Eq. (123) as an integral relation between the transverse part of the local field in the medium under consideration and the total induced current density, $\vec{J}^{i}(\vec{r};\omega)$. As discussed in Ref. 9 one finds

$$\vec{E}_{\rm T}(\vec{r};\omega) = \vec{E}^{e}_{\rm T}(\vec{r};\omega) - i\mu_0\omega\int_V\left[\overleftrightarrow{D}_0^{\rm T}(\vec{r}-\vec{r}\,';\omega) + \overleftrightarrow{g}_{\rm T}(\vec{r}-\vec{r}\,';\omega)\right]\cdot\vec{J}^{i}(\vec{r}\,';\omega)d^3r', \tag{126}$$

for $\vec{r}\in V_{\rm T}$, where $\overleftrightarrow{D}_0^{\rm T}(\vec{r}-\vec{r}\,';\omega)$ is the so-called *transverse* free-space propagator, and

$$\overleftrightarrow{g}_{\rm T}(\vec{r}-\vec{r}\,';\omega) = \frac{1}{3q_0^2}\overleftrightarrow{\delta}_{\rm T}(\vec{r}-\vec{r}\,') \tag{127}$$

is the transverse self-field propagator. Although the integration in Eq. (126) extends over the domain V where the induced current density itself is different from zero the integral relation has to be used whenever the observation point $\vec{r}$ is located inside $V_{\rm T}$, as mentioned. The transverse propagator is related to the free-space propagator itself via

$$\overleftrightarrow{D}_0(\vec{r}-\vec{r}\,';\omega) = \overleftrightarrow{D}_0^{\rm T}(\vec{r}-\vec{r}\,';\omega) + \overleftrightarrow{g}_{\rm L}(\vec{r}-\vec{r}\,';\omega), \tag{128}$$

where

$$\overleftrightarrow{g}_{\rm L}(\vec{r}-\vec{r}\,';\omega) = \frac{1}{q_0^2}\overleftrightarrow{\delta}_{\rm L}(\vec{r}-\vec{r}\,') \tag{129}$$

is the longitudinal self-field propagator. Since it is known (see Ref. 9) that the transverse free-space propagator has a singularity of the form $|\vec{r}-\vec{r}\,'|^{-1}$ at the point where $\vec{E}_{\mathrm{T}}(\vec{r};\omega)$ is to be calculated the integral in Eq. (126) is absolute convergent at $\vec{r}$. In turn, there is no need to cut out an infinitesimal spherical exclusion volume. Using the full free-space propagator $\overleftrightarrow{D}_0(\vec{r}-\vec{r}\,';\omega)$ as in Eq. (123) the exclusion volume is necessary since $\overleftrightarrow{D}_0(\vec{r}-\vec{r}\,';\omega)$ exhibits a singularity of the form $|\vec{r}-\vec{r}\,'|^{-3}$ at $\vec{r}$. Such a singularity makes the integral only conditionally convergent.

By means of the longitudinal self-field propagator in Eq. (129), the formal relation in Eq. (122), and the relation between $\vec{E}_{\mathrm{L}}(\vec{r};\omega)$ and $\vec{J}^{i}_{\mathrm{L}}(\vec{r};\omega)$ given in Eq. (118), the longitudinal part of the local field in our medium can be related to the total induced current density by means of the following integral relation:

$$\vec{E}_{\mathrm{L}}(\vec{r};\omega) = -i\mu_0\omega \int_V \overleftrightarrow{g}_{\mathrm{L}}(\vec{r}-\vec{r}\,';\omega)\cdot\vec{J}^{i}(\vec{r};\omega)d^3r', \quad \vec{r}\in V_{\mathrm{T}}. \tag{130}$$

By addition of Eqs. (126) and (130), it appears that the local field, $\vec{E}(\vec{r};\omega)$, may in compact form be related to the induced current density via a single integral relation, viz.

$$\vec{E}(\vec{r};\omega) = \vec{E}^{e}_{\mathrm{T}}(\vec{r};\omega) - i\mu_0\omega \int_V \overleftrightarrow{G}_0(\vec{r}-\vec{r}\,';\omega)\cdot\vec{J}^{i}(\vec{r}\,';\omega)d^3r', \tag{131}$$

where the dyadic electromagnetic propagator $\overleftrightarrow{G}_0(\vec{r}-\vec{r}\,';\omega)$ is given by

$$\overleftrightarrow{G}_0(\vec{r}-\vec{r}\,';\omega) = \overleftrightarrow{D}^{\mathrm{T}}_0(\vec{r}-\vec{r}\,';\omega) + \overleftrightarrow{g}_{\mathrm{T}}(\vec{r}-\vec{r}\,';\omega) + \overleftrightarrow{g}_{\mathrm{L}}(\vec{r}-\vec{r}\,';\omega). \tag{132}$$

Since the external sources was assumed to be located outside the transverse (or equivalent longitudinal) current-density domain of the medium under study, the transverse external field in Eq. (131) necessarily is equal to the total external field, i.e. $\vec{E}^{e}_{\mathrm{T}}(\vec{r};\omega) = \vec{E}^{e}(\vec{r};\omega)$.

It appears from the preceding analysis that the medium and source regions can be considered as spatially separated provided the transverse current-density domains of the source particles (V^{e}_{T}) and medium (V_{T}) under study do not overlap.

3.2. PRESCRIBED CURRENT SOURCES LOCATED INSIDE THE MEDIUM

It may happen that the source and medium regions overlap. In such cases it is still possible to establish an integral relation between the local field and the induced current density. To realize this let us for simplicity assume that the "external" sources are located completely inside V_{T}. In this case, the field $\vec{E}^{e}(\vec{r};\omega)$, appearing in Eq. (131), and steming from sources outside V_{T}, is zero. Furthermore, since the prevailing current density inside the medium now is given by the distribution $\vec{J}^{e}(\vec{r},t) + \vec{J}^{i}(\vec{r},t)$, cf. Eq. (17), the relevant equation replacing Eq. (131) is

$$\vec{E}(\vec{r};\omega) = -i\mu_0\omega \int_V \overleftrightarrow{G}_0(\vec{r}-\vec{r}\,';\omega)\cdot\left(\vec{J}^{e}(\vec{r}\,';\omega) + \vec{J}^{i}(\vec{r}\,';\omega)\right)d^3r'. \tag{133}$$

If one assumes that the dynamics hidden in $\vec{J}^{e}(\vec{r}\,';\omega)$ evolves independently of the induced fields one may consider the field

$$\vec{E}(\vec{r};\omega) \equiv -i\mu_0\omega \int_V \overleftrightarrow{G}_0(\vec{r}-\vec{r}\,';\omega)\cdot\vec{J}^{e}(\vec{r}\,';\omega)d^3r' \tag{134}$$

as a *prescribed* driving field for the system. By combining Eqs. (133) and (134) one obtains an integral relation

$$\vec{E}(\vec{r};\omega) = \vec{E}^{e}(\vec{r};\omega) - i\mu_0\omega \int_V \overleftrightarrow{G}_0(\vec{r}-\vec{r}';\omega)\cdot\vec{J}^{i}(\vec{r}';\omega)d^3r', \tag{135}$$

which in form is almost identical to that of Eq. (131). The driving field, which one in a wide sense may call the external field, now, besides the transverse part

$$\vec{E}^{e}_{\mathrm{T}}(\vec{r};\omega) = \frac{1}{3i\varepsilon_0\omega}\vec{J}^{e}_{\mathrm{T}}(\vec{r};\omega) - i\mu_0\omega \int_V \overleftrightarrow{D}^{\mathrm{T}}_0(\vec{r}-\vec{r}';\omega)\cdot\vec{J}^{e}(\vec{r}';\omega)d^3r', \tag{136}$$

has a longitudinal part

$$\vec{E}^{e}_{\mathrm{L}}(\vec{r};\omega) = \frac{1}{i\varepsilon_0\omega}\vec{J}^{e}_{\mathrm{L}}(\vec{r};\omega). \tag{137}$$

Contributions from external sources located outside V_{T} in a domain V^e can be incorporated in the scheme of Eq. (135) by adding a term $-i\mu_0\omega\int_{V^e}\overleftrightarrow{D}^{\mathrm{T}}_0(\vec{r}-\vec{r}';\omega)\cdot\vec{J}^{e}(\vec{r}';\omega)d^3r'$ to Eq. (136). In the general case the integral relation between the local field and the induced current density thus may be written in the form given in Eq. (135), inserting for the longitudinal and transverse parts of the prescribed driving field, $\vec{E}^{e}(\vec{r};\omega) = \vec{E}^{e}_{\mathrm{T}}(\vec{r};\omega) + \vec{E}^{e}_{\mathrm{L}}(\vec{r};\omega)$, the expression in Eq. (137) and

$$\vec{E}^{e}_{\mathrm{T}}(\vec{r};\omega) = \frac{1}{3i\varepsilon_0\omega}\vec{J}^{e}_{\mathrm{T}}(\vec{r};\omega) - i\mu_0\omega \int_{V+V^e} \overleftrightarrow{D}^{\mathrm{T}}_0(\vec{r}-\vec{r}';\omega)\cdot\vec{J}^{e}(\vec{r}';\omega)d^3r'. \tag{138}$$

In retrospective it appears that the full propagator $\overleftrightarrow{D}_0(\vec{r}-\vec{r}';\omega)$ $(\equiv \overleftrightarrow{D}^{\mathrm{OUT}}_0(\vec{r}-\vec{r}';\omega))$ were used in Eq. (39) (with Eq. (38) inserted) when calculating the field from the sources spatially separated from our system, whereas the transverse propagator $\overleftrightarrow{D}^{\mathrm{T}}_0(\vec{r}-\vec{r}';\omega)$ is used in Eq. (138). This is of course consistent since the domain V^e in Eq. (138) is spatially separated from V (in the sense that V^e_{T} and V_{T} do not overlap). Outside the transverse current density domain V^e_{T}, $\overleftrightarrow{D}_0(\vec{r}-\vec{r}';\omega) = \overleftrightarrow{D}^{\mathrm{T}}_0(\vec{r}-\vec{r}';\omega) + \overleftrightarrow{g}_{\mathrm{L}}(\vec{r}-\vec{r}';\omega)$ and $\overleftrightarrow{D}^{\mathrm{T}}_0(\vec{r}-\vec{r}';\omega)$ are identical, since $\overleftrightarrow{g}_{\mathrm{L}}(\vec{r}-\vec{r}';\omega)$ is zero here.

4. Many-body constitutive relation and loop equation

4.1. RESPONSE TO THE TRANSVERSE LOCAL FIELD PLUS THE LONGITUDINAL EXTERNAL FIELD

To determine the electrodynamic behaviour of a system of charged particles subjected to a prescribed external field it was realized in chapter 2 that the microscopic Maxwell-Lorentz equation had to be supplemented by a material (constitutive) equation of one sort or another in order to close the set of equations needed for a complete description. Within the framework of the standard linear response theory this is done by introducing a linear but nonlocal relation between the induced current density and either the selfconsistent electric field (see Eq. (44)) or the external field (see Eq. (50)). In chapter 2, the constitutive relations were introduced in a phenomenological manner and it is certainly not obvious how these may appear from a rigorous theory. To establish the material relation (in one form or another) within the framework of a nonrelativistic description one has to invoke the many-body Schrödinger equation.

As described in Ref. 9 one may start from the Pauli Hamiltonian of the particles in the system. In this Hamiltonian appears a vector, $\vec{A}$, and a scalar potential, ϕ. These potentials have a contribution $(\vec{A}^e, \phi^e)$ from the external field driving the system. In the present case when one assumes (response theory) that the external fields are *prescribed* quantities, $\vec{A}^e$ and ϕ^e do not play a role of dynamical variables for the electromagnetic field in the system. Contrary to this, the induced vector, $\vec{A}^i$, and scalar, ϕ^i, potentials (plus the time derivative of these) are indeed dynamical variables, and in fact, *the* dynamical variables of the field. To describe the electromagnetics of the system in an echonomic manner it is adequate to write the Pauli Hamiltonian in the Coulomb gauge as far as the dynamical field variables are concerned. Doing so the redundant dynamical part of the scalar potential, ϕ^i, is eliminated in favour of the particle coordinate variables. In the Coulomb gauge the induced vector potential is transverse, i.e. $\vec{A}^i = \vec{A}^i_{\mathrm{T}}$. To describe the field originating in the broad-sense external sources there is no need to choose *a priori* the Coulomb gauge. Altogether we thus may use the following vector and scalar potentials in the Pauli Hamiltonian:

$$\vec{A} = \vec{A}^e + \vec{A}^i_{\mathrm{T}}, \tag{139}$$

and

$$\phi = \phi^e. \tag{140}$$

For calculational purposes it is often convenient to choose an external-field gauge in which $\phi^e = 0$. Doing so, the particle-field interaction enters the Pauli Hamiltonian only through a vector potential

$$\vec{A} = \vec{A}^e_{\mathrm{T}} + \vec{A}^e_{\mathrm{L}} + \vec{A}^i_{\mathrm{T}}. \tag{141}$$

Since the transverse part of the prevailing local field is $\vec{A} = \vec{A}^e_{\mathrm{T}} + \vec{A}^i_{\mathrm{T}}$, Eq. (141) may be written

$$\vec{A} = \vec{A}_{\mathrm{T}} + \vec{A}^e_{\mathrm{L}}. \tag{142}$$

In the frequency-space domain the vector potential in Eq. (142) is related to the electric field via

$$\vec{A}(\vec{r};\omega) = (i\omega)^{-1}\left(\vec{E}_{\mathrm{T}}(\vec{r};\omega) + \vec{E}^e_{\mathrm{L}}(\vec{r};\omega)\right). \tag{143}$$

Starting from the Liouville equation for the many-body density-matrix operator ρ, i.e.

$$i\hbar\frac{\partial\rho}{\partial t} = [\mathcal{H}, \rho], \tag{144}$$

where $\mathcal{H}$ is the Pauli Hamiltonian, and the general expression

$$\vec{J} = \mathrm{Tr}\{\rho\vec{j}\} \tag{145}$$

for the current density in the system, $\vec{j}$ being the current-density operator, it is possible upon linearization in the electric field (or equivalently the vector potential) to establish in the frequency-space domain a constitutive relation of the form[9]

$$\vec{J}^i(\vec{r};\omega) = \int \overleftrightarrow{\sigma}^{\mathrm{MB}}(\vec{r},\vec{r}\,';\omega)\cdot\left[\vec{E}_{\mathrm{T}}(\vec{r}\,';\omega) + \vec{E}^e_{\mathrm{L}}(\vec{r}\,';\omega)\right] d^3r', \tag{146}$$

where $\overleftrightarrow{\sigma}^{\mathrm{MB}}(\vec{r},\vec{r}\,';\omega)$ is the linear (and nonlocal) many-body (MB) conductivity response tensor of the system. In the absense of spin-orbit coupling effects $\overleftrightarrow{\sigma}^{\mathrm{MB}}$ is the sum of a space and a spin part, i.e.

$$\overleftrightarrow{\sigma}^{\mathrm{MB}}(\vec{r},\vec{r}\,';\omega) = \overleftrightarrow{\sigma}^{\mathrm{SPACE}}(\vec{r},\vec{r}\,';\omega) + \overleftrightarrow{\sigma}^{\mathrm{SPIN}}(\vec{r},\vec{r}\,';\omega). \tag{147}$$

In Eq. (146) the integral extends over the domain of the system under study, and for notational simplicity we omit indicating this on the integral sign here and on those appearing in the subsequent part of the article.

Although the constitutive relation in Eq. (146) was established on the basis of an external-field gauge in which $\phi^e = 0$, it is possible to prove that the gauge invariance of the induced current-density response is maintained after linearization of the electrodynamic response[9].

4.2. LOOP EQUATION FOR THE TRANSVERSE PART OF THE LOCAL FIELD

We have realized in chapter 3 that the microscopic Maxwell-Lorentz equations leads to the integral relation in Eq. (135) between the local electric field and the induced current density, with the broad-sense externale (driving) field given by the sum of Eqs. (137) and (138). The constitutive relation between the induced current density and the sum of the transverse local field and the longitudinal part of the external field (Eq. (146)), which was derived from a field-linearization of the Schrödinger-field electrodynamics closes the system of equations needed to establish the electrodynamics of our system in a prescribed external field.

Let us now take a closer look at the two basic relations between the induced current density and the prevailing field (Eqs. (135) and (146)). We begin by a split of Eq. (135) into its divergence-free and rotational-free parts, viz.

$$\vec{E}_{\mathrm{T}}(\vec{r}) = \vec{E}^e_{\mathrm{T}}(\vec{r}) - i\mu_0\omega \int \left[\overleftrightarrow{D}^{\mathrm{T}}_0(\vec{r}-\vec{r}\,') + \overleftrightarrow{g}_{\mathrm{T}}(\vec{r}-\vec{r}\,')\right] \cdot \vec{J}^i_{\mathrm{T}}(\vec{r}\,')d^3r', \tag{148}$$

and

$$\vec{E}_{\mathrm{L}}(\vec{r}) = \vec{E}^e_{\mathrm{L}}(\vec{r}) - i\mu_0\omega \int \overleftrightarrow{g}_{\mathrm{L}}(\vec{r}-\vec{r}\,') \cdot \vec{J}^i_{\mathrm{L}}(\vec{r}\,')d^3r', \tag{149}$$

omitting here, and in the following, for brevity the reference to ω from the notation.

For conceptual purposes it is convenient next to introduce the so-called TT, LL, TL, and LT many-body conductivity response functions via

$$\overleftrightarrow{\sigma}^{\mathrm{MB}}_{\mathrm{AB}}(\vec{r}-\vec{r}\,') \equiv \int \overleftrightarrow{\delta}_{\mathrm{A}}(\vec{r}-\vec{r}\,'') \cdot \overleftrightarrow{\sigma}^{\mathrm{MB}}(\vec{r}\,'',\vec{r}\,''') \cdot \overleftrightarrow{\delta}_{\mathrm{B}}(\vec{r}\,'''-\vec{r}\,')d^3r'''d^3r'', \tag{150}$$

where A, B = T or L. In terms of these response tensors the many-body conductivity tensor can be written

$$\overleftrightarrow{\sigma}^{\mathrm{MB}}(\vec{r},\vec{r}\,') = \overleftrightarrow{\sigma}^{\mathrm{MB}}_{\mathrm{TT}}(\vec{r},\vec{r}\,') + \overleftrightarrow{\sigma}^{\mathrm{MB}}_{\mathrm{TL}}(\vec{r},\vec{r}\,') + \overleftrightarrow{\sigma}^{\mathrm{MB}}_{\mathrm{LT}}(\vec{r},\vec{r}\,') + \overleftrightarrow{\sigma}^{\mathrm{MB}}_{\mathrm{LL}}(\vec{r},\vec{r}\,'). \tag{151}$$

The division in Eq. (151) allows one to obtain from Eq. (146) the following expressions for the transverse and longitudinal parts of the induced current density:

$$\vec{J}^i_{\mathrm{T}}(\vec{r}) = \int \overleftrightarrow{\sigma}^{\mathrm{MB}}_{\mathrm{TT}}(\vec{r},\vec{r}\,') \cdot \vec{E}_{\mathrm{T}}(\vec{r}\,')d^3r' + \int \overleftrightarrow{\sigma}^{\mathrm{MB}}_{\mathrm{TL}}(\vec{r},\vec{r}\,') \cdot \vec{E}^e_{\mathrm{L}}(\vec{r}\,')d^3r', \tag{152}$$

and

$$\vec{J}^i_{\mathrm{L}}(\vec{r}) = \int \overleftrightarrow{\sigma}^{\mathrm{MB}}_{\mathrm{LT}}(\vec{r},\vec{r}\,') \cdot \vec{E}_{\mathrm{T}}(\vec{r}\,')d^3r' + \int \overleftrightarrow{\sigma}^{\mathrm{MB}}_{\mathrm{LL}}(\vec{r},\vec{r}\,') \cdot \vec{E}^e_{\mathrm{L}}(\vec{r}\,')d^3r'. \tag{153}$$

By combining Eqs. (148) and (152) it is realized that the transverse part of the local field satisfies the integral equation

$$\vec{E}_{\mathrm{T}}(\vec{r}) = \vec{E}^{\mathrm{drive}}_{\mathrm{T}}(\vec{r}) + \int \overleftrightarrow{K}_{\mathrm{TT}}(\vec{r},\vec{r}\,') \cdot \vec{E}_{\mathrm{T}}(\vec{r}\,')d^3r', \tag{154}$$

where the kernel $\overleftrightarrow{K}_{\mathrm{TT}}$ is given by

$$\overleftrightarrow{K}_{\mathrm{TT}}(\vec{r},\vec{r}') = -i\mu_0\omega \int \left[\overleftrightarrow{D}_0^{\mathrm{T}}(\vec{r}-\vec{r}'') + \overleftrightarrow{g}_{\mathrm{T}}(\vec{r}-\vec{r}'')\right] \cdot \overleftrightarrow{\sigma}_{\mathrm{TT}}^{\mathrm{MB}}(\vec{r}'',\vec{r}')d^3r'', \tag{155}$$

and the effective transverse driving (drive) field by

$$\vec{E}_{\mathrm{T}}^{\mathrm{drive}}(\vec{r}) = \vec{E}_{\mathrm{T}}^{e}(\vec{r})$$
$$-i\mu_0\omega \int \left[\overleftrightarrow{D}_0^{\mathrm{T}}(\vec{r}-\vec{r}'') + \overleftrightarrow{g}_{\mathrm{T}}(\vec{r}-\vec{r}'')\right] \cdot \overleftrightarrow{\sigma}_{\mathrm{TL}}^{\mathrm{MB}}(\vec{r}'',\vec{r}') \cdot \vec{E}_{\mathrm{L}}^{e}(\vec{r}')d^3r''d^3r'. \tag{156}$$

It appears from Eq. (156) that the transverse part of the local field is driven not only by the transverse part of the external field but also indirectly by the longitudinal part of the external field. In a sense this is obvious since $\vec{E}_{\mathrm{L}}^{e}$ gives rise to a *prescribed* contribution to the transverse part of the induced current density via $\overleftrightarrow{\sigma}_{\mathrm{TL}}^{\mathrm{MB}}$. In turn, this prescribed current density causes a prescribed transverse electromagnetic field, which one may calculate using the transverse propagator $\overleftrightarrow{D}_0^{\mathrm{T}} + \overleftrightarrow{g}_{\mathrm{T}}$.

Formally, the loop equation in (154) for the transverse part of the local field has the solution

$$\vec{E}_{\mathrm{T}}(\vec{r}) = \int \overleftrightarrow{\Gamma}_{\mathrm{TT}}(\vec{r},\vec{r}') \cdot \vec{E}_{\mathrm{T}}^{\mathrm{drive}}(\vec{r}')d^3r', \tag{157}$$

where one may call $\overleftrightarrow{\Gamma}_{\mathrm{TT}}(\vec{r},\vec{r}')$ the transverse field-field response tensor. This tensor fulfils the dyadic integral equation

$$\overleftrightarrow{\Gamma}_{\mathrm{TT}}(\vec{r},\vec{r}') = \overleftrightarrow{\delta}_{\mathrm{T}}(\vec{r}-\vec{r}') + \int \overleftrightarrow{K}_{\mathrm{TT}}(\vec{r},\vec{r}'') \cdot \overleftrightarrow{\Gamma}_{\mathrm{TT}}(\vec{r}'',\vec{r}')d^3r''. \tag{158}$$

By inserting Eq. (153) into Eq. (149) it readily appears that there is no fundamental loop equation for the longitudinal part of the local field. Instead one has an integral relation between the $\vec{E}_{\mathrm{L}}$ and $\vec{E}_{\mathrm{T}}$. Let us assume that the solution for the transverse part of the local field has been obtained, in practice an approximate one, of course. Then, the longitudinal part of the local field is determined by direct integration from

$$\vec{E}_{\mathrm{L}}(\vec{r}) = \vec{E}_{\mathrm{L}}^{e}(\vec{r}) - i\mu_0\omega \int \overleftrightarrow{g}_{\mathrm{L}}(\vec{r}-\vec{r}'') \cdot \overleftrightarrow{\sigma}_{\mathrm{LL}}^{\mathrm{MB}}(\vec{r}'',\vec{r}') \cdot \vec{E}_{\mathrm{L}}^{e}(\vec{r}')d^3r''d^3r'$$
$$-i\mu_0\omega \int \overleftrightarrow{g}_{\mathrm{L}}(\vec{r}-\vec{r}''') \cdot \overleftrightarrow{\sigma}_{\mathrm{LT}}^{\mathrm{MB}}(\vec{r}''',\vec{r}'') \cdot \overleftrightarrow{\Gamma}_{\mathrm{TT}}(\vec{r}'',\vec{r}') \cdot \vec{E}_{\mathrm{T}}^{\mathrm{drive}}(\vec{r}')d^3r'''d^3r''d^3r' \tag{159}$$

4.3. RELATIONS BETWEEN THE MANY-BODY AND IMPIRICAL CONDUCTIVITY RESPONSE FUNCTIONS

In the preceding part of this paper I have introduced three different linear material relations, viz., the one rigorously derived from many-body response theory (Eq. (146)), and the two impirical ones relating the induced current density to the local field (Eq. (44)) and the external field (Eq. (50)), respectively. For systems exhibiting translational invariance in time the two impirical ones take the forms

$$\vec{J}^{i}(\vec{r};\omega) = \int \overleftrightarrow{\sigma}(\vec{r},\vec{r}';\omega) \cdot \vec{E}(\vec{r}';\omega)d^3r', \tag{160}$$

and

$$\vec{J}^i(\vec{r};\omega) = \int \overleftrightarrow{\sigma}^e(\vec{r},\vec{r}';\omega)\cdot\vec{E}^e(\vec{r}';\omega)d^3r', \tag{161}$$

in the frequency-space domain. The three conductivities $\overleftrightarrow{\sigma}^{\mathrm{MB}}(\vec{r},\vec{r}';\omega)$, $\overleftrightarrow{\sigma}(\vec{r},\vec{r}';\omega)$, and $\overleftrightarrow{\sigma}^e(\vec{r},\vec{r}';\omega)$, associated with Eqs. (146), (160), and (161) are not independent. In fact we have already established a relation between $\overleftrightarrow{\sigma}(\vec{r},\vec{r}',t,t')$ and $\overleftrightarrow{\sigma}^e(\vec{r},\vec{r}',t,t')$ (see Eq. (58)).

In cases where the external sources are located outside the transverse current density domain of the system, the relations between the many-body conductivity and each of the two impirical conductivities are formally simple. Thus, for $\vec{E}^e_{\mathrm{L}} = 0$, Eq. (146) may be written

$$\vec{J}^i(\vec{r}) = \int \overleftrightarrow{\sigma}^{\mathrm{MB}}(\vec{r},\vec{r}'')\cdot\overleftrightarrow{\delta}_{\mathrm{T}}(\vec{r}''-\vec{r}')\cdot\vec{E}(\vec{r}'')d^3r''d^3r', \tag{162}$$

and by means of Eq. (150) and comparison to Eq. (160), it is seen that

$$\overleftrightarrow{\sigma}(\vec{r},\vec{r}') = \overleftrightarrow{\sigma}^{\mathrm{MB}}_{\mathrm{LT}}(\vec{r},\vec{r}') + \overleftrightarrow{\sigma}^{\mathrm{MB}}_{\mathrm{TT}}(\vec{r},\vec{r}'). \tag{163}$$

Also, the relation between $\overleftrightarrow{\sigma}^{\mathrm{MB}}$ and $\overleftrightarrow{\sigma}^e$ may readily be obtained when $\vec{E}^e_{\mathrm{L}} = 0$, since $\vec{E}^{\mathrm{drive}}_{\mathrm{T}} = \vec{E}^e_{\mathrm{T}}$ in this case. Hence, by combining Eqs. (146) and (157) one gets

$$\vec{J}^i(\vec{r}) = \int \overleftrightarrow{\sigma}^{\mathrm{MB}}(\vec{r},\vec{r}'')\cdot\overleftrightarrow{\Gamma}_{\mathrm{TT}}(\vec{r}'',\vec{r}')\cdot\vec{E}^e_{\mathrm{T}}(\vec{r}')d^3r''d^3r', \tag{164}$$

and upon comparison to Eq. (161)

$$\overleftrightarrow{\sigma}^e(\vec{r},\vec{r}') = \int \overleftrightarrow{\sigma}^{\mathrm{MB}}(\vec{r},\vec{r}'')\cdot\overleftrightarrow{\Gamma}_{\mathrm{TT}}(\vec{r}'',\vec{r}''')\cdot\overleftrightarrow{\delta}_{\mathrm{T}}(\vec{r}'''-\vec{r}')d^3r'''d^3r''. \tag{165}$$

In the general case, where $\vec{E}^e_{\mathrm{L}} \neq 0$ inside the medium being investigated, the relation between $\overleftrightarrow{\sigma}^{\mathrm{MB}}$ and $\overleftrightarrow{\sigma}^e$ (or $\overleftrightarrow{\sigma}$) is more complicated, though still easy to derive. Thus, by inserting the expression for $\vec{E}^{\mathrm{drive}}_{\mathrm{T}}$ given in Eq. (156) into Eq. (157), one can express the transverse part of the local field in terms of the transverse and longitudinal parts of the external field. Inserting next the equation obtained for $\vec{E}_{\mathrm{T}}$ into Eq. (146) one may derive the relation between $\vec{J}^i$ and $\vec{E}^e$. Upon comparison to Eq. (161) one then finds the following result:

$$\begin{aligned}\overleftrightarrow{\sigma}^e(\vec{r},\vec{r}') = &\int \overleftrightarrow{\sigma}^{\mathrm{MB}}(\vec{r},\vec{r}_2)\cdot\Big\{\overleftrightarrow{\Gamma}_{\mathrm{TT}}(\vec{r}_2,\vec{r}_3)\cdot\overleftrightarrow{\delta}_{\mathrm{T}}(\vec{r}_3-\vec{r}')\delta(\vec{r}_4)\delta(\vec{r}_5) + \overleftrightarrow{\delta}_{\mathrm{L}}(\vec{r}_2-\vec{r}')\delta(\vec{r}_3)\delta(\vec{r}_4)\delta(\vec{r}_5)\\ &-i\mu_0\omega\overleftrightarrow{\Gamma}_{\mathrm{TT}}(\vec{r}_2,\vec{r}_3)\cdot\Big[\overleftrightarrow{D}^{\mathrm{T}}_0(\vec{r}_3-\vec{r}_4)+\overleftrightarrow{g}_{\mathrm{T}}(\vec{r}_3-\vec{r}_4)\Big]\cdot\overleftrightarrow{\sigma}^{\mathrm{MB}}_{\mathrm{TL}}(\vec{r}_4,\vec{r}_5)\cdot\overleftrightarrow{\delta}_{\mathrm{L}}(\vec{r}_5-\vec{r}')\Big\}\\ &\times d^3r_5d^3r_4d^3r_3d^3r_2.\end{aligned} \tag{166}$$

Neglecting the terms in Eq. (166) containing $\overleftrightarrow{\delta}_{\mathrm{L}}$ one regains Eq. (165), as required.

5. Coupled-antenna theory

5.1. MANY-BODY SPACE CONDUCTIVITY AND ITS TENSOR-PRODUCT STRUCTURE

We have realized in section 4.2 that in order to describe the electrodynamics of a particle system subjected to a prescribed external field the main challenge from a conceptual point of view lies in solving the loop equation (Eq. (154)) for the transverse part of the local field inside the transverse current density domain. In the heart of a further analysis hence it is needed to study the kernel $\overleftrightarrow{K}_{\rm TT}(\vec{r},\vec{r}')$ of the loop equation. The kernel contains the transverse vacuum propagator and the TT-part of the many-body conductivity tensor, cf. Eq. (155). Since, we have already examined the electromagnetic propagator it appears natural by now to focus our attention on the many-body conductivity.

Though the method of analysis presented in this chapter can be adopted using the complete many-body conductivity, I shall for simplicity restrict myself to situations where the spin-part of the conductivity tensor is negligible. Thus, I start by the following well known expression for the spatial part of the many-body response tensor:

$$\overleftrightarrow{\sigma}^{\rm SPACE}(\vec{r},\vec{r}';\omega) = \overleftrightarrow{\sigma}^{\rm dia}(\vec{r},\vec{r}';\omega) + \overleftrightarrow{\sigma}^{\rm para}(\vec{r},\vec{r}';\omega), \tag{167}$$

where

$$\overleftrightarrow{\sigma}^{\rm dia}(\vec{r},\vec{r}';\omega) = \frac{ie^2}{m\omega} N_0(\vec{r})\delta(\vec{r}-\vec{r}')\overleftrightarrow{U} \tag{168}$$

is the diamagnetic part of the conductivity, and

$$\overleftrightarrow{\sigma}^{\rm para}(\vec{r},\vec{r}';\omega) = \frac{2i}{\omega}\sum_{I,J}\frac{P_J - P_I}{\hbar\omega + E_J - E_I}\vec{J}_{I\to J}(\vec{r})\vec{J}_{J\to I}(\vec{r}') \tag{169}$$

is its paramagnetic part. In Eq. (168), $N_0(\vec{r})$ is the many-body electron density in the field-unperturbed system. The summation in Eq. (169) runs over the various many-body energy eigenstates of our system, these being denoted by capital letters I, J, etc. For an energy eigenstate A, the many-body wave function is named Ψ_A, the energy E_A, and the probability that this state is occupied P_A. In Eq. (169) also appears the many-body transition current densities $\vec{J}_{I\to J}(\vec{r})$ and $\vec{J}_{J\to I}(\vec{r}')$ calculated at space points $\vec{r}$ and $\vec{r}'$. In explicit form, the transition current density from state A to state B is given by

$$\vec{J}_{A\to B}(\vec{r}) = -\frac{e\hbar}{2mi}\sum_{\alpha}\int \delta(\vec{r}-\vec{r}_\alpha)\left[\Psi_B^*\vec{\nabla}_\alpha\Psi_A - \Psi_A\vec{\nabla}\Psi_B^*\right]\prod_{\beta} d^3r_\beta, \tag{170}$$

where $\vec{r}_\alpha$ (and $\vec{r}_\beta$) denotes the different electron coordinates. The summation ($\sum_\alpha$) and product ($\prod_\beta$) run over all electron coordinates.

It is adequate at this stage to seek to rewrite the diamagnetic conductivity in a tensor product form analogous to the one for the paramagnetic response function (see Eq. (169)). To achieve this goal, I take as a starting point the relation

$$\chi^{\rm MB}(\vec{r},\vec{r}';\omega) = \frac{i}{e\omega}\vec{\nabla}'\cdot\vec{\nabla}\cdot\overleftrightarrow{\sigma}^{\rm SPACE}(\vec{r},\vec{r}';\omega), \tag{171}$$

where $\chi^{\rm MB}(\vec{r},\vec{r}';\omega)$ is the well known many-body density-density (scalar) response function, relating in a nonlocal fashion the induced particle density at $\vec{r}$ to the external scalar potential at $\vec{r}'$. For a derivation of Eq. (171) the reader is referred to Ref. 9, where also the explicit expression for $\chi^{\rm MB}(\vec{r},\vec{r}';\omega)$ can be found. It readily appears from this expression that $\chi^{\rm MB}(\vec{r},\vec{r}';\omega)$ approaches a finite value for $\omega\to 0$. By a comparison to Eq. (171) it thus appears that $\vec{\nabla}'\cdot\vec{\nabla}\cdot\overleftrightarrow{\sigma}^{\rm SPACE}(\vec{r},\vec{r}';\omega)$ necessarily has to approach zero when $\omega\to 0$. In turn, this implies that the LL-part of the response tensor must vanish in the static limit, i.e.

$$\overleftrightarrow{\sigma}^{\rm SPACE}_{\rm LL}(\vec{r},\vec{r}';\omega\to 0) = 0. \tag{172}$$

Let us now be a bit more restrictive in our demands by requiring that

$$\overleftrightarrow{\sigma}^{\mathrm{SPACE}}(\vec{r},\vec{r}\,';\omega\rightarrow 0)=0. \tag{173}$$

The condition in Eq. (174) is sufficient if one requires that the current density induced in our system must vanish identically. However, it appears from the fundamental constitutive relation in Eq. (146), decomposed possibly as indicated in Eqs. (152) and (153), that since the Maxwell-Lorentz equation in Eq. (1) implies that any transverse field $\vec{E}_{\mathrm{T}}(\vec{r};\omega)$ must vanish in the electrostatic limit, there is no need *a priori* for $\overleftrightarrow{\sigma}^{\mathrm{MB}}_{\mathrm{LT}}$ and $\overleftrightarrow{\sigma}^{\mathrm{MB}}_{\mathrm{TT}}$ to vanish in the limit $\omega \rightarrow 0$, even if one requires the induced current density to be zero at $\omega = 0$. Before proceeding let me stress that a particular analysis is needed in the case of a superconducting system. For such a system, the paramagnetic response tensor vanishes for frequencies below the superconducting gap in the low temperature limit. The infinite conductivity of a superconductor thus is associated with the diamagnetic effect inevitably present.

From the strong requirement in Eq. (173) follows the identity

$$\frac{e^2}{m}N_0(\vec{r})\delta(\vec{r}-\vec{r}\,')\overleftrightarrow{U}+2\sum_{I,J}\frac{P_J-P_I}{E_J-E_I}\vec{J}_{I\rightarrow J}(\vec{r})\vec{J}_{J\rightarrow I}(\vec{r}\,')=0, \tag{174}$$

which one may consider as a selfconsistency requirement on the many-body states of the field-unperturbed system. The identity in Eq. (174) immediately allows one to write the diamagnetic conductivity in the tensor product form sought for, i.e.

$$\overleftrightarrow{\sigma}^{\mathrm{dia}}(\vec{r},\vec{r}\,';\omega)=\frac{2}{i\omega}\sum_{I,J}\frac{P_J-P_I}{E_J-E_I}\vec{J}_{I\rightarrow J}(\vec{r})\vec{J}_{J\rightarrow I}(\vec{r}\,'). \tag{175}$$

One may wonder what happens to the factor $(P_J-P_I)/(E_J-E_I)$ when the many-body states I and J become identical. Using the fact that $P_A = C\exp[-E_A/(kT)]$ in thermal equilibrium, where C is a constant, it readily follows that

$$\lim_{E_J\rightarrow E_I}\frac{P_J-P_I}{E_J-E_I}=-\frac{C}{kT}\exp\left(\frac{-E_I}{kT}\right), \tag{176}$$

so that the afore-mentioned ratio is finite for $J = I$.

By adding the expressions in Eqs. (169) and (175), we finally obtain the following formula for the total space conductivity:

$$\overleftrightarrow{\sigma}^{\mathrm{SPACE}}(\vec{r},\vec{r}\,';\omega)=\frac{2\hbar}{i}\sum_{I,J}\frac{P_J-P_I}{(E_J-E_I)(\hbar\omega+E_J-E_I)}\vec{J}_{I\rightarrow J}(\vec{r})\vec{J}_{J\rightarrow I}(\vec{r}\,'). \tag{177}$$

It is easy to check that the conductivity in Eq. (177) indeed does vanish in the static limit. Thus, starting from

$$\overleftrightarrow{\sigma}^{\mathrm{SPACE}}(\vec{r},\vec{r}\,';\omega\rightarrow 0)=\frac{2\hbar}{i}\sum_{I,J}\frac{P_J-P_I}{(E_J-E_I)^2}\vec{J}_{I\rightarrow J}(\vec{r})\vec{J}_{J\rightarrow I}(\vec{r}\,'), \tag{178}$$

and from the relation

$$\vec{J}_{B\rightarrow A}(\vec{r})=-\vec{J}_{A\rightarrow B}(\vec{r}), \tag{179}$$

valid if real many-body wave functions are used, one just needs to add the two terms in Eq. (178) for which $(I, J) = (A, B)$ and $(I, J) = (B, A)$. Hence, since it appears that

$$\frac{2\hbar}{i}\left[\frac{P_B - P_A}{(E_B - E_A)^2}\vec{J}_{A\to B}(\vec{r})\vec{J}_{B\to A}(\vec{r}\,') + \frac{P_A - P_B}{(E_A - E_B)^2}\vec{J}_{B\to A}(\vec{r})\vec{J}_{A\to B}(\vec{r}\,')\right] = 0 \tag{180}$$

if one uses Eq. (179), it follows that the contributions to $\overleftrightarrow{\sigma}^{\mathrm{SPACE}}(\vec{r}, \vec{r}\,'; \omega \to 0)$ vanish pairwise.

To display the relative role of the para- and diamagnetic contributions to the space conductivity, let us rewrite Eq. (177) in the form

$$\overleftrightarrow{\sigma}^{\mathrm{SPACE}}(\vec{r}, \vec{r}\,'; \omega) = \frac{2i}{\omega}\sum_{I,J}\left(\frac{\hbar\omega}{E_I - E_J}\right)\frac{P_J - P_I}{\hbar\omega + E_J - E_I}\vec{J}_{I\to J}(\vec{r})\vec{J}_{J\to I}(\vec{r}\,'). \tag{181}$$

By comparison to Eq. (169) it is seen that each term in the summation for $\overleftrightarrow{\sigma}^{\mathrm{SPACE}}$ deviates with a factor $\hbar\omega/(E_I - E_J)$ from the corresponding term in the paramagnetic response. Thus, if one chooses a frequency close to a given many-body transition, say $A \to B$, so that $\hbar\omega \approx E_B - E_A$, the factor $\hbar\omega/(E_B - E_A)$ is almost one. A resonant contribution to $\overleftrightarrow{\sigma}^{\mathrm{SPACE}}$ hence is dominated by the paramagnetic response.

5.2. ON THE SOLUTION OF THE LOOP EQUATION FOR THE TRANSVERSE LOCAL FIELD

If one denotes the transverse parts of the various transition current densities by a superscript T, so that e.g. $\vec{J}^{\mathrm{T}}_{I\to J}(\vec{r})$ stands for the transverse part of $\vec{J}_{I\to J}(\vec{r})$, the TT-part of the space conductivity tensor in Eq. (177) is given by

$$\overleftrightarrow{\sigma}^{\mathrm{SPACE}}_{\mathrm{TT}}(\vec{r}, \vec{r}\,'; \omega) = \sum_{I,J} A_{IJ}(\omega)\vec{J}^{\mathrm{T}}_{I\to J}(\vec{r})\vec{J}^{\mathrm{T}}_{J\to I}(\vec{r}\,'), \tag{182}$$

with the abbreviation

$$A_{IJ}(\omega) = \frac{2\hbar}{i}\frac{P_J - P_I}{(E_J - E_I)(\hbar\omega + E_J - E_I)}. \tag{183}$$

By inserting the expression for $\overleftrightarrow{\sigma}^{\mathrm{SPACE}}_{\mathrm{TT}}(\vec{r}, \vec{r}\,'; \omega)$ in Eq. (182) into the formula for the kernel $\overleftrightarrow{K}_{\mathrm{TT}}$ (Eq. (155)) this becomes

$$\overleftrightarrow{K}_{\mathrm{TT}}(\vec{r}, \vec{r}\,') = \sum_{I,J}\vec{F}^{\mathrm{T}}_{I\to J}(\vec{r})\vec{J}^{\mathrm{T}}_{J\to I}(\vec{r}\,'), \tag{184}$$

where

$$\vec{F}^{\mathrm{T}}_{I\to J}(\vec{r}) = -i\mu_0\omega A_{IJ}(\omega)\int\left[\overleftrightarrow{D}^{\mathrm{T}}_0(\vec{r} - \vec{r}\,'') + \overleftrightarrow{g}_{\mathrm{T}}(\vec{r} - \vec{r}\,'')\right]\cdot\vec{J}^{\mathrm{T}}_{I\to J}(\vec{r}\,'')d^3r'' \tag{185}$$

is a known transverse vector field of $\vec{r}$. The fact that $\overleftrightarrow{K}_{\mathrm{TT}}(\vec{r}, \vec{r}\,')$ consists of a sum of tensor products in which the first and second vector is a function only of $\vec{r}$ and $\vec{r}\,'$, respectively, makes it possible to transform the integral equation problem for the transverse part of the local field into a matrix equation problem. Thus, by inserting the expression for $\overleftrightarrow{K}_{\mathrm{TT}}(\vec{r},\vec{r}\,')$ given into Eq. (184) into the loop equation in (154) one obtains

$$\vec{E}_{\mathrm{T}}(\vec{r}) = \vec{E}_{\mathrm{T}}^{\mathrm{drive}}(\vec{r}) + \sum_{I,J} \beta_{J\to I} \vec{F}_{I\to J}^{\mathrm{T}}(\vec{r}), \tag{186}$$

with the yet unknown $\beta_{J\to I}$ numbers given by

$$\beta_{J\to I} = \int \vec{J}_{J\to I}^{\mathrm{T}}(\vec{r}) \cdot \vec{E}_{\mathrm{T}}(\vec{r}) d^3r. \tag{187}$$

The physical interpretation of Eq. (186) is as follows. The transverse local field at $\vec{r}$, $\vec{E}_{\mathrm{T}}(\vec{r})$, is equal to the transverse driving field, $\vec{E}_{\mathrm{T}}^{\mathrm{drive}}(\vec{r})$, at $\vec{r}$, plus the sum of the field radiated from the various $I \to J$ transitions. The relevant current density distribution associated with the transition $I \to J$ is $\vec{J}_{I\to J}^{\mathrm{T}}(\vec{r}\,'')$, and the field propagation from this distribution is given in terms of the sum of the transverse vacuum ($\overleftrightarrow{D}_0^{\mathrm{T}}(\vec{r}-\vec{r}\,'')$) and self-field ($\overleftrightarrow{g}_{\mathrm{T}}(\vec{r}-\vec{r}\,'')$) propagators. The strength of the radiation from the individual transitions are given by the numbers $\beta_{J\to I}$ appearing in Eq. (186). It is obvious that the strength $\beta_{J\to I}$ is determined by the work carried out by the yet unknown local field on the current density distribution $\vec{J}_{J\to I}^{\mathrm{T}}(\vec{r})$ associated with the transition $J \to I$, i.e. the excitation process.

Since the radiation (and hence absorption) associated with the various transitions are coupled, the strengths, $\beta_{J\to I}$, have to be determined in a selfconsistent manner. This is done by inserting the ansatz in Eq. (186) back into Eq. (187). This leads to the equation

$$\beta_{J\to I} - \sum_{K,L} N_{K\to L}^{J\to I} \beta_{L\to K} = H_{J\to I}, \tag{188}$$

where

$$N_{K\to L}^{J\to I} = \int \vec{J}_{J\to I}^{\mathrm{T}}(\vec{r}) \cdot \vec{F}_{K\to L}^{\mathrm{T}}(\vec{r}) d^3r, \tag{189}$$

and

$$H_{J\to I} = \int \vec{J}_{J\to I}^{\mathrm{T}}(\vec{r}) \cdot \vec{E}_{\mathrm{T}}^{\mathrm{drive}}(\vec{r}) d^3r \tag{190}$$

in principle are known numbers. Letting the $J \to I$-transition run through all possible (relevant) combinations, Eq. (188) gives a set of inhomogeneous algebraic equations among the unknown $\beta_{J\to I}$'s. The solutions to the (homogeneous) set of equations, appearing in the absence of the driving field $\vec{E}_{\mathrm{T}}^{\mathrm{drive}}$, are the possible local-field resonances of the system. Thus, from the condition that the determinant (Det$\{\cdots\}$) of the system of homogeneous equations vanishes, i.e.

$$\mathrm{Det}\left\{\delta_{J\to I,K\to L} - N_{K\to L}^{J\to I}(\omega^{\mathrm{RES}})\right\} = 0, \tag{191}$$

where $\delta_{J\to I,K\to L}$ is the Kronecker delta, the local-field resonance frequencies ω^{RES} may be obtained. An elaborated discussion of the coupled-antenna picture described above can be found in Ref. 9.

6. Three examples from the physics of mesoscopic systems

6.1. LINEAR ELECTRODYNAMICS OF QUANTUM WELLS

Let us briefly review the basic microscopic formalism describing the linear electrodynamics of a quantum well placed on top of a dielectric substrate occupying, in a Cartesian

xyz-coordinate system, the half-space $z \geq 0$. Inside the quantum well the local field, $\vec{E}(z;\vec{q}_{\|},\omega) \equiv \vec{E}(z)$, satisfies the integral equation

$$\vec{E}(z) = \vec{E}^B(z) + \int \overleftrightarrow{K}(z,z') \cdot \vec{E}(z')dz', \tag{192}$$

in the random-phase-approximation (RPA) approach, under the assumption that the entire system exhibits translational invariance parallel to the plane of the quantum well and is excited by a plane, monochromatic wave having a wave vector $\vec{q}_{\|}$ along the surface. The kernel in Eq. (192) is given by

$$\overleftrightarrow{K}(z,z') = -i\mu_0\omega \int \overleftrightarrow{G}^B_{<<}(z,z') \cdot \overleftrightarrow{\sigma}(z'',z')dz''. \tag{193}$$

In Eq. (192), $\vec{E}^B(z)$ is the background field consisting of the incident field plus the field reflected from the substrate in the absence of the quantum well. The associated electromagnetic propagator is denoted by $\overleftrightarrow{G}^B_{<<}(z,z')$, cf. Ref. 8. The quantity $\overleftrightarrow{\sigma}(z'',z')$ is the conductivity tensor in the RPA approach. Formally, the solution of Eq. (192) may be written

$$\vec{E}(z) = \int \overleftrightarrow{\Gamma}(z,z') \cdot \vec{E}^B(z')dz', \tag{194}$$

where $\overleftrightarrow{\Gamma}(z,z')$ is the field-field response tensor. Once the field-field response function has been obtained, the induced current density, $\vec{J}(z)$, in the quantum well can be calculated from

$$\vec{J}(z) = \int \overleftrightarrow{\sigma}^B(z,z') \cdot \vec{E}^B(z')dz', \tag{195}$$

where

$$\overleftrightarrow{\sigma}^B(z,z') = \int \overleftrightarrow{\sigma}(z,z'') \cdot \overleftrightarrow{\Gamma}(z'',z')dz'' \tag{196}$$

is the so-called background (or external) conductivity response tensor.

For observation points outside the quantum well in the vacuum domain the prevailing electric field will be given by

$$\vec{E}(z) = \vec{E}^B(z) - i\mu_0\omega \int \overleftrightarrow{P}^B_{<<}(z,z') \cdot \overleftrightarrow{\sigma}^B(z',z'') \cdot \vec{E}^B(z'')dz''dz', \tag{197}$$

where the propagator

$$\overleftrightarrow{P}^B_{<<}(z,z') = \overleftrightarrow{G}^B_{<<}(z,z') - \left(\frac{c_0}{\omega}\right)^2 \delta(z-z')\vec{e}_z\vec{e}_z \tag{198}$$

is obtained from $\overleftrightarrow{G}^B_{<<}(z,z')$ by subtraction of the self-field part. In Eq. (198), $\vec{e}_z$ is a unit vector in the z-direction. The expression in Eq. (197) can be rewritten in a physically appealing form by use of the displacement matrix $\overleftrightarrow{N}(z'',z_0)$ given via

$$\vec{E}^B(z'') = \overleftrightarrow{N}(z'',z_0) \cdot \vec{E}^B(z_0). \tag{199}$$

Hence,

$$\vec{E}(z) = \vec{E}^B(z) - i\mu_0\omega\overleftrightarrow{P}^B_{<<}(z,z_0) \cdot \vec{J}^S(z_0), \tag{200}$$

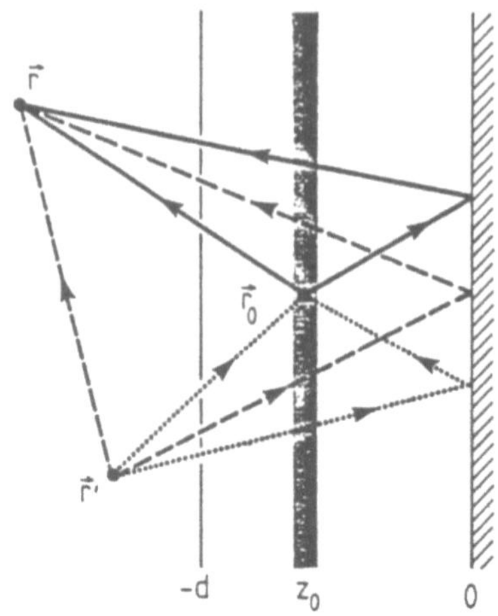

Figure 1. Schematic illustration of the sheet-model radiation channels. The quantum well occupies the region $-d \leq z \leq 0$, and the equivalent current-density sheet is placed at $z = z_0$. Radiation from a source point ($\vec{r}'$) can reach the point of observation ($\vec{r}$) by various channels. The contribution from the background field (dashed lines) consists of a direct term plus an indirect one (reaching $\vec{r}$ upon reflection from the substrate). As indicated by the dotted line, a secondary source is created at the point $\vec{r}_0$ in the sheet by the primary radiation from $\vec{r}'$, reaching $\vec{r}_0$ by the direct and indirect paths. The radiation from $\vec{r}_0$ in turn reaches $\vec{r}$ by direct and indirect propagation, as indicated by the solid lines.

where

$$\vec{J}^S(z_0) = \overleftrightarrow{S}(z_0) \cdot \vec{E}^B(z_0), \tag{201}$$

with

$$\overleftrightarrow{S}(z_0) = \int \overleftrightarrow{N}(z', z_0) \cdot \overleftrightarrow{\sigma}^B(z', z'') \cdot \overleftrightarrow{N}(z'', z_0) dz'' dz'. \tag{202}$$

The result in Eq. (200) may be interpreted as follows. The prevailing field outside the quantum well at the plane z consists of a sum of the background field and the field radiated from a current sheet carrying a sheet (superscript S) current density $\vec{J}^S(z_0)$ and placed at an arbitrary chosen position z_0 inside the quantum well domain (see also Fig. 1). The sheet conductivity tensor, $\overleftrightarrow{S}(z_0)$, is given by the expression in Eq. (202). Though the sheet conductivity depends on z_0, the field radiated from the sheet into the vacuum domain is independent of z_0. If the sheet is so thin that the phase shift of the electromagnetic background field across the sheet is negligible, the sheet conductivity tensor is given by

$$\overleftrightarrow{S}^{\text{ED-ED}} = \int \overleftrightarrow{\sigma}^B(z', z'') dz'' dz'. \tag{203}$$

In the above-mentioned case the sheet acts as an electric dipole receiver and radiator, hence the superscript ED-ED.

In macroscopic electrodynamics the presence of so-called surface currents and charges, assumed to be present only in a layer of infinitesimal thickness located at the boundary, affects the electromagnetic boundary (jump) conditions at a vacuum-medium interface[10]. The ED-ED sheet thus may be considered as the microscopic analogy of a boundary carrying induced currents and charges. If one assumes that the quantum well extends from $z = 0^-$ to $z = 0^+$, the jump conditions for the electromagnetic field across the sheet are, cf. Ref. 12

$$\vec{e}_x \cdot \left[\vec{E}(0^+) - \vec{E}(0^-)\right] = \frac{1}{\varepsilon_0}\frac{q_\|}{\omega}\vec{e}_z \cdot \overleftrightarrow{S}_0 \cdot \vec{E}(0^-), \tag{204}$$

$$\vec{e}_z \cdot \left[\vec{D}(0^+) - \vec{D}(0^-)\right] = \frac{1}{\varepsilon_0}\frac{q_\|}{\omega}\vec{e}_x \cdot \overleftrightarrow{S}_0 \cdot \vec{D}(0^-), \tag{205}$$

and

$$\vec{e}_z \times \left[\vec{B}(0^+) - \vec{B}(0^-)\right] = \mu_0 \left(\overleftrightarrow{U} - \vec{e}_z\vec{e}_z\right) \cdot \overleftrightarrow{S}_0 \cdot \vec{E}(0^-), \tag{206}$$

where

$$\overleftrightarrow{S}_0 = \overleftrightarrow{S}^{\text{ED-ED}} \cdot \left[\overleftrightarrow{U} - i\mu_0\omega\overleftrightarrow{P}^B_{<<}(0^-,0^-) \cdot \overleftrightarrow{S}^{\text{ED-ED}}\right]^{-1}. \tag{207}$$

As discussed in detail in Ref. 12, the correct jump conditions are different from those derived in the macroscopic standard description[10]. This difference stems from the fact that oscillating surface currents flowing perpendicular to the surface plane are neglected in the traditional approach. Only by keeping the normal component or the surface current does consistency exist among the various choices of jump conditions that one uses to derive the Fresnel reflection and transmission coeffiecients. Although the surface currents and charges are induced by the background field, the question of how to relate these quantities to the prevailing field is left untouched in the textbook descriptions. The analysis outlined in this section demonstrates that to derive the jump conditions for the electromagnetic field inevitably one has to deal with the local-field problem of the quantum well (sheet). A detailed account of the sheet-model description, including an analysis of the dynamics inside the sheet can be found in Refs. 12 and 13, where also references to the work of other authors as well as comparisons to available experimental data are given.

6.2. QUANTUM DOTS OF LIGHT

Let us imagine that a mesoscopic particle is excited by a monochromatic external field $\vec{E}^e(\vec{r};\omega) \equiv \vec{E}^e(\vec{r})$. In the RPA approach, the field inside the particle fulfils the integral equation

$$\vec{E}(\vec{r}) = \vec{E}^e(\vec{r}) + \int \overleftrightarrow{K}(\vec{r},\vec{r}\,') \cdot \vec{E}(\vec{r}\,')d^3r', \tag{208}$$

with a kernel given by

$$\overleftrightarrow{K}(\vec{r},\vec{r}\,') = -i\mu_0\omega \int \overleftrightarrow{G}_0(\vec{r},\vec{r}\,'') \cdot \overleftrightarrow{\sigma}(\vec{r}\,'',\vec{r}\,')d^3r''. \tag{209}$$

The formal solution of Eq. (208) is as before

$$\vec{E}(\vec{r}) = \int \overleftrightarrow{\Gamma}(\vec{r},\vec{r}\,') \cdot \vec{E}^e(\vec{r}\,')d^3r', \tag{210}$$

where $\overleftrightarrow{\Gamma}(\vec{r},\vec{r}\,')$ is the field-field response tensor in the RPA approximation. In terms of $\overleftrightarrow{\sigma}(\vec{r},\vec{r}\,')$ and $\overleftrightarrow{\Gamma}(\vec{r}\,',\vec{r}\,'')$ the induced current density inside the particle takes the form

$$\vec{J}(\vec{r}) = \int \overleftrightarrow{\sigma}(\vec{r},\vec{r}\,') \cdot \overleftrightarrow{\Gamma}(\vec{r}\,',\vec{r}\,'') \cdot \vec{E}^e(\vec{r}\,'')d^3r''d^3r'. \tag{211}$$

Since a mesoscopic particle has an extension much less than the wavelength of the external field it is usually correct to assume that the external field is constant across the particle, i.e. $\vec{E}^e(\vec{r}'') \simeq \vec{E}^e(\vec{0})$, if the center of the particle is located at $\vec{r}'' = \vec{0}$. Thus, because the electric dipole moment of the current density distribution is

$$\vec{p}(\omega) = \frac{i}{\omega}\int \vec{J}(\vec{r})d^3r, \tag{212}$$

one obtains

$$\vec{p}(\omega) = \overleftrightarrow{\alpha}(\omega)\cdot\vec{E}^e(\vec{0};\omega), \tag{213}$$

where

$$\overleftrightarrow{\alpha} = \frac{i}{\omega}\int \overleftrightarrow{\sigma}(\vec{r},\vec{r}')\cdot\overleftrightarrow{\Gamma}(\vec{r}',\vec{r}'')d^3r''d^3r'd^3r \tag{214}$$

is the dyadic ED-ED polarizability of the mesoscopic particle. Due to the fact that the field-field response tensor satisfies the integral equation

$$\overleftrightarrow{\Gamma}(\vec{r},\vec{r}') = \overleftrightarrow{U}\delta(\vec{r}-\vec{r}') + \int \overleftrightarrow{K}(\vec{r},\vec{r}'')\cdot\overleftrightarrow{\Gamma}(\vec{r}'',\vec{r}')d^3r'', \tag{215}$$

the RPA-polarizability may be written in the appealing form

$$\overleftrightarrow{\alpha}(\omega) = \overleftrightarrow{\alpha}_0(\omega) + \overleftrightarrow{\Delta}(\omega), \tag{216}$$

where

$$\overleftrightarrow{\alpha}_0(\omega) = \frac{i}{\omega}\int \overleftrightarrow{\sigma}(\vec{r},\vec{r}')d^3r'd^3r \tag{217}$$

is the polarizability in the absence of local-field corrections, and

$$\overleftrightarrow{\Delta}(\omega) = \frac{i}{\omega}\int \overleftrightarrow{\sigma}(\vec{r},\vec{r}')\cdot\overleftrightarrow{K}(\vec{r}',\vec{r}'')\cdot\overleftrightarrow{\Gamma}(\vec{r}'',\vec{r}''')d^3r'''d^3r''d^3r'd^3r \tag{218}$$

is the local-field correction. In Figs. 2–4 are shown two examples underlining the role of local-field corrections for the ED-ED polarizability of mesoscopic metal (Fig. 2) and semiconductor (Figs. 3 and 4) particles. The data in these figures are taken from Refs. 14 and 15. The polarizability discussed above determines among other things the ability of the mesoscopic particle to radiate. To which extend is it possible to confine the radiated (or absorbed) electromagnetic field in space then? In a qualitative approach the spatial confinement problem of the source field may be discussed within the framework of a self-field approximation. Neglecting retardation effects the local field will be given by

$$\vec{E}(\vec{r}) = \vec{E}^e(\vec{r}) + \frac{1}{3i\varepsilon_0\omega}\vec{J}_T(\vec{r}) + \frac{1}{i\varepsilon_0\omega}\vec{J}_L(\vec{r}). \tag{219}$$

It appears from Eq. (219) that we cannot hope to confine the local field to a volume smaller than that occupied by the transverse (or longitudinal) part of the induced current density, i.e. V_T. The reason that the confinement volume of the electromagnetic field does not coincide with the region in space where the induced current density itself is different from zero originates in the fact that the near-field electrodynamics is different for transverse and longitudinal fields, cf. the "extra" factor 3^{-1} in front of the $\vec{J}_T$-term in Eq. (219).

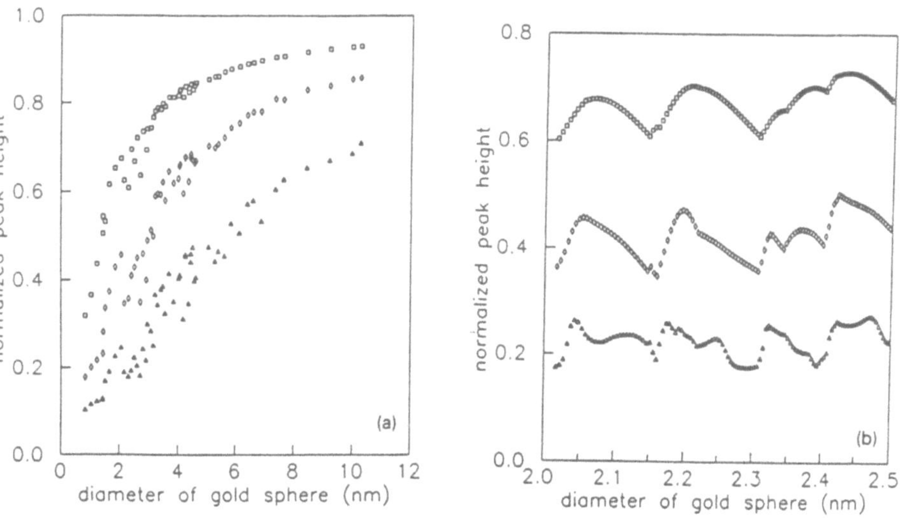

Figure 2. (a) Normalized to the Lorentz-Mie peak value, the height of the resonance peak of the imaginary part of the polarizability of a gold sphere is plotted as a function of the sphere diameter for three different damping rates, viz. $0.1\omega_p$ (□), $0.03\omega_p$ (◇), and $0.01\omega_p$ (△), ω_p being the unscreened plasmon frequency of the homogeneous electron gas. (b) A detailed view of the normalized height of the resonance peak of (a) in the range 2.0nm≤sphere diameter≤2.5nm. Going from the left to the right each new point corresponds to an addition of two electrons, one with spin-up and one with spin-down.

To illustrate the confinement problem let us consider two extremely simplified examples. In the first example, we assume complete confinement of the induced current density, at say $\vec{r} = \vec{0}$. The model current density thus is

$$\vec{J}(\vec{r}) = \vec{J}_0 \delta(\vec{r}). \tag{220}$$

The assumed complete electron confinement leads to a singularity at $\vec{r} = 0$ in $\vec{J}_{\mathrm{T}}$ (and $\vec{J}_{\mathrm{L}}$ as well). For $\vec{r} \neq 0$, the forms of $\vec{J}_{\mathrm{T}}$ and $\vec{J}_{\mathrm{L}}$ are given by

$$\vec{J}_{\mathrm{L}}(\vec{r}) = -\vec{J}_{\mathrm{T}}(\vec{r}) = \frac{1}{4\pi r^3}\left(\vec{U} - \frac{3\vec{r}\vec{r}}{r^2}\right) \cdot \vec{J}_0. \tag{221}$$

For a system with perfect electronic confinement the electrodynamic field confinement exhibits characteristic wings of the form r^{-3}. This was to be expected since the longitudinal current density distribution from an oscillating point dipole is identical to the ED-field distribution in the quasi-static limit. In the second example, I remove the unphysical assumption that the electron motion is completely confined in space and consider a hydrogen(-like) atom undergoing a prescribed motion between the $1s$ and $2p_z$ energy eigenstates. Described in a spherical (r, θ, ϕ)-coordinate system with unit vectors $\vec{e}_r$, $\vec{e}_\theta$, and $\vec{e}_\phi$ the $1s \rightarrow 2p_z$ transition current density is given by

$$\vec{J}_{1s\rightarrow 2p_z}(\vec{r}) = A\left[\vec{e}_r\left(2 + \frac{r}{a_0}\right)\cos\theta - 2\vec{e}_\theta \sin\theta\right] e^{-br}, \tag{222}$$

where A is a constant and $b = 3/(2a_0)$, a_0 being the effective Bohr radius. The transverse

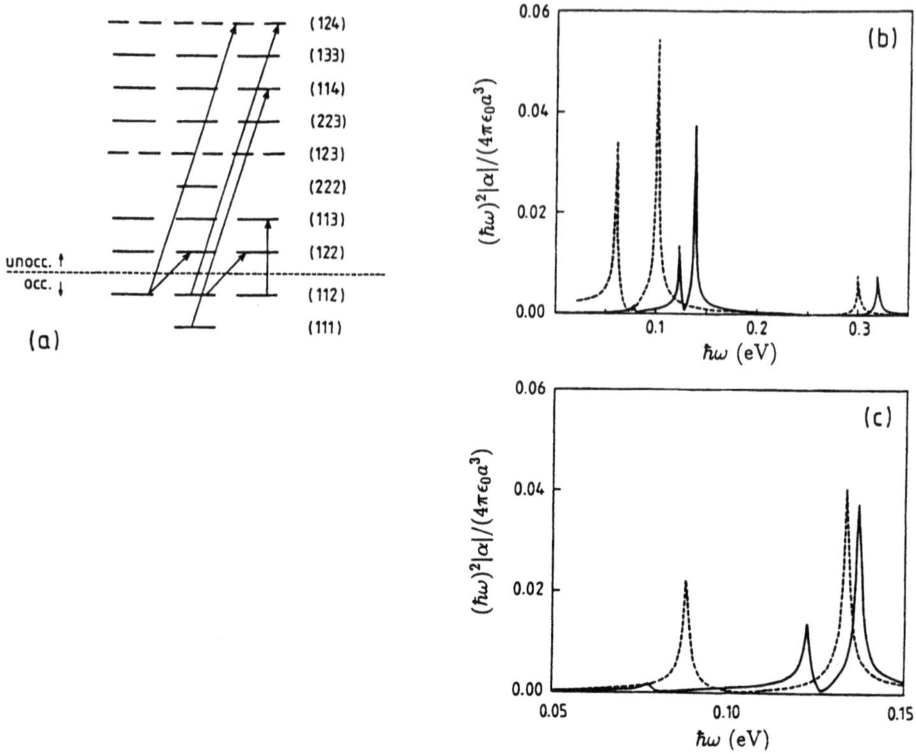

Figure 3. Numerical magnitude of the normalized polarizability of a cubic GaAs particle (side length $a = 200\text{Å}$) as a function of the photon energy calculated with (solid line in (b) and (c)) and without the local-field correction (dashed line in (b)), as well as within the so-called self-coupling approximation (dashed line in (c)). In the self-coupling approximation only the radiative coupling between the up- and down-transitions of a given pair of levels is included. The appropriate transitions and energy eigenstates are shown in (a). Readers who want to understand the level classification used are referred to Ref. 14.

and longitudinal parts of the current density in Eq. (222) are given by

$$\vec{J}^{\mathrm{T(L)}}_{1s\to 2p_z}(\vec{r}) = \vec{J}^{\mathrm{T(L)}}_{\theta}\vec{e}_\theta + \vec{J}^{\mathrm{T(L)}}_{r}\vec{e}_r, \tag{223}$$

where

$$\vec{J}^{\mathrm{T}}_{\theta} = -\frac{8A}{3}\sin\theta\left\{e^{-br}\left[1+\frac{1}{br}+\frac{2}{(br)^2}+\frac{2}{(br)^3}\right]-\frac{2}{(br)^3}\right\}, \tag{224}$$

$$\vec{J}^{\mathrm{L}}_{\theta} = \frac{2A}{3}\sin\theta\left\{e^{-br}\left[1+\frac{4}{br}+\frac{8}{(br)^2}+\frac{8}{(br)^3}\right]-\frac{8}{(br)^3}\right\}, \tag{225}$$

$$\vec{J}^{\mathrm{T}}_{r} = -\frac{16A}{3}\cos\theta\left\{e^{-br}\left[\frac{1}{br}+\frac{2}{(br)^2}+\frac{2}{(br)^3}\right]-\frac{2}{(br)^3}\right\}, \tag{226}$$

and

$$\vec{J}^{\mathrm{L}}_{r} = \frac{4A}{3}\cos\theta\left\{e^{-br}\left[\frac{3+br}{2}+\frac{4}{br}+\frac{8}{(br)^2}+\frac{8}{(br)^3}\right]-\frac{8}{(br)^3}\right\}. \tag{227}$$

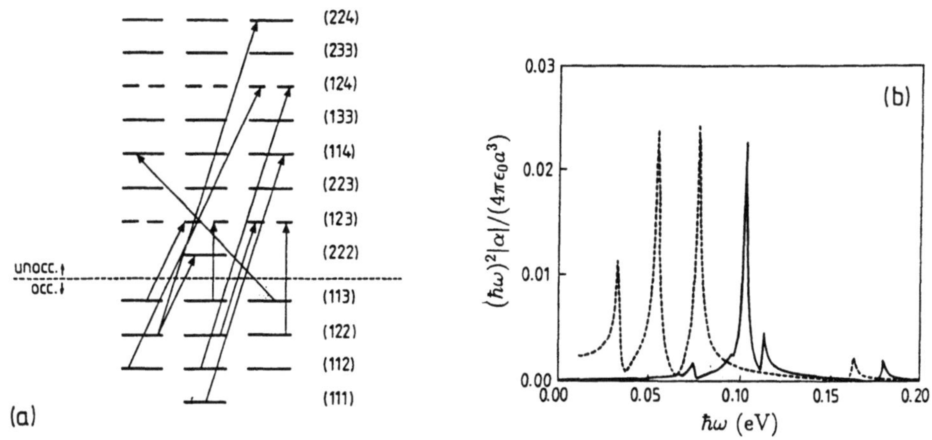

Figure 4. Numerical magnitude of the normalized polarizability of a cubic GaAs particle (side length $a = 271Å$) as a function of the photon energy calculated with (solid line in (b)) and without the local-field correction (dashed line in (b)). The appropriate transitions and energy eigenstates are shown in (a) [classification as in Fig. 3].

The tails of the two current densities exhibit a characteristic $(br)^{-3}$-dependence. This means that although the spatial confinement region for the transition current density itself essentially has the exponential form e^{-br} the confinement region for the electromagnetic field is determined by the much slower decaying $(br)^{-3}$-function. In passing we note that despite it appears at a first sight as if the transverse and longitudinal current densities are singular at $r = 0$, this is not so, as one can easily verify.

The considerations put forward above indicate that it might be possible to focus electromagnetic fields to beam waists considerably less in spatial extension than predicted by the classical far-field diffraction theory. A new route for achieving confinement of electromagnetic fields beyond the classical diffraction limit is based on the phase conjugation of the middle- and near-field radiated by the tip of an optical near-field microscope. For a description of the various theoretical and experimental aspects of light confinement by phase conjugation the reader is referred to Refs. 16 and 17. Light foci smaller than those possible within the scope of classical diffraction theory are called quantum dots of light[16].

6.3. PHOTON DRAG IN MESOSCOPIC RINGS

In a nonlinear medium containing mobile charged particles the excitation of even order higher harmonics is accompanied by a generation of a DC current in the particle system. This so-called photon-drag effect, which has been reviewed by Mulser[18] recently, has been studied for many years in semiconductors and metals, where a net momentum is transferred from the electromagnetic field to the conduction electron system. Since the momentum exchange between the field and particles requires that the finite value of the wave vector of the electromagnetic field is taken into account, from the outset a nonlocal response theory is needed *a priori*. In mesoscopic media the photon-drag phenomenon until quite recently has been studied only in quantum well systems[19–24]. A short while ago the present author proposed that one should draw attention to the possibility of generating photon-drag currents in mesoscopic semiconductor and metal rings[9]. When the mesoscopic medium forms a closed ring a photon-drag current may be excited transferring only angular momentum to the conduction electrons of the ring. To create a photon-drag current in a closed ring it is thus necessary that the incident electromagnetic field is elliptically or circularly polarized.

If the intensity of the electromagnetic field is not too high the photon-drag current is dominated by contributions which are of second (lowest) order in the field amplitude. To calculate the photon-drag current density, $\vec{J}_0(\vec{r})$, one hence may start from the expression

$$\vec{J}_0(\vec{r}) = \mathrm{Tr}\left\{\rho_0 \vec{j}_F\right\} + \tfrac{1}{4}\mathrm{Tr}\left\{\rho_1 \vec{j}_1^{\dagger}\right\} + \tfrac{1}{4}\mathrm{Tr}\left\{\rho_1^{\dagger} \vec{j}_1\right\}, \tag{228}$$

where ρ_0 and ρ_1 are the nonlinear DC and first harmonic parts of the density-matrix operator, respectively, and $\vec{j}_F$ and $\vec{j}_1$ are the free (F) and first harmonic components of the current-density operator. The dagger ($\dagger$) stands for Hermitian conjugation.

A particularly simple and beautiful analysis of the photon-drag phenomenon can be carried out for a circular mesoscopic ring with cross-sectional dimensions so small that only transitions between quantum states along the ring are allowed. If the ring is excited by a right- (plus sign) or left- (minus sign) hand circular polarized external field the associated photon-drag current along the ring, $I_0^{\pm}$, is to be calculated from

$$I_0^{\pm} = \vec{e}_\theta \cdot \int \vec{J}_0^{\pm}(\vec{r})dA, \tag{229}$$

where $\vec{e}_\theta$ is a unit vector along the ring. The integration in Eq. (229) extends over the cross section of the ring. If one assumes that the external plane-wave field propagates in the direction perpendicular to the plane of the circular ring the photon-drag current becomes in the RPA approach[9]

$$I_0^{\pm} = -\frac{\hbar e^3 |E_0|^2}{32\pi m^2 \omega^2 r_0^2} \sum_{\alpha} \frac{(f_\alpha - f_{\alpha\pm1})(2\alpha \pm 1)}{\hbar(\omega + i/\tau) + \varepsilon_\alpha - \varepsilon_{\alpha\pm1}} + \text{cc.}, \tag{230}$$

where E_0 is the amplitude of the circular polarized external field, r_0 is the radius of the ring, and τ is a phenomenologically introduced electronic relaxation time. The summation in Eq. (230) extends over the various energy eigenstates along the ring, i.e. $\alpha = \pm 1, \pm 2, \cdots$, and since it has been assumed that the electron dynamics is free-electron-like along the ring, $\varepsilon_\alpha - \varepsilon_{\alpha\pm1} = -\hbar^2(1 \pm 2\alpha)/(2mr_0^2)$. The quantities f_α and $f_{\alpha\pm1}$ are Fermi-Dirac distribution factors. Although not indicated in the notation, these functions also depend on the two remaining spatial quantum numbers needed for a complete characterization of the various electronic energy eigenstates of the ring.

Let us end our brief introduction to the photon-drag phenomena in mesoscopic rings by presenting a few numerical data for a gold ring of radius $r_0 = 0.41\mu$m, and let us assume that the cross section of the ring is rectangular with the dimensions $a = 49$Å and $b = 38$Å. For simplicity, we adopt the free-electon infinite-barrier model, and assume that the energy eigenvalues are close to those of a simple box model, i.e.

$$\varepsilon_\alpha = \frac{\pi\hbar^2}{2m}\left[\frac{p^2}{a^2} + \frac{q^2}{b^2} + \frac{\alpha^2}{\pi^2 r_0^2}\right], \tag{231}$$

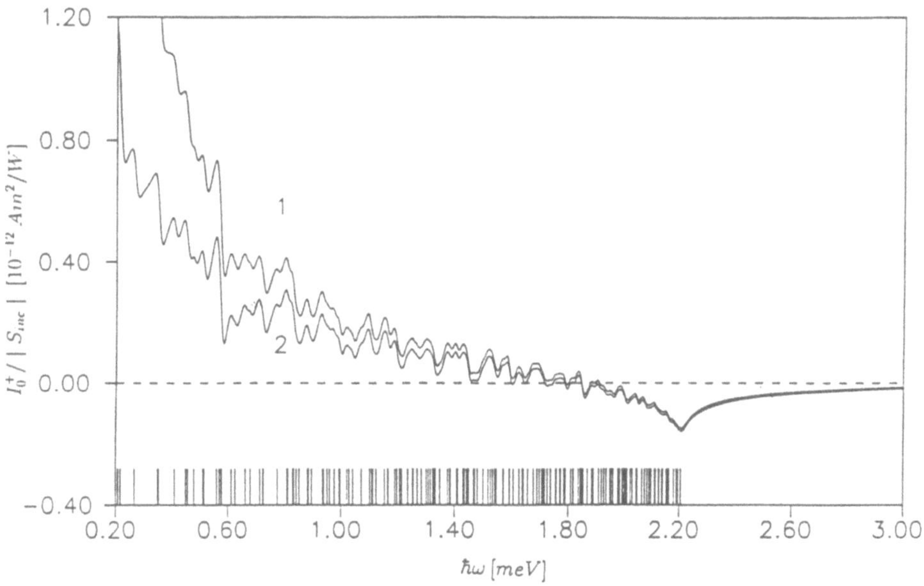

Figure 5. Normalized photon-drag current, I_0^+/S_{inc}, as a function of the photon energy for a mesoscopic Au ring (dimensions given in the main text) with an electronic relaxation time $\tau = 5 \times 10^{-11}$s. For curve 1, nonresonant (counter-rotating) terms have been omitted from the calculation. Both resonant and non-resonant terms are kept for curve 2.

where $p, q, \alpha = \pm 1, \pm 2, \cdots$. In a low-temperature (T=0K) calculation all electronic states below the Fermi energy ($\varepsilon_F \approx 5.51$eV in Au) are occupied, and all states above are empty. This means that for a chosen set of (p,q)-quantum numbers among the say positive α-values only the one satisfying the inequalities

$$\varepsilon_{p,q,\alpha} < \varepsilon_F < \varepsilon_{p,q,\alpha\pm1} \tag{232}$$

can contribute to the photon-drag effect. Letting the quantum numbers p and q run through all the relevant combinations satisfying the inequality $\pi^2\hbar^2(p^2/a^2+q^2/b^2)/(2m) < \varepsilon_F$, all the relevant positive α-values can be determined. For the present Au ring the actual number of α values is 199. For each positive α-transition $\alpha \Rightarrow \alpha+1$, there is of course an associated transition $-\alpha \Rightarrow -(\alpha+1)$ between negative quantum-number states. Since the photon-drag current in the present (low intensity) approach is proportional to the square of the amplitude of the external field, the ratio between $I_0^\pm$ and the magnitude, S_{inc}, of the Poynting vector of the incident (external) electromagnetic field is independent of the field intensity. In Figs. 5–7 we show the results of numerical calculations[25] of the normalized photon-drag current I_0^+/S_{inc} as a function of the photon energy, $\hbar\omega$, for different relaxation times. The small vertical lines inserted in the various figures mark the electronic transition energies which contribute to the photon-drag current along the ring. In Fig. 5 a relaxation time of $\tau = 5 \times 10^{-11}$s was chosen, and it appears from this figure that I_0^+ goes rapidly towards zero with increasing photon energy once $\hbar\omega$ is larger than the largest of the relevant transition energies, as one would expect. At low frequencies, $I_0^+ > 0$ due to the

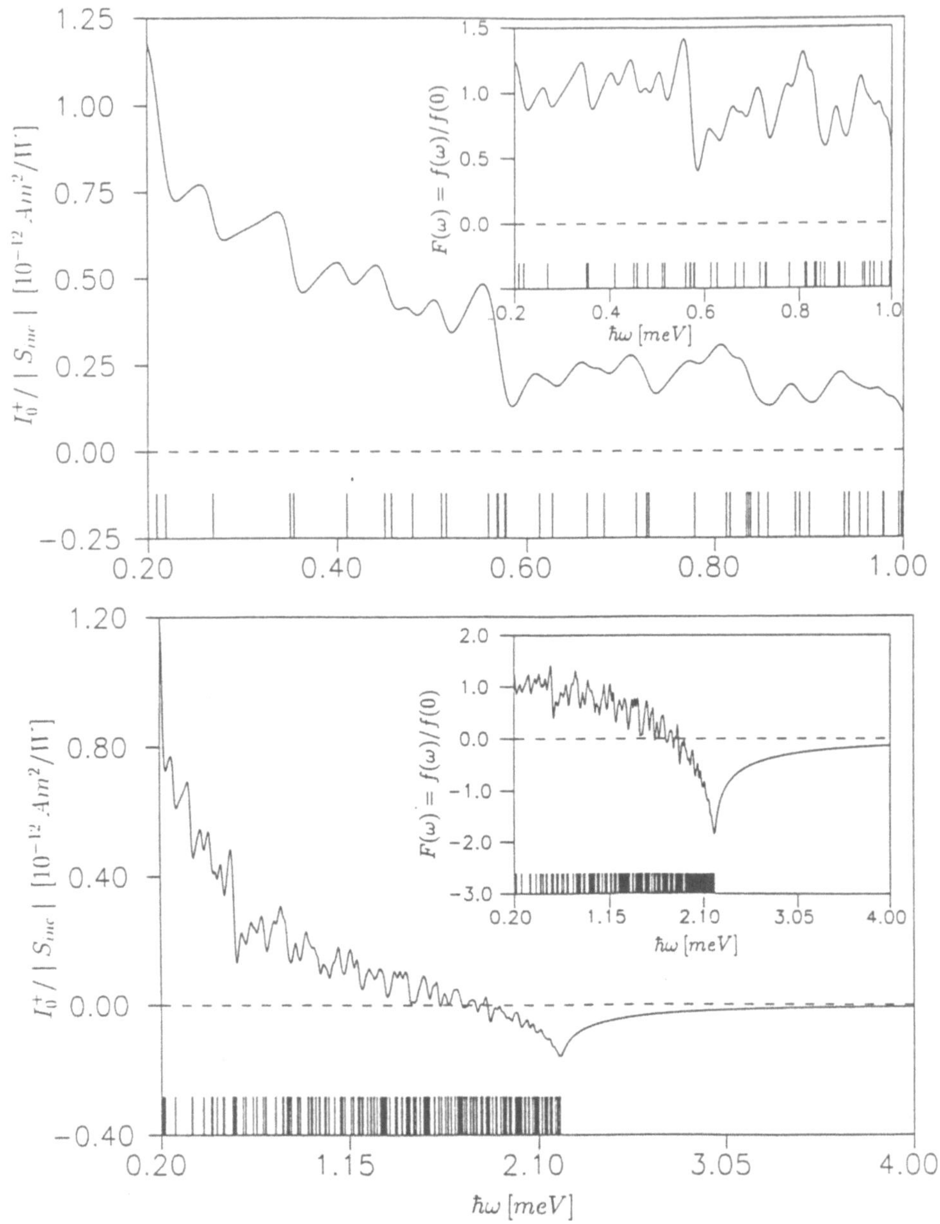

Figure 6. Normalized photon-drag current, $I_0^+/S_{\rm inc}$, as a function of the photon energy for four different relaxation times, viz. (1) 4×10^{-11}s, (2) 3×10^{-11}s, (3) 2×10^{-11}s and (4) 1×10^{-11}s (same Au ring as in Fig. 5).

fact that $\hbar\omega+\varepsilon_\alpha<\varepsilon_{\alpha+1}$ for all α, cf. Eq. (230). As the frequency is increased more and

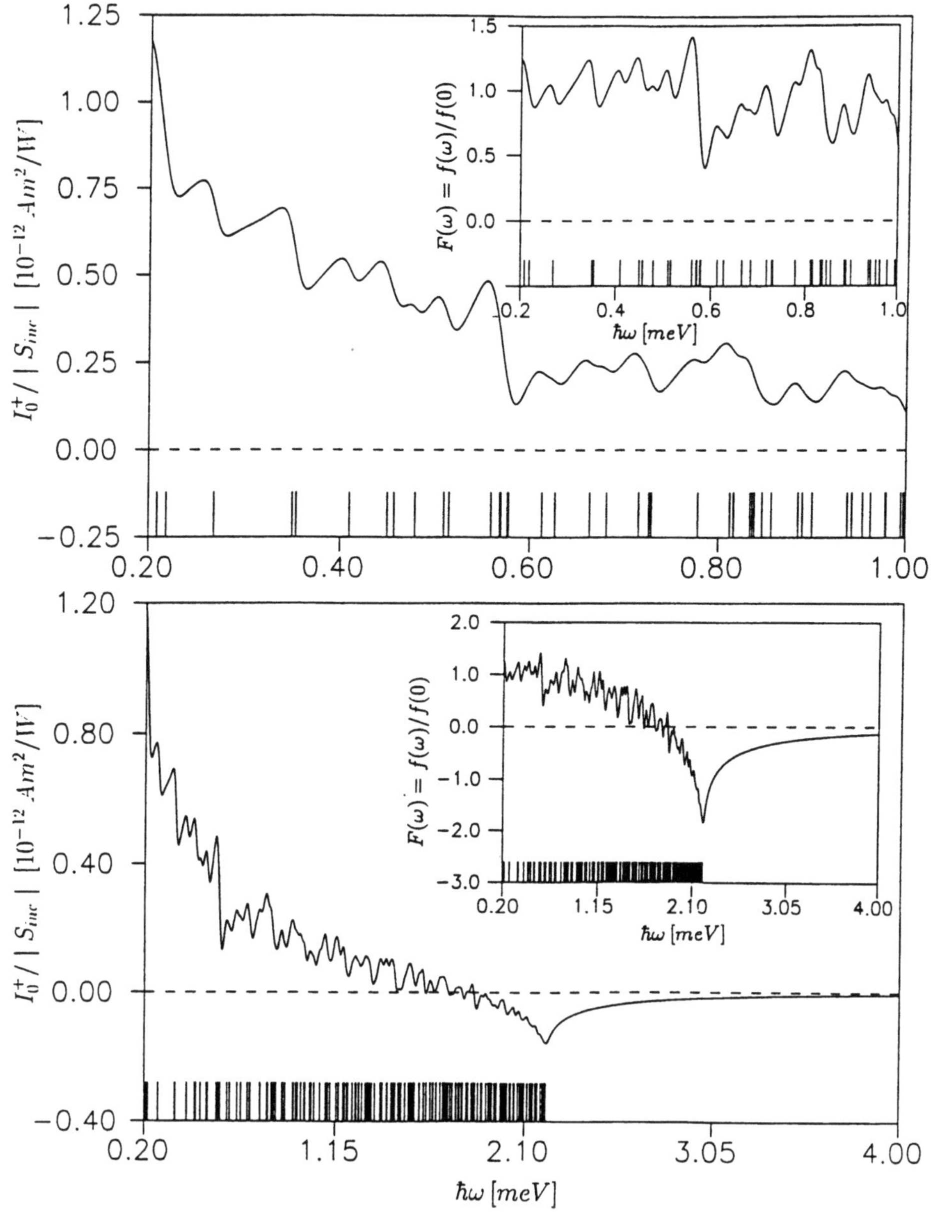

Figure 7. Normalized photon-drag current, I_0^+/S_{inc}, as a function of the photon energy for a relaxation time $\tau = 1 \times 10^{-10}$s. The upper part of the figure shows a detailed view of a part of the spectrum plotted in the bottom figure. In the inserts the photon-drag current has been normalized in such a manner that it approaches unity in the low-frequency limit.

more transitions give opposite contributions to I_0^+ ($\hbar\omega + \varepsilon_\alpha > \varepsilon_{\alpha+1}$ for these transitions), and hence the current exhibits an overall but slow decrease with increasing photon energy. At a certain photon energy the current, I_0^+, changes sign. Curve 1 in Fig. 5 is obtained from a calculation where non-resonant (counter-rotating) terms are neglected, and curve 2 stems from a calculation where both resonant and non-resonant terms are retained. The "noisy" structure appearing in the frequency range below the highest electronic transition frequency is a more or less resolved fingerprint of the individual electronic transitions. For large electronic collision rates, only groups of narrow-lying transitions reflect themselves in I_0^+, cf. the results of Fig. 6. A good fingerprint of most of the transitions is obtained when the collision time is as long as $\tau = 1 \times 10^{-10}$s, see Fig 7. In all figures (5 to 7) an overall ω^{-2} dependence, originating in the ω^{-2}-factor in front of the summation sign in Eq. (230), is present. This factor essentially has no particular relation to the photon-drag phenomenon and is removed in the plots shown in the inserts of Fig. 7. Also, the curves in the inserts have been further normalized so that the ratio $I_0^+/S_{\rm inc}$ approaches one in the limit $\omega \to 0$.

References

1. A.O. Barut and J. Kraus, Found. Phys. 13, 189 (1983).
2. A.O. Barut and J.F. Van Huele, Phys. Rev. A32, 3187 (1985).
3. A.O. Barut and J.P. Dowling, Phys. Rev. A36, 649 (1987).
4. A.O. Barut, J.P. Dowling, and J.F. Van Huele, Phys. Rev. A38, 4405 (1988).
5. B. Blaive, A.O. Barut, and R. Boudet, J. Phys. B: At. Mol. Opt. Phys. 24, 3121 (1991).
6. A.O. Barut and B. Blaive, Phys. Rev. A45, 2810 (1992).
7. O. Keller, Phys. Rev. B34, 3883 (1986).
8. O. Keller, Phys. Rev. B37, 10588 (1988).
9. O. Keller, Phys. Rep. (in press).
10. J.D. Jackson, *Classical Electrodynamics* (Wiley, New York, 1975).
11. L.V. Keldysh, D.A. Kirznitz, and A.A. Maradudin (eds.), *The Dielectric Function of Condensed Systems, Vol. 24 in Modern Problems in Condensed Matter Sciences* (North-Holland, Amsterdam, 1989).
12. O. Keller, J. Opt. Soc. Amer. 12, 987 (1995).
13. O. Keller, J. Opt. Soc. Amer. 12, 997 (1995).
14. O. Keller, M. Xiao, and S. Bozhevolnyi, Opt. Commun. 102, 238 (1993).
15. O. Keller, M. Xiao, and S. Bozhevolnyi, Opt. Commun. 114, 491 (1995).
16. O. Keller, J. Nonlinear Optical Physics and Materials, Vol. 5, No. 1 and 2 (in press).
17. S.I. Bozhevolnyi, O. Keller, and I.I. Smolyaninov, Opt. Lett. 19, 1601 (1994).
18. P. Mulser, J. Opt. Soc. Amer. B2, 1814 (1985).
19. F.T. Vas'ko, Sov. Phys. Semicond. 19, 808 (1985).
20. S. Luryi, Phys. Rev. Lett. 58, 2263 (1987).
21. A.A. Grinberg and S. Luryi, Phys. Rev. B38, 87 (1988).
22. M.I. Stockman, L.N. Pandey, and T.F. George, Phys. Rev. Lett. 65, 3433 (1990).
23. O. Keller, in O. Keller (ed.) *Studies in Classical and Quantum Nonlinear Optics* (Nova Science Publishers, New York, 1995) p. 269.
24. O. Keller and G. Wang (unpublished).

PERSISTENT CURRENT AND PERSISTENT CHARGE IN NANOSTRUCTURES

I.O.KULIK[1]
Department of Physics, Bilkent University
Bilkent 06533, Ankara, Turkey

Abstract. Quantúm effects in the interaction of electromagnetic field with small, but not microscopically small, metallic particles and rings are investigated. The aspects of mesoscopic systems related to time-independent (or magnetic) Aharonov-Bohm effect, time-dependent (electric) Aharonov-Bohm effect, and to quantum high-frequency effects in a coupled system: optical fiber+ mesoscopic conducting loop are discussed.

1. Introduction

Electromagnetic field is primarily a wave. The quantum mechanics tell us that the field exists in a form of quanta, the photons.

Electrons are believed to be primarily particles, the point-like objects. Quantum physics then introduces a wave aspect of electron through the notion of particle-wave dualism. The electrons and photons are to be considered on same footing.

Unlike photons, electrons are charged, which brings a new aspect to the wave mechanics of electrons interacting with the electromagnetic field. This was first recognized by Aharonov and Bohm [1] who have shown that electromagnetic potentials $\mathbf{A}$, ϕ are of primary importance, rather than electromagnetic fields $\mathbf{H}$, $\mathbf{E}$ themselves, and that at certain topology of space (or space-time) it may appear that the effects related to vector potential $\mathbf{A}$ alone (with $\mathbf{H}$ and $\mathbf{E}$ identically equal to zero), or that of scalar potential ϕ alone (again, with $\mathbf{H} \equiv 0, \mathbf{E} \equiv 0$) may exist. These effects are: the persistent currents in metallic loops [2,3], persistent charges in metallic granules [4] and resistance oscillations in mesoscopic rings and networks [5,6].

We will consider the wave phenomena in collection of electrons which are large in number (say, $N \sim 10^{10}$ in a metallic granule of size $\sim 1\mu m$) but at certain condition may behave similar to a single atom. The condition is specified by the requirement that electron in a granule does not suffer inelastic collision which acts as a measuring event, resulting in the reduction of the wave packet and elimination of coherent electron phase. [The elastic scattering, on contrary, preserves the electron phase.] The phase-breaking length of electron

[1]Also at: B. Verkin Institute for Low Temperature Physics and Engineering, Acad. Sci. of Ukraine, Lenin ave. 47, Kharkov 310164, Ukraine.

T. Hakioğlu and A. S. Shumovsky (eds.), Quantum Optics and the Spectroscopy of Solids, 45–56.

l_φ is related to the electron-electron and electron-phonon scattering as

$$l_\varphi^{-1} \sim l_{inel}^{-1} \sim \frac{1}{v_F}(\lambda_{e-e}\frac{T^2}{\varepsilon_F} + \lambda_{e-ph}\frac{T^3}{\omega_D^2}) \tag{1}$$

in a clean metal ($l_{elast} \gg L$), and

$$l_\varphi \sim (l_{inel} l_{elast})^{1/2} \tag{2}$$

in a dirty metal ($l_{elast} \ll L$) where l_{inel}, l_{elast} are inelastic and elastic mean free paths, L is the typical size of the system, ω_D the Debye energy, and v_F and ε_F are the electron Fermi velocity and Fermi energy, respectively. λ_{e-e} and λ_{e-ph} are dimensionless electron-electron and electron-phonon coupling constants which can be crudely taken as quantities of order 1. Typically, $\varepsilon_F \sim 1 - 10eV$ and $\omega_D \sim 10 - 100meV$ which means that mesoscopic behavior appears in system of micron size at temperature T below $1K$.

2. Persistent current

Persistent current is a dissipationless non-decaying current in a mesoscopic loop induced by a static magnetic flux piercing the loop (Fig.1). The current was first predicted for a normal-metal loop in the paper [2] following the pioneering work of Imry and Gunther [7] who have shown that in a one-dimensional *superconducting* loop with a superconductivity destroyed by quantum fluctuations, a non-decaying, or "persistent" current (according to later terminology [3]) reappears . As stated in Ref.[2], in a normal metal "...the current

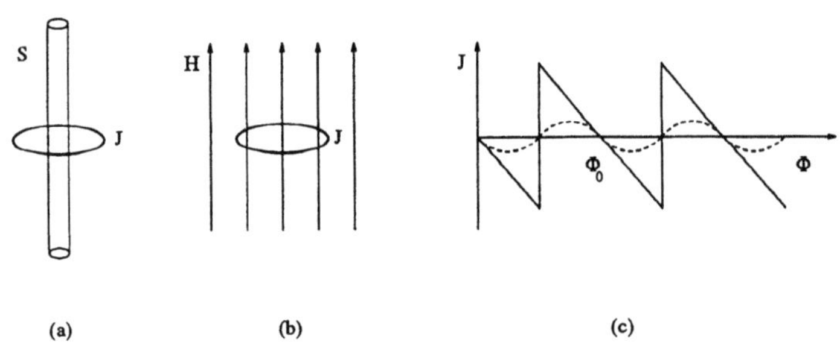

Figure 1. (a) Schematic of the Aharonov-Bohm effect in the ring with a solenoid, and, (b) in an homogeneous magnetic field; (c) The dependence of current on magnetic flux. Solid line corresponds to $T = 0$, dotted line to $T > 0$.

state corresponds to a minimum of free energy, so that allowance for dissipation does not lead to its decay".

The origin of the current in a ring can be understood with a generic Hamiltonian

$$H = -t\sum_{n=1}^{N}(a_n^+ a_{n+1} e^{i\alpha} + a_{n+1}^+ a_n e^{-i\alpha}). \tag{3}$$

where a_n is an electron annihilation operator at site n, t is the transmission amplitude between the nearest sites, and α is the phase related to the magnetic flux in the ring

$$\alpha = 2\pi\frac{\Phi}{N\Phi_0}, \quad \Phi_0 = \frac{hc}{e}. \tag{4}$$

The quantity $\Phi_0 = 4 \cdot 10^{-7} G \cdot cm^2$ is the flux quantum of a normal metal. The flux in the ring determined as

$$\Phi = \int \mathbf{A} dl = \int \mathbf{H} d\mathbf{S}, \tag{5}$$

can be produced either by a solenoid inserted inside the ring (Fig.1(a)), or by an externally applied magnetic field (Fig.1(b)). In the first case, no classical effect of magnetic field is expected because for the electrons confined within the ring, the magnetic field appears to be identically zero. The effect of vector potential is introduced by the phase factor $\exp(i\alpha)$ in the hopping amplitude in Eq.(3) where $\alpha = \int_n^{n+1} \mathbf{A} dl$. Calculating the eigenstates of the electron within the Hamiltonian of Eq.(3)-

$$\varepsilon_k = -2t\cos(k+\alpha), \quad k = \frac{2\pi}{N} m, \; m = 0, 1, 2 ... N-1 \tag{6}$$

and calculating the thermodynamic potential

$$\Omega = -T \sum_k \ln(1 + e^{-\beta(\varepsilon_k - \mu)}) \tag{7}$$

where $\beta = 1/T$ and μ is the chemical potential, we evaluate a current J as a derivative

$$J = -c\frac{\partial \Omega}{\partial \Phi}. \tag{8}$$

The dependence $\Omega(\Phi)$ vanishes in the limit $N \to \infty$ consistent with the van Leuven theorem stating that in classical mechanics thermodynamic parameters are independent of $\mathbf{A}$ and ϕ. Calculation shows that J is periodic in Φ with a period of flux quantum Φ_0, see Fig.1(c).

The oscillatory dependence is a hallmark of the persistent-current effect and serves as an experimental indication that the effect is there [8,9]. The magnitude of the persistent current

$$J_c \sim ev_F/L \sim 10^2 nA \quad \text{at} \quad T = 0 \text{ and } \quad L \sim 1\mu m \tag{9}$$

where $L = Na$ is the circumference of the ring (a is atomic period). Estimate (9) corresponds to a current produced by *one* electron orbiting around the ring with velocity of order of Fermi velocity $v_F \sim 10^8 cm/s$.

The magnetic moment corresponding to this current $M = \frac{1}{c} JS$ where S is the cross section of the ring, is much larger than microscopic Bohr magneton

$$M/\mu_B \sim k_F L \sim 10^4. \tag{10}$$

However, the magnetic energy associated with the moment, $E_1 \sim M^2/L^3$, is much smaller than electron level spacing $\Delta\varepsilon \sim \hbar v_F/L$,

$$E_1/\Delta\varepsilon \sim \frac{e^2/a}{mc^2} \sim 10^{-5} \tag{11}$$

which means that we may neglect the self-action of the current on the magnetic field within the ring.

The current remains nonzero if scattering (both elastic and inelastic), $3d$ lattice effects and finite temperature are taken into account. However, the magnitude of critical current reduces.

Inelastic scattering reduces the magnitude of J_c according to an exponential factor $\exp(-L/l_\varphi)$. The $3d$ effects *increase* the value of J_c by a factor of $\sqrt{N_\perp}$ where $N_\perp = k_F^2 S/4\pi^2$ is the number of "perpendicular channels" (number of components of electron momenta perpendicular to the direction of current). The effect of finite temperature is to mix the electron states around the Fermi energy and therefore to reduce the value of J_c because due to k-dependence of the energy of individual electron states (6), the α-dependence of Ω flattens at increasing T. The effect is accounted for by a factor $\exp(-L/\xi_T)$ [2] where ξ_T is the temperature dependent "coherence length" of normal metal

$$\xi_T = \hbar v_F / 2\pi T. \tag{12}$$

[In a dirty metal, ξ_T is given by $\xi_T = (\hbar v_F l_{elast}/2\pi T)^{1/2}$.]

The effect of elastic collision on magnitude of the persistent current is more subtle. It seems at first that elastic scattering does not influence J_c since it preserves the coherent phase of electron. The other extreme is to introduce the uncertainty of the energy due to scattering $\delta\varepsilon \sim \hbar v_F/l_{elast}$ and to compare that with the energy spacing in presence of vector potential $\Delta\varepsilon \sim \hbar v_F/L$. The reduction factor of order l_{elast}/L may follow [10] which however is not correct. Kirczenow [11] showed that smaller reduction does in fact emerges in a specific model of potential scattering.

If we introduce some barriers V_i representing impurities in ring than at large V_i the Φ-dependent part of energy will be proportional to $1/|V_i|$, whereas resistivity due to these barriers will increase as $|V_i|^2$. This means that $J_c \sim \Delta\Omega$ should scale with $1/\sqrt{R}$ and not with $1/R$ as follows from the reduction factor l_{elast}/L.

Consider perfect $1d$ ring interrupted at some point with a δ-functional barrier of height V. Eigenvalue problem in the tight-binding approximation (3) is easily solved giving for the current from a particular eigenstate (6) labeled with m, an expression

$$j_m = (-1)^m \frac{em^2}{2\pi V} \sin 2\pi \frac{\Phi}{\Phi_0}. \tag{13}$$

Summing over all k_m up to $|k_m| = k_F$ gives an estimate of J_c

$$J_c \sim \frac{ev_F}{L} \cdot |\frac{t'}{t}| \tag{14}$$

where t' is the transmission amplitude in the ring with a barrier.

Considering more general case of many barriers and combining the efect of many transverse channels we receive an estimate of the maximal current in dirty metal

$$J_c \sim \frac{ev_F}{L} (\frac{R_0}{R})^{1/2} \tag{15}$$

where R_0 is quantum of resistance $R_0 = h/2e^2 \simeq 12.9k\Omega$ and R is the reristance along the ring (resistance of a rod of length L and cross section S received by cutting the ring at some point). The formula (15) is in qualitative agreement with the experimentally measured magnitude of persistent current [8].

3. Persistent charge

Consider two small (mesoscopic) pieces of metal P_1, P_2 separated by distance d and placed either in the field of time-dependent scalar potential $\phi(t)$, with electric field confined within a capacitor (Fig.2(a)), or in an external electric field producing the scalar potential difference between the particles $\phi = Ed$ (Fig.2(b)). In both cases, the quantity in question

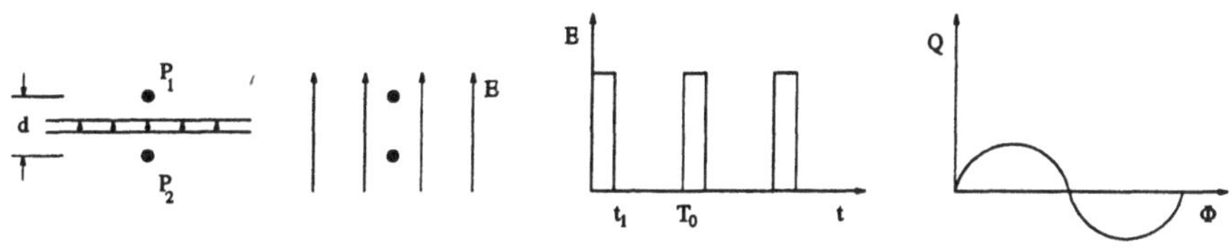

Figure 2. Schematic of the electric Aharonov-Bohm effect with (a) a thin capacitor , and, (b) in an homogeneous electric field. (c) and (d) represent $E(t)$ and $Q(\Phi)$ dependences respectively.

which determines the Aharonov-Bohm effect is the "electric flux" Φ' defined as

$$\Phi' = \int E dx dt. \tag{16}$$

Integral is taken over the period (T_0) of the electric field variation. If, for instance, we choose the dependence $E(t)$ in the form of the Kronig-Penney barrier (Fig.2(c))

$$E(t) = \begin{cases} E_0, & \text{if } |t - nT_0| < t_1; \\ 0, & \text{if } |t - nT_0| > t_1 \end{cases} \tag{17}$$

then we receive $\Phi' = E_0 t_1 d$. The generic Hamiltonian of the problem

$$H = \frac{e\phi(t)}{2}\sigma_z + T_{12}\sigma_x \tag{18}$$

where σ_i are Pauli matrices and T_{12} is the hopping amplitude between P_1 and P_2. Suppose that at $t = 0$ the system acquired amplitudes u_0, v_0 in the upper and lower states of (18). Then, at a later time, it may occupy any of two states with amplitudes u, v thus creating the dipole moment $P = ed(|u|^2 - |v|^2)$. Solution of the time-dependent Schrödinger equation gives

$$P(t) = 4e\mathrm{Re}(u_0^* v_0)\mathrm{Im}\zeta(t), \tag{19}$$

where

$$\zeta(t) = \frac{e^{2\pi i\nu} - e^{2\pi i N\nu}}{1 - e^{2\pi i\nu}} + Fe^{2\pi i N\nu}, \quad \nu = \frac{eV_0}{h} = \frac{\Phi'}{\Phi_0}, \tag{20}$$

and $N = [t/T_0]$, $F = \{t/T_0\}$ where $[x]$ and $\{x\}$ are the integer and the fractional parts of x, respectively. The quantity ν is the electric flux Φ' in units of flux quantum (4). The charge $Q = P/e$ is accumulated between the granules. The value of the charge is an oscillating function of Φ'/Φ_0, analogous to the oscillation of persistent current vs Φ/Φ_0 in a static Aharonov-Bohm effect.

If the electric flux Φ' is slowly varying in time (with period T_1 much larger than T_0) then we will observe the periodic variation of Q with Φ' provided that T_1 is much smaller than the phase-breaking time $\tau_\varphi = l_\varphi/v_F$. The oscillating behavior will persist to temperature of order of $\hbar/T_0$. No such experiment have sofar been performed.

4. Resistance oscillation

Suppose that mesoscopic loop of Fig.1(a) is connected by conducting wires with two thermal reservoirs R_1, R_2 held at different voltages V_1, V_2 (Fig.3(a)). Then the transport

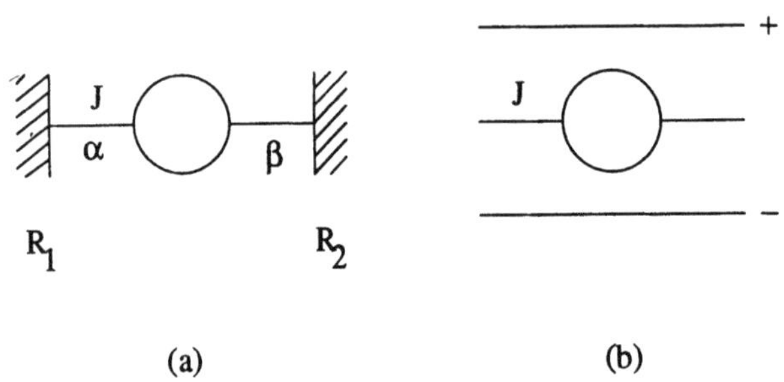

Figure 3. (a)Mesoscopic loop connected to incoherent voltage sources (thermal reservoirs) R_1, R_2. (b)A "quantum mechanical transistor" [13].

current flowing through the loop will be superimposed over the persistent current produced by the external flux. The electrical resistance of the loop $R = (V_2 - V_1)/J_{tr}$ can be calculated with the Landauer formula [6,12]

$$R = \frac{2e^2}{h} \sum_{\alpha\beta} |t_{\alpha\beta}|^2 \tag{21}$$

where $t_{\alpha\beta}$ is the transmission amplitude between one of perpendicular channels of electron eigenstate to the left of the ring (α) and to the right of one (β). This formula is applicable when electron state α enters the ring from the left equilibrium reservoir, and emerges to the state β in the right reservoir. It is assumed that energy conserves within the ring, i.e. motion of electron is "ballistic" in energy. Magnetic flux Φ piercing the ring will affect electronic states α, β and therefore will alter the resistance of the ring.
Resistance oscillation are found in many electric measurements on mesoscopic rings (see [5] and refererences therein). The temperature, size, and purity dependence of resistance oscillation are similar to those of a persistent current. Typical magnitude of conductance variation is of order of e^2/h. An example of $R(\Phi)$ dependence is shown in Fig.4.

In a loop configuration shown in Fig.3(b), an additional electric field was applied [13] perpendicular to the direction of the transport current (the "control gate" of "quantum mechanical transistor") resulting in the shift of oscillation pattern of Fig.4 proportional to the gate voltage. The possible explanation of the experiment may be in considering various classical paths of electron within the ring (Fig.5).

In the dirty ("diffusive") regime $l_{elast} \ll L$, various local loops of electron trajectory may be formed inside the ring. These loops change their energy in a random way when

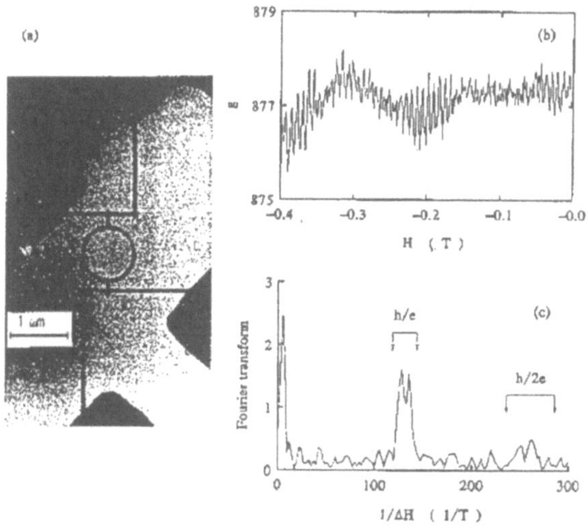

Figure 4. Resistance oscillation in the loop of gold. (a)Loop configuration; (b)R vs H dependence; (c)Fourier transform of $R(H)$ showing maximum at flux quantum periodicity. From Ref.[5].

magnetic flux Φ changes, causing the electron redistribution between the loops and, as a result, the shift in the oscillation pattern $R(\Phi)$. The effect is reduced in magnitude because of screening of electric field inside the metal. Quite large voltages on the control gate ($v_G \sim 1V$) are required to see the substantial shift in the interference pattern $R(\Phi)$, consistent with this reduction.

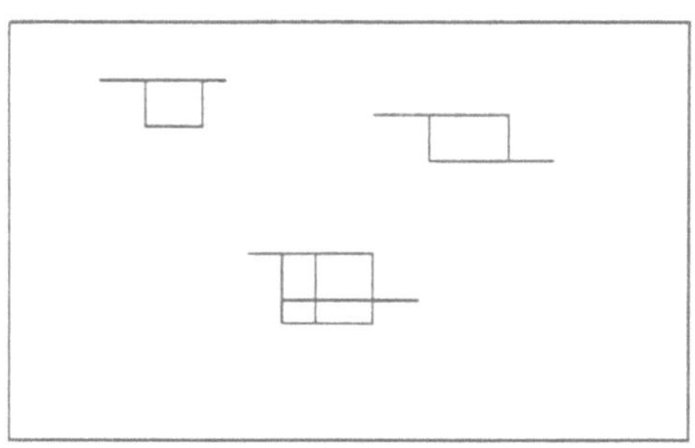

Figure 5. Random network of local loops inside the disordered conductor. Changing of magnetic field causes electron redistribution between the loops.

5. Quantum interference in high frequency field

We now turn to another configuration of the Aharonov-Bohm experiment with mesoscopic rings [14]. Suppose that nanoscale loop encloses an optical fiber with the high frequency field pumped through it (Fig.6). If the static magnetic field H_{dc} is simultaneously applied

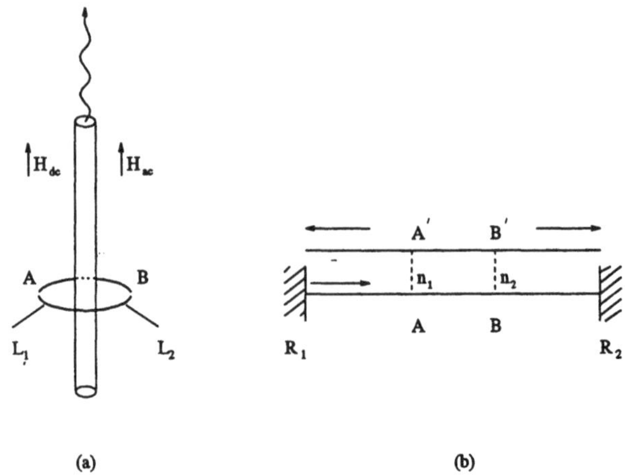

Figure 6. (a)The mesoscopic loop (L) enclosing an optical fiber. In the TE_{01} mode, the magnetic field H_{ac} is aligned parallel to external field H_{dc} and to fiber axis.
(b)Schematic of weakly coupled loop with weak links at points A, B coupled to thermal reservoirs R_1, R_2.

along the fiber axis, resistance of the loop will periodically change not only with the static flux $\Phi = H_{dc}S$ but also with the amplitude of a.c. power in the fiber. Most strong effect is expected in the TE_{01} mode of the fiber in which magnetic field H_{ac} has component along the fiber axis (Fig.7). We assume that loop is "weakly connected", i.e. has narrow regions

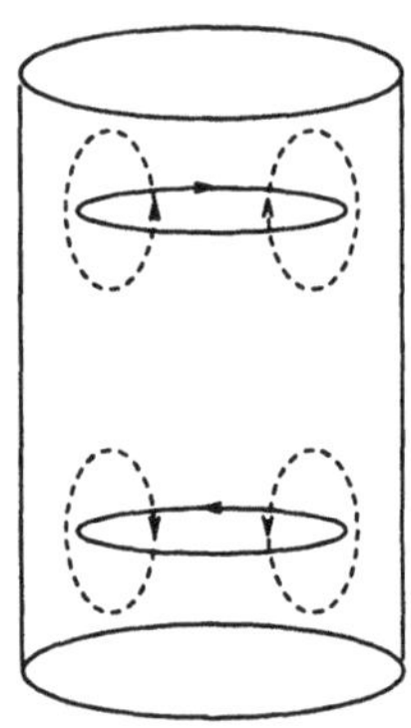

Figure 7. Field configuration in the TE_{01} mode of optical fiber. Solid lines are lines of force of magnetic field, dotted lines the lines of force of electric field.

A, B such that an a.c. field is concentrating near the latter. The size of the loop should be smaller than few wavelengths of optical field to ensure that total a.c. flux within the loop is not equal to zero.
The generic Hamiltonian corresponding to the setup of Fig.6(b) can be represented in the

form

$$H = -t_0 \sum_n (a_n^+ a_{n+1} + b_n^+ b_{n+1}) + H_{int},$$
$$H_{int} = -t_1 a_{n_1}^+ b_{n_1} e^{i\alpha_1} - t_2 a_{n_2}^+ b_{n_2} e^{i\alpha_2} + h.c. \tag{22}$$

where the first term accounts for the hopping along two independent chains (a_n, b_n are the electron annihilation operators in the chains), whereas the second one ascribes the interaction between the chains. The lower chain is connected to two thermal reservoirs R_1, R_2 for measurement of the resistance variation with the a.c. and d.c. fields.
Hopping amplitudes between the chains are modulated in phase due to static and alternating fields according to $t_i e^{i\alpha_i}$ with

$$\alpha_i = \alpha_i^0 + A_i \sin(\omega t + \delta_i). \tag{23}$$

The phase difference $\alpha_1^0 - \alpha_2^0$ is controlled by a static flux

$$\alpha_1^0 - \alpha_2^0 = \frac{2\pi\Phi_{dc}}{\Phi_0}, \qquad \Phi_{dc} = \int \mathbf{B} d\mathbf{S}, \qquad \Phi_0 = \frac{hc}{e} \tag{24}$$

whereas the time-dependent parts relate to an a.c. power in fiber.
Solution of Hamiltonian (22) can be achieved by perturbation over t_1, t_2 in the frequency domain $\hbar\omega > 4|t_0|$ in which inelastic transitions corresponding to change in energy $n\hbar\omega$ are forbidden. Hamiltonian (22) does not have unperturbed states outside the bandwidth of the one-dimensional metal $4|t_0|$.
Employing the identity

$$e^{iz\sin\varphi} = \sum_{n=-\infty}^{\infty} J_n(z) e^{in\varphi}$$

where $J_n(z)$ are Bessel functions of order n, we can decompose H_{int} into the Fourier series

$$H_{int} = \sum_{n=-\infty}^{\infty} H_{int}^{(n)} e^{in\omega t}, \tag{25}$$

with

$$H_{int}^{(n)} \equiv -t_1 e^{i\alpha_1^0} J_n(A_1) a_{n_1}^+ b_{n_1} - t_2 e^{i\alpha_2^0} J_n(A_2) a_{n_2}^+ b_{n_2}\,. \tag{26}$$

Coupling between the plane-wave states of the unperturbed Hamiltonian

$$\psi_k = \sum_n e^{ikn} a_n^+ |0\rangle, \qquad \phi_k = \sum_n e^{ikn} b_n^+ |0\rangle \tag{27}$$

results in transition between the chains with probability

$$W_\pm(\psi_k \leftrightarrow \phi_{\pm k}) = |\langle \phi_{\pm k} | H_{int}^{(0)} | \psi_k \rangle|^2 / |v_k| \tag{28}$$

where subscript "+" refers to transition $\psi_k \to \phi_{-k}$ and "-" to transition $\phi_k \to \psi_{-k}$. Since momentum $|k|$ is conserved at transition, we may introduce kinetic equations for the populations $f_1^{\pm}$ of electrons with momenta k and $-k$ in the lower chain, and similarly $f_2^{\pm}$ in the upper chain. Taking into account symmetry between the transitions $\psi_k \to \phi_{\pm k}$ and $\phi_k \to \psi_{\pm k}$ we receive

$$\frac{df_2^-}{dt} = -W_0 f_2^-(1-f_1^-) - W_- f_2^-(1-f_1^+) + W_0 f_1^-(1-f_2^-) + W_- f_1^+(1-f_2^-),$$
$$\frac{df_2^+}{dt} = -W_0 f_2^+(1-f_1^-) - W_0 f_2^+(1-f_1^+) + W_+ f_1^-(1-f_2^+) + W_0 f_1^+(1-f_2^+) \quad (29)$$

in the upper chain, and

$$\frac{df_1^-}{dt} = Q_+ - W_0 f_1^-(1-f_2^-) - W_+ f_1^-(1-f_2^+) + W_0 f_2^-(1-f_1^-) + W_+ f_2^+(1-f_1^-),$$
$$\frac{df_1^+}{dt} = Q_- - W_- f_1^+(1-f_2^-) - W_0 f_1^+(1-f_2^+) + W_- f_2^-(1-f_1^+) + W_0 f_2^+(1-f_1^+) \quad (30)$$

in the lower chain where $Q_{\pm}$ is the influx of electrons from the reservoirs

$$Q_{\pm} = \frac{v_k}{L}(f_1^{\pm} - f_0(\varepsilon_k \pm \frac{eV}{2})), \quad (31)$$

and $f_0(\varepsilon) = (\exp(\beta(\varepsilon-\mu))+1)^{-1}$ is the equilibrium Fermi distribution.
In a steady-state, $df/dt = 0$, Eq.(29) gives

$$f_2^- = \frac{W_0 f_1^- + W_- f_1^+}{W_0 + W_-}, \quad f_2^+ = \frac{W_0 f_1^+ + W_+ f_1^-}{W_0 + W_+}. \quad (32)$$

In the lower chain, solution of Eq.(30) at small W's gives

$$f_1^- \simeq f_0(\varepsilon_k - eV/2) + [W_0 f_2^- + W_+ f_2^+ - (W_0 + W_+) f_0(\varepsilon_k - eV/2)]/|v_k|,$$
$$f_1^+ \simeq f_0(\varepsilon_k + eV/2) + [W_0 f_2^+ + W_- f_2^- - (W_0 + W_-) f_0(\varepsilon_k + eV/2)]/|v_k|. \quad (33)$$

The current flowing between the reservoirs is

$$J = \int_0^{\pi} v_k (f_1^+ - f_1^-)\frac{dk}{2\pi}. \quad (34)$$

After simple manipulation we receive

$$J = \int_0^{\pi} \frac{dk}{2\pi} W(0)[\frac{W(k)}{W(0)+W(k)} + \frac{W(-k)}{W(0)+W(-k)}][f_0(\varepsilon_k - \frac{eV}{2}) - f_0(\varepsilon_k + \frac{eV}{2})]. \quad (35)$$

where

$$W(k) = t_1^2 J_0^2(A_1) + t_2^2 J_0^2(A_2) + 2t_1 t_2 J_0(A_1) J_0(A_2)\cos(\alpha + 2Nk). \quad (36)$$

Conductance of the system $G = dJ/dV$ is represented as $G_0 + G_1$ where G_1 is the interference term proportional to $|t_i|^2$

$$G_1 = \frac{e^2 t^2 X_0}{2hT}\int_0^{\pi} \frac{X_0[X_+(k) + X_-(k)] + X_+(k)X_-(k)}{[X_0 + X_+(k)][X_0 + X_-(k)]}\frac{dk}{|v_k|\cosh^2(\varepsilon_k - \mu)/2T}. \quad (37)$$

In this expression

$$X_s(k) = A + B\cos(\alpha + skL), \qquad s = -1, 0, +1$$
$$A = t_1^2 J_0^2(A_1) + t_2^2 J_0^2(A_2),\ B = 2t_1 t_2 J_0(A_1) J_0(A_2),\ t = \sqrt{t_1^2 + t_2^2}. \tag{38}$$

where $L = 2Na$ is the total length of the ring.
Conductance of the ring is a function of both static magnetic flux Φ_{dc} and the amplitude of an a.c. field $A_{ac} = \text{const} \cdot \sqrt{P}$ where P is an optical power in a fiber.
The dependence $G(\Phi_{dc})$ is similar to oscillations of static resistance discussed in Sect.4. The dependence $G(P)$ shown in Fig.8 is similar to the amplitude oscillation in the a.c. Josephson effect in superconductors [15,16]. The temperature variation of both oscillation

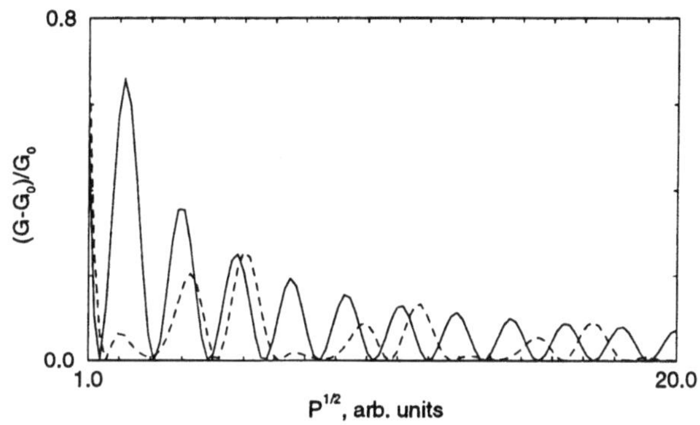

Figure 8. d.c. conductance of the loop as a function of an a.c. power. Solid line corresponds to $A_1 = A_2$, the dashed line to $A_1 = 0.5A_2$.

dependences is somewhat different from the temperature dependence of the amplitude of static oscillations. Formerly, the oscillation resulted from the Φ dependence of the electron states (6). Integration with respect to k eliminated the oscillating component. In case of an a.c. interference, the transition probability (a coefficient before $f_0^+ - f_0^-$ in Eq. (35)) does not vanish after the integration with respect to k. Therefore the oscillation will have a non-exponential small amplitude at temperature larger than the level spacing $\Delta\varepsilon = \hbar v_F/L$.

6. Other Aharonov-Bohm effects

In recent years, there has been an interest in extension of the Aharonov-Bohm effects to systems other than metals [17,18], to solid cylinders [19,20] and antidots [21], to unhomogeneous magnetic fields with radial [22] or azimuthal [23] components. Aharonov and Casher [24] have considered the interaction of electron spin with electrically charged rods. The effect results in the shift of AB oscillation. Being quite small, Aharonov-Casher effect is enhanced in semiconductors with strong spin-orbit coupling and small effective mass of electrons [25]. Actually, such effects are nothing else than the manifestation of spin-orbit interaction [26] which shifts the $R(\Phi)$ dependence but do not reveal the full period of oscillation like in conventional (static) AB experiment. Interaction of particle spin with electrically charged body is can be considered in context of "Berry phase" in quantum mechanics [27]. These and similar AB effects are reviewed in a paper [28].

References

1. Y. Aharonov and D. Bohm, *Phys. Rev.* **115**, 485 (1959).
2. I. O. Kulik, *JETP Lett.* **11**, 275 (1970).
3. M. Buttiker, Y. Imry, and R. Landauer, *Phys. Lett. A* **96**, 365 (1983).
4. I. O. Kulik, *Physica C, Cond. Matt. (Amsterdam)* **218**, 252 (1996).
5. S. Washburn, in *Mesoscopic Phenomena in Solids, p.1.* Eds. B.L. Altshuler, P.A. Lee, and R.A. Webb (North-Holland, Amsterdam, 1991).
6. Y. Imry, in *Directions in Condensed Matter Physics, p.101.* Eds G. Grinstein and E. Mazenko (World Scientific, Singapore, 1986).
7. L. Gunther, and Y. Imry, *Sol. St. Communs.* **7**, 1391 (1969).
8. V. Chandrasekhar, R.A. Webb, M.J. Brady, M.B. Ketchen, W.J. Gallagher, and A. Kleinsasser, *Phys. Rev. Lett.* **67**, 3578 (1991).
9. D. Mally, C. Chapelier, and A. Benoit, *Phys. Rev. Lett.* **70**, 2020 (1993).
10. H. F. Cheung, E. K. Riedel, and Y. Gefen, *Phys. Rev. Lett.* **62**, 587 (1989).
11. G. Kirczenow, *J. Phys.: Condens. Matter* **7**, 2021 (1995).
12. R. Landauer, *Phil. Mag.* **21**, 863 (1970).
13. S. Washburn, H. Schmid, D. Kern, and R. A. Webb, *Phys. Rev. Lett.* **59**, 1791 (1987).
14. I. O. Kulik, and A. S. Shumovsky, *Appl. Phys. Lett.*, to appear Nov. 1996.
15. I. O. Kulik, and I. K. Yanson, *Josephson Effect in Superconductive Tunneling Structures* (Israel Program for Scientific Translations, Jerusalem, 1972).
16. A. Barone, and G. Paterno, *Physics and Applications ot the Josephson Effect* (Wiley, N.Y., 1982).
17. I. O. Kulik, A. S. Rozhavsky, and E. N. Bogachek, *JETP Lett.*, **47**, 302 (1988).
18. E. N. Bogachek, I. V. Krive, I. O. Kulik, and A. S. Rozhavsky, *Sov. Phys. JETP* **70**, 863 (1990).
19. E. N. Bogachek, and G. A. Gogadze, *Sov. Phys. JETP* **36**, 973 (1973).
20. N. B. Brandt, E. N. Bogachek, D. V. Gitsu, G. A. Gogadze, I. O. Kulik, A. A. Nikolaeva, and Ya. G. Ponomarev, *Sov. J. Low Temp. Phys.* **8**, 358 (1982).
21. E. N. Bogachek, and U. Landman, *Phys. Rev. B* **52**, 14067 (1995).
22. D. Loss, P. Goldbart, and A. V. Balatsky, *Phys. Rev. Lett.* **65**, 1655 (1990).
23. A. Stern, *Phys. Rev. Lett.* **68**, 1022 (1992).
24. Y. Aharonov, and A. Casher, *Phys. Rev. Lett.* **53**, 319 (1984).
25. E. N. Bogachek, and U. Landman, *Phys. Rev. B* **50**, 2678 (1994).
26. Y. Meir, Y. Gefen, and O. Entin-Wohllman, *Phys. Rev. Lett.* **63**, 798 (1989).
27. E. N. Bogachek, I. V. Krive, I. O. Kulik, and A. S. Rozhavsky, *Mod. Phys. Lett. B* **5**, 1607 (1991).
28. I. V. Krive, and A. S. Rozhavsky, *Int. J. Mod. Phys. B* **6**, 1255 (1992).

QUANTUM OPTICS AND SOLID STATE SPECTROSCOPY

ALEXANDER S. SHUMOVSKY
Department of Physics, Bilkent University
Bilkent 06533, Ankara, Turkey

Abstract. Possibility of formation of strongly correlated states of the bosonic excitations in solids similar to the squeezed states of photons is considered. The possibility of observation of these states in optical spectroscopy is discussed.

1. Introduction

By definition, the Quantum Optics is considered as a branch of physics studying the quantum statistical properties of the optical fields, in other words, the quantum state of light and the picture of its quantum fluctuations. Since the experimental information of the quantum statistical properties of light is provided by the correlation measurements , the quantum correlation functions are the principal objects of this branch of modern physics. Some specific effects with no classical analogies such as antibunching, sub-Poissonian statistics of photons and squeezing of quantum fluctuations are stipulated by the nonlinear processes.

We note that within the framework of quantum field theory the photons are described by the Bose-operators which obey the commutation relations of the Weyl-Heisenberg algebra. In solid state physics we know a number of quasi-particles or collective excitations such as phonons, polaritons, excitons, Cooper pairs (approximately) which are also bosons. In many physically important cases they take part in various kinds of nonlinear processes or interactions. Therefore, in analogy to the photons, it is quite natural to expect some unconventional quantum statistical properties of the Bose-type excitations in solids which could be important for deeper understanding of a number of physical phenomena.

Unfortunately, we usually cannot measure directly the high-order correlation functions for the collective excitations in solids. However, in many important cases the light can be scattered by these excitations. In this case, the photon states are correlated with the degrees of freedom of the excitations inside the solid. Therefore, the scattered light is naturally expected to reflect the quantum statistical properties of the excitations. In turn, the correlation measurements for the scattered light could give information about the quantum statistics of the excitations.

Therefore, there are two important questions. 1) What kind of the statistical properties we can expect for the bosonic excitations in solids? 2) How to observe these properties with the aid of optical correlation measurements? For all that, it is usually necessary to

T. Hakioğlu and A. S. Shumovsky (eds.), Quantum Optics and the Spectroscopy of Solids, 57–68.

take into account that boson excitations in condensed matter ought to be considered at thermal equilibrium.

The main purpose of this paper is to discuss briefly the present state of the problems mentioned above.

2. Squeezed States

The definition of the squeezed states of bosons [1, 2, 3] is based on the Bogoliubov's canonical transformation. This transformation has been introduced in connection with the problem of superfluidity of liquid ^4He [4]. The original form of this transformation is

$$U^+ a_k U = u_k a_k + v_k a_{-k}^+ \equiv \mathbf{a}_k, \qquad |u_k|^2 - |v_k|^2 = 1 \tag{1}$$

where the operators $a_{\pm k}$, $a_{\pm k}^+$ describe the over-condensate excitations ($k \neq 0$) at low temperatures when almost all atoms of the ^4He are in the Bose-Einstein condensate state. The unitary operator U has the form

$$U = \prod_k U_k, \qquad U_k = \exp(\xi_k a_k^+ a_{-k}^+/2 - \xi_k^* a_{-k} a_k/2) \tag{2}$$

with

$$|\xi_k|^2 = \cosh^{-1} u_k, \qquad \arg \xi_k = \arg v_k - \pi$$

and diagonalizes the Hamiltonian of over-condensate excitations

$$H = \frac{1}{2} \sum_{k \neq 0} [\epsilon_k (a_k^+ a_k + a_{-k}^+ a_{-k}) + \lambda_k a_k^+ a_{-k}^+ + \lambda_k^* a_{-k} a_k]. \tag{3}$$

Hence, the operators $\mathbf{a}_k, \mathbf{a}_k^+$ describe the simultaneous creation and annihilation of the bosons with opposite momenta. Here

$$u_k = \frac{|\lambda_k|}{\sqrt{2\epsilon_k(\epsilon_k - E_k)}}, \qquad v_k = \frac{\epsilon_k - E_k}{\sqrt{2\epsilon_k(\epsilon_k - E_k)}} e^{i \arg \lambda_k}$$

and $E_k = \sqrt{\epsilon_k^2 - |\lambda_k|^2}$ represents the spectrum of pair over-condensate excitations

$$H = \frac{1}{2} \sum_{k \neq 0} E_k (\mathbf{a}_k^+ \mathbf{a}_k + \mathbf{a}_{-k}^+ \mathbf{a}_{-k}) \tag{4}$$

One can see that U_k in the Eq.. (2) formally coincides with the operator corresponding to the so-called two-mode coherent state in quantum optics [5, 6, 7]. The vacuum state of the system described by the Hamiltonian (4) is

$$|\Phi\rangle = \prod_{k \neq 0} |\psi_k\rangle, \qquad |\psi_k\rangle = U_k |0\rangle_k |0\rangle_{-k} \tag{5}$$

where $|0\rangle_{\pm k}$ denotes the vacuum state defined by the ground state stability condition $a_{\pm k}|0\rangle_{\pm k} = 0$. Then

$$\langle a_k^+ a_k \rangle_\Phi \equiv \overline{n}_k = |v_k|^2,$$
$$\langle (a_k^+ a_k)^2 \rangle_\Phi = |v_k|^2 (u_k^2 + |v_k|^2) = \overline{n}_k (2\overline{n}_k + 1)$$

where $\langle ... \rangle_\Phi$ denotes the expectation value with respect to the state (5). The number variance is

$$V_\Phi(a_k^+ a_k) \equiv \langle (a_k^+ a_k)^2 \rangle_\Phi - \langle a_k^+ a_k \rangle_\Phi^2 = \overline{n}_k^2 + \overline{n}_k \neq 0 \tag{6}$$

while in the case of ordinary vacuum state $V(a_k^+ a_k) = 0$. It means that there are very strong quantum fluctuations of the number of ^4He atoms in the over-condensate state of a superfluid system [8, 9]. It should be stressed that this result is valid also for the total number of particles since

$$N = \sum_{k \neq 0} a_k^+ a_k, \qquad \overline{N} = \sum_{k \neq 0} |v_k|^2, \qquad V_\Phi(N) = \sum_{k \neq 0} u_k^2 |v_k|^2.$$

Let us note that using the terminology of quantum optics (see, e.g. [10]) one can say now that the vacuum state (5) of the system (3) and (4) corresponds to a super-Poissonian number distribution as far as

$$V_\Phi(N) > \overline{N}.$$

Although the investigation of the *squeezing phenomenon* is not the aim of this paper, let us note that there is no squeezing in the mode k since

$$V_\Phi([a_k + a_k^+]/2) = (2u_k^2 - 1)/4 > 1/4$$

although at

$$\frac{\pi}{6} + l\pi < \arg \lambda_k < \frac{5\pi}{6} + l\pi, \qquad l = 0, 1, ...$$

squeezing is present in the mixture of two modes with the opposite k.

The investigation of the quantum number fluctuations in the superfluid liquid ^4He is not a subject of the optical spectroscopy and can be done with the aid of neutron scattering [8, 9]. Therefore, let us consider now a different example of manifestation of squeezed states in condensed matter physics which is the polariton in an ionic crystal. This system is usually described by the Hamiltonian [11, 12]

$$\begin{aligned} H = & \sum_{k \geq 0} [\omega_k (a_k^+ a_k + a_{-k}^+ a_{-k}) + \Omega_L (b_k^+ b_k + b_{-k}^+ b_{-k})] \\ & - \sum_{k \geq 0} D_k [(a_k^+ - a_{-k})(b_k + b_{-k}^+) + (a_{-k}^+ - a_k)(b_{-k} + b_k^+)]. \end{aligned} \tag{7}$$

Here the operators a describe photons with the frequency $\omega_k = kc$, the operators b describe the transversal optical (TO) phonons, $\Omega_L = \Omega_T \sqrt{\epsilon_0 \epsilon_\infty}$ is the frequency of longitudinal vibrations of ions and

$$D_k = -\frac{i}{2} \sqrt{\Omega_L \omega_k \eta}, \qquad \eta = \frac{\epsilon_0 - \epsilon_\infty}{\epsilon_0} > 0.$$

One can see that the operator structure of the Hamiltonian (7) is somewhat different in comparison with (3) since it contains the supplementary terms $a_{\pm k}^+ b_{\pm k}$ and $b_{\pm k}^+ a_{\pm k}$. The Bogoliubov's canonical transformations in this case have the form

$$c_{kj} = u_{1j}(k) a_k + u_{2j}(k) b_k - v_{1j}^*(k) a_{-k}^+ - v_{2j}^*(k) b_{-k}^+, \quad j = 1, 2 \tag{8}$$

and obey the conditions

$$\sum_l (u_{lj}u^*_{lj'} - v^*_{lj}v_{lj'}) = \delta_{jj'}.$$

The explicit form of the coefficients in (8) is

$$\begin{aligned}
u_{1j}(k) &= (i)^{j-1}\frac{(E_{kj}+\omega_k)|D_k|\sqrt{\Omega_L}}{\sqrt{\Phi_{kj}}},\\
v_{1j}(k) &= (-i)^{j-1}\frac{(E_{kj}-\omega_k)|D_k|\sqrt{\Omega_L}}{\sqrt{\Phi_{kj}}},\\
u_{2j}(k) &= i^j\frac{(E^2_{kj}-\omega^2_k)(E_{kj}+\Omega_L)}{2\sqrt{\Omega_L\Phi_{kj}}},\\
v_{2j}(k) &= (-1)^{j-1}i^j\frac{(E^2_{kj}-\omega^2_k)(E_{kj}-\Omega_L)}{2\sqrt{\Omega_L\Phi_{kj}}}
\end{aligned} \tag{9}$$

where

$$\Phi_{kj} \equiv E_{kj}[(E^2_{kj}-\omega^2_k)^2 + 4\omega_k\Omega_L|D_k|^2],$$

and

$$E^2_{kj} = \frac{\omega^2_k+\Omega^2_L}{2} \pm \frac{1}{2}\sqrt{(\omega^2_k-\Omega^2_L)^2+16\omega_k\Omega_L|D_k|^2} \tag{10}$$

are the two branches of the polariton spectrum [11, 12] (see Fig. 1). In the polariton

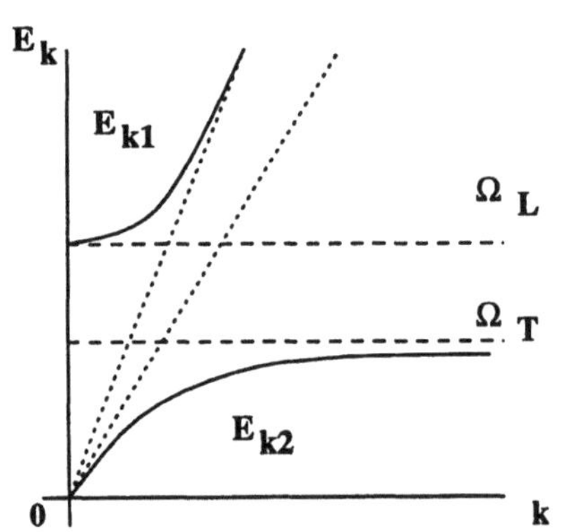

Figure 1. Two branches of the polariton excitations versus k.

(diagonal) representation, the Hamiltonian (7) has the following form

$$H_{pol} = \sum_{k\geq 0}\sum_j E_{kj}(c^+_{kj}c_{kj} + c^+_{-kj}c_{-kj}) + E_0 \times \mathbf{1} \tag{11}$$

where

$$E_0 = -\sum_k \sum_{l,j} E_{kj} |v_{lj}|^2.$$

The eigenstates of the Hamiltonian (11) which are the Fock number states of polaritons are squeezed states in terms of the photon and phonon degrees of freedom. It means, that its structure corresponds to the simultaneous creation of pairs of photons and phonons and also of a photon and phonon simultaneously. The corresponding squeezing operator has the form [9]

$$S_k = \exp(\xi_k a_k^{+2} + \eta_k b_k^{+2} + \zeta a_k^+ b_{-k}^+ - h.c)$$

where ξ_k, η_k, and ζ_k are known but quite complicated functions of the coefficients u and v. One can see that S_k represents some generalization of the two-mode squeezing operator (2).

Let us examine statistical properties of phonons in the system (7). As in standard quantum optics, these properties can be characterized by the degree of coherence [10]

$$G_\Theta^{(2)} = \frac{\langle b_k^{+2} b_k^2 \rangle}{\langle b_k^+ b_k \rangle^2} \tag{12}$$

Without going into straightforward calculations let us examine the temperature dependence of (12) given in Fig. 2 (also see [13]). Taking into account that in the case of Gaussian distribution of phonons $G^{(2)} = 2$ at $\Theta = 0$, one can see from the Fig. 2 that there is a *super-Gaussian* distribution of phonons in the polariton system. It means that at low

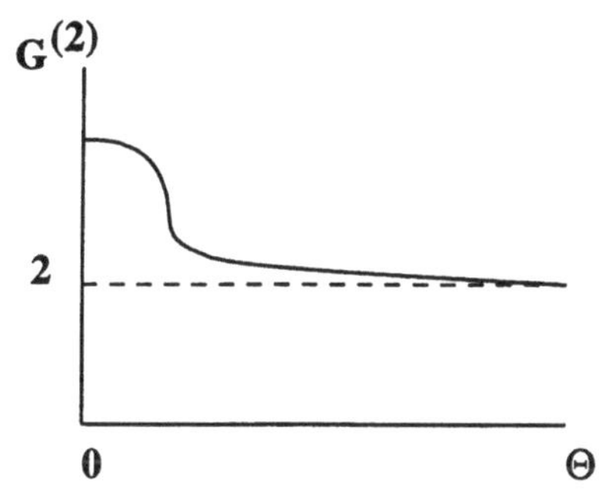

Figure 2. Temperature dependence of the degree of coherence (12) at fixed k.

temperature, there are quite strong quantum number fluctuations of TO phonons in an ionic crystal. Increase of temperature amplifies the thermal number fluctuations and make the quantum fluctuations unobservable. The estimation of the threshold of observation of the quantum number fluctuations gives for a typical ionic crystal $T_{th} \leq 30K$ [13] which is not too low temperature for the experimental study. A similar behavior can be observed also in the photon sub-system [14].

Thus, the above considered examples show that realization of a squeezed state in material systems (quantum liquid, solid state etc.) leads to a quite strong quantum number fluctuations of bosons which can be observed at relatively low temperatures.

Let us discuss briefly some other sources of squeezed states in solids. Since the origin of the emergence of squeezed states is the simultaneous creation or annihilation of two (or more) bosons, the simplest process in a real crystal which leads to a squeezed state is connected with the third-order phonon anharmonicity. In this case, we have for example the scattering of two phonons into one or decay of one phonon into two phonons (see Fig. 3). These processes are realized under the condition of the conservation of energy but it is not necessary to have conservation of quasi-momentum [11].

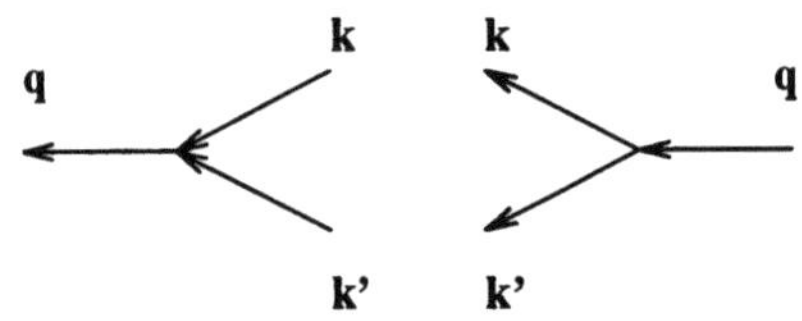

Figure 3. Scattering of two phonons into one and decay of a phonon into two phonons. The processes obey the energy conservation $\Omega(q) = \Omega(k) + \Omega(k')$ and one of the relations $\vec{q} = \vec{k} + \vec{k}'$ or $\vec{q} = \vec{k} + \vec{k}' + \vec{g}$ where $\vec{g}$ is the vector of reciprocal lattice.

Another example is provided by the photon-phonon interaction responsible for the processes of emission and absorption of electromagnetic waves by a crystal. The energy of interaction in the dipole approximation in this case has the form

$$H_{int} = -\sum_f \vec{d}_{f\alpha} \cdot \vec{E}(f) \tag{13}$$

where

$$\vec{d}_{f\alpha} = \sum_k \sqrt{\frac{\hbar}{2\Omega_\alpha(k)N}}\vec{d}_\alpha(k)e^{ikf}(b_{k\alpha} + b^+_{-k\alpha})$$

is the electric dipole moment of the primitive cell of the crystal, index α marks the branch of the phonon oscillations of the frequency Ω_α, b and b^+ are the phonon operators, and

$$\vec{E}(f) = i\sqrt{\frac{2\pi\hbar}{N\epsilon}}\sum_{q,\sigma}\vec{e}_\sigma(q)\sqrt{\omega_q}(a_{q\sigma} - a^+_{-q\sigma})e^{iqf}$$

is the electric field on the site f. Here ϵ is the permittivity of the crystal, $\vec{e}$ describes the photon polarization, and the operators a and a^+ describe the photons with the frequency ω_q corresponding to the region of transparency of the crystal. One can see that the Hamiltonian (13) formally coincides with the polariton model (7).

Besides the phonons, the processes related to the excitons in semiconductors and molecular crystals can be also considered as the sources of squeezed states. For example, in the case of a molecular crystal the exciton-photon interaction can lead to the formation of the exciton-polaritons [15]. The interaction energy corresponding to the above process is

$$H_{EP} = -\frac{e}{mc}\sum_f \vec{A}(f) \cdot \vec{P}_f + \frac{e^2 S}{2mc^2}\sum_f A^2(f) \tag{14}$$

where $\vec{A}(f)$ describes the vector potential of electromagnetic field in the site f and $\vec{P}_f$ is the total momentum of all S optically active electrons of a neutral molecule in the lattice site f. Going over to the quantization in the reciprocal space (the second quantization representation), one can rewrite (14) in the form

$$H_{EP} = \sum_{k,\alpha,f} \mathcal{D}(k,\alpha,f)(B^+_{kf} - B_{-kf}(a_{k\alpha} + a^+_{-k\alpha}\\ + \sum_{k,\alpha} \frac{\hbar\Omega_p^2 S}{4c|k|}((a^+_{k\alpha} + a_{-k\alpha})(a_{k\alpha} + a^+_{-k\alpha}) \qquad (15)$$

where Ω_p^2 is the squared plasma frequency and $\mathcal{D}$ is the coupling constant [15]. Since the exciton operators B^+, B obey the boson commutation relations we again have the Hamiltonian of the form of (7). The exciton-phonon interaction can be also described in a similar way [15].

Let us mention here an important and interesting problem of the exciton-polaritons in a semiconducting micro-cavity formed by a multiple-quantum-well structure placed between the distributed Bragg reflectors (DBR) [16, 17, 18, 19]. In this case, the two branches of the polariton formed by coupled modes of the radiation field and collective excitations of semiconductor provide a two-level system. Unlike the Jaynes-Cummings model [20, 21, 22], the occupation number of any level of that two-level system can take an arbitrary integer value due to the bosonic nature of the polariton. Nevertheless, the vacuum-field induced Rabi oscillations can be observed in such a system [16, 17]. Let us stress that the study of Rabi oscillations in various physical systems is now an active field of research due to its importance in the investigation of spontaneous emission rate in a cavity and plays a crucial role in the development of new light emitting devices with special properties, as for instance narrow line-width, low-threshold pump-rate, and low noise level [23]. In the case of exciton-polariton system, the properties of radiation are determined by the specific correlation between the processes of simultaneous creation and annihilation of the photons and excitons described in terms of states quite similar to (5) [19].

Another example of great interest is provided by the competition of direct and indirect radiative transitions in the GaAs/AlAs superlattices and multiple-quantum-well structures [24, 25]. The band structure of this system is represented in Fig. 4. In this superlattice, the lowest electronic states of the conducting band are located in the AlAs layers, whereas the lowest electronic states of the valence band are confined in the GaAs layers. After photoexcitation due to the quite short laser pulse close to the resonance with the transition $|0\rangle \leftrightarrow |1\rangle$ electrons scatter from the Γ states in the GaAs to the lower lying X states located in the AlAs. This Γ-X transfer is either dominated by the LO-phonon emission [24] or is due to the interface scattering [25]. After that, the level $|2\rangle$ is populated and indirect (phonon-assisted) transition can occur (type II transition in Fig. 4). This transition leads to an additional peak in the luminescence spectrum [24, 25] which is shifted downwards from the type I peak by the energy $\sim 0.05eV$ corresponding to the phonon energy. In this case, we again have the simultaneous creation of a photon and phonon. Thus, we can expect also here some kind of squeezed state.

The above examples provide the case when the squeezed states result from the interactions of bosons (phonon-photon, phonon-phonon, exciton-phonon, etc.). However, they can also occur as a result of interaction between bosons and some other particles such as electrons, protons of the hydrogen bonds and so on. An important example is provided by

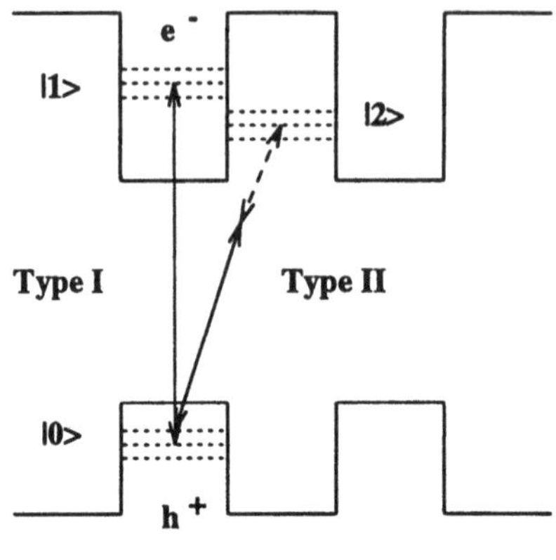

Figure 4. Band structure of a GaAs/AlAs superlattice. Solid lines correspond to the photons while the dotted line corresponds to a phonon.

the phonon squeezing via correlations in the superconducting electron-phonon interaction [26, 27, 28]. Similar process takes place in a polaron system [29, 30].

Thus, it is shown in this section that many different mechanisms can lead to the formation of squeezed states (or states with correlated processes of creation and annihilation) of the boson-type excitations in solids. It is manifested in the change of quantum statistical properties of these excitations, first of all in the increase of quantum number fluctuations with respect to non-squeezed states.

Let us now turn our attention to the problem of how to measure the quantum statistical properties of the boson excitations in solids.

3. Quantum Spectroscopy of Solids

It was shown in the previous section that the type of bose statistics can be qualitatively determined by the normalized correlation function (12). This means that it is necessary to measure the number correlation or, using the terminology of quantum optics, the intensity correlation. For the photons emitted by some source this can be done with the aid of the Hanbury-Brown a nd Twiss correlation interferometer [10]. In this device, the photon beam is splitted into two equivalent beams which are then detected by two independent photodetectors. To determine the photon correlations photon counters and digital correlations are used. Unfortunately, it is almost impossible to use the similar scheme of correlation measurements for phonons, excitons and other boson excitations in solids.

At the same time, these excitations can interact with photons in the process of emission or scattering of light by solids. In this case, it is reasonable to think that the quantum statistical properties of photons should contain some information about the quantum statistical properties of bosons in solids.

A simple illustrative example of this process is provided by the photoluminescence in the system shown in Fig. 4 and briefly discussed in the previous section. In fact, the process of photoluminescence in the type II transition is characterized by a correlation between creation of photons and phonons and thus strongly depend on the type of statistics of the initial phonon state.

In many cases, bosonic excitations in solids are Raman active [31]. The process of Raman scattering with generation of the Stokes and anti-Stokes photons is shown in Fig. 5. If we suppose the monochromatic coherent laser field and scattering on an optical

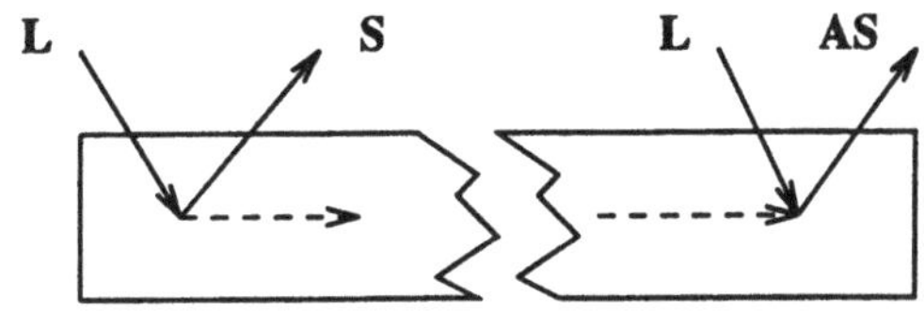

Figure 5. Scheme of the Raman scattering with generation of Stokes (A) and anti -Stokes (AS) components. The Raman active excitations of a medium are shown by the dotted lines.

phonon, the Hamiltonian of the process in question is

$$H = \omega_L a_L^+ a_L + \Omega b^+ b + \sum_k (\omega_{Sk} a_{Sk}^+ a_{Sk} + \omega_{Ak} a_{Ak}^+ a_{Ak}) + \sum_k [\gamma_{Sk} a_L^+ a_{Sk} b + \gamma_{Ak} b^+ a_L^+ a_{Ak} + h.c.]. \quad (16)$$

Here the operators a_L describe the laser (pump) field with the frequency ω_L, the phonons of frequency Ω are described by the operators b. Due to the Manley-Rowe relations [31] $\omega_{Sk} - \omega_{Ak} \sim 2\Omega$. The operators a_{Sk} and a_{Ak} correspond to the Stokes and anti-Stokes photons respectively and γSk, γ_{Ak} are the coupling constants.

The investigation of quantum properties of this Hamiltonian is the subject of a number of works (e.g., see Refs. [32, 33, 34] and references there). However, due to the cubic nonlinearity, the problem is not solved exactly except the case of $\gamma_{Ak} = 0$ [34, 35, 36].

The Hamiltonian (16) has the following conservation laws

$$\begin{cases} n_L + \sum_k (n_{Sk} + n_{Ak}) \equiv N = const \times 1 \\ n_b - \sum_k (n_{Sk} - n_{Ak}) \equiv M = const \times 1 \end{cases} \quad (17)$$

Here n_x denotes the number operator of the field x.

To examine the problem with the Hamiltonian (16) let us use the method of elimination of Bose variables [37, 38]. In view of our main aim, we follow here the case considered in the Ref. [39]. We should eliminate from the equations of motion the phonon field dynamics of which is described by the following formal expression

$$b(t) = b_f(t) - iB(t), \qquad b_f(t) = b(t_0) e^{-i\Omega(t-t_0)},$$

$$B(t) = \int_{t_0}^{t} d\tau e^{-i\Omega(t-\tau)} \sum_k [\gamma_{Sk} a_{Sk}^+(\tau) a_L(\tau) + \gamma_{Ak} a_L^+(\tau) a_{Ak}(\tau)]. \quad (18)$$

Consider now the time evolution of the expectation values $\langle n_{Sk} \rangle_t$ and $\langle n_{Ak} \rangle_t$ describing the process of emission of the Stokes and anti-Stokes photons respectively. Here $\langle ... \rangle_t$ denotes the average of corresponding operator in the Heisenberg representation with respect to the density matrix of the initial state of the complete system [38]. We have

$$i\frac{d}{dt}\langle n_{Sk} \rangle_t = -\gamma_{Sk}(\langle a_L^+ a_{Sk} b \rangle - \langle b^+ a_{Sk}^+ a_L \rangle),$$

$$i\frac{d}{dt}\langle n_{Ak} \rangle_t = -\gamma_{Ak}(\langle b^+ a_L^+ a_{Ak} \rangle - \langle a_{Ak}^+ a_L b \rangle).$$

We have to substitute here the expression (18) and conjugated expression for b^+. It leads to the expectation values containing the mixed averages with participation of $b(t_0)$ and $b^+(t_0)$. Depending on the choice of the initial state, these averages have different values. In the case of thermal harmonic phonons we get [39]

$$\langle \mathcal{A} b_f \rangle_t = i\overline{n}_b \langle [B, \mathcal{A}] \rangle_t \tag{19}$$

where $\mathcal{A}$ is an arbitrary operator and $\overline{n}_b$ is the initial number of phonons depending on the temperature, while in the case of squeezed thermal phonons

$$\langle \mathcal{A} b_f \rangle_t = i[|v|^2 + (u^2 + |v|^2)\overline{n}_b] \langle [B, \mathcal{A}] \rangle_t - iuv(2\overline{n}_b + 1) \langle [B^+, \mathcal{A}] \rangle_t. \tag{20}$$

Here u and v are the parameters describing the squeezed state. It follows from these exact relations that the mixed correlation functions of thermal harmonic phonons do not contribute into the equations of motion at $T \sim 0$ when $\overline{n}_b \sim 0$ while in the case of squeezed phonons, they give a non-vanishing contribution even at $T = 0$. Thus, there is a principal difference between these two cases.

Consider, for simplicity, just the case of low temperatures when $\overline{n}_b \sim 0$ and use the Markov approximation. Then, the equations of motion in the case of harmonic phonons have the form

$$\begin{aligned} \frac{d}{dt}\langle n_{Sk} \rangle_t = 2\pi\gamma_S^2 \langle a_L^+ a_L \rangle_t \delta_{kk_S} + \pi\gamma_{Sk}\gamma_S(\langle a_L^+ a_L a_S^+ a_{Sk} \rangle_t + \langle a_{Sk}^+ a_S a_L^+ a_L \rangle_t) \\ \pi\gamma_{Sk}\gamma_A(\langle a_L^{+2} a_{Sk} a_A \rangle_t + \langle a_A^+ a_{Sk}^+ a_L^2 \rangle_t) \end{aligned} \tag{21}$$

and

$$\begin{aligned} \frac{d}{dt}\langle n_{Ak} \rangle_t = -\pi\gamma_{Ak}\gamma_A(\langle {}_A^+ a_{Ak} \rangle_t + \langle a_{Ak}^+ a_A \rangle_t) - \pi\gamma_{Ak}\gamma_A(\langle a_L^+ a_L a_A^+ a_{Ak} \rangle_t + \langle a_{Ak}^+ a_A a_L^+ a_L \rangle_t) \\ -\pi\gamma_{Ak}\gamma_S(\langle a_L^{+2} a_S a_{Ak} \rangle_t + \langle a_{Ak}^+ a_S^+ a_L^2 \rangle_t). \end{aligned} \tag{22}$$

Here a_S, γ_S and a_A, γ_A denote the resonance values defined by the conditions

$$\omega_{Sk} = \omega_L - \Omega, \qquad \omega_{Ak} = \omega_L + \Omega$$

respectively.

One can see that unlike the Eq. (21), the right-hand side of (22) does not contain any direct dependence on the mean number of photons in the laser mode $\langle a_L^+ a_L \rangle_t$. Using the short-time approximation, it is not hard to see that this property leads to a delay of the anti-Stokes radiation with respect to the Stokes radiation (Fig. 6). In the case of the squeezed phonons, the right-hand side of the Eq. (21) should be added by the terms

$$\begin{aligned} 2\pi|v|^2\gamma_S^2 \langle a_L^+ a_L \rangle_t \delta_{kk_S} - \pi|v|^2\gamma_{Sk}\gamma_S(\langle a_s^+ a_{Sk} \rangle_t + \langle a_{Sk}^+ a_S \rangle_t) \\ +\pi\gamma_{Sk}\gamma_A(uv\langle a_A^+ a_{Sk} \rangle_t + uv^* \langle a_{Sk}^+ a_A \rangle_t) \end{aligned} \tag{23}$$

which shows the increase of dependence on $\langle a_L^+ a_L \rangle_t$ and at the same time, the increase of correlation between the Stokes and anti-Stokes photons. Similar additional terms for (22) have the form

$$\begin{aligned} 2\pi|v|^2\gamma_A^2 \langle a_L^+ a_L \rangle_t \delta_{kk_A} - |v|^2\pi\gamma_{Ak}\gamma_A(\langle a_A^+ a_{Ak} \rangle_t + \langle a_{Ak}^+ a_A \rangle_t) \\ +\pi\gamma_{Ak}\gamma_S(uv\langle a_{Ak}^+ a_S \rangle_t + uv^* \langle a_S^+ a_{Ak} \rangle_t). \end{aligned} \tag{24}$$

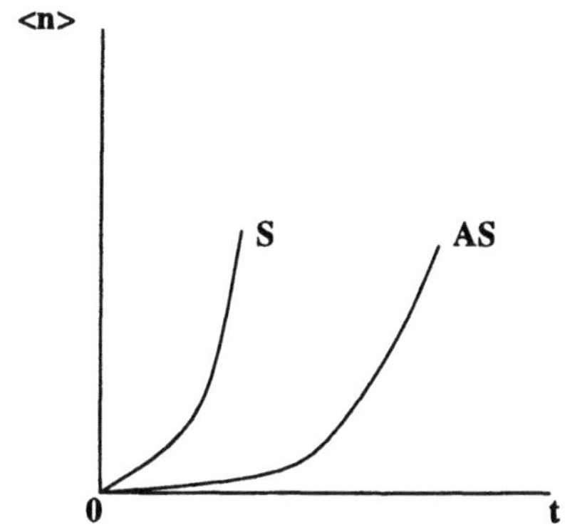

Figure 6. Short-time behavior of the mean photon number for the Stokes and anti -Stokes components described by the Eqs. (21) and (22).

One can see from (24) that now there is a direct dependence of the right-hand side of the equation of motion for the average number of anti-Stokes photons on the number of photons in the laser mode. Therefore, in the case of squeezed phonons there is no delay between the Stokes and anti-Stokes radiation. Thus, the change of state of the phonons leads to some qualitative change of the spectra of scattered light in the Raman scattering process. Some other changes are discussed in the Refs. [9, 33, 34, 35].

Of course, the analysis of the high-order intensity correlations of the Stokes and anti-Stokes components can give richer picture of qualitative and quantitative effects of the dependence of photon statistics of scattered light on the quantum statistical properties of the scatterers [9, 36]. These high-order intensity correlations can be measured by the methods of correlation spectroscopy [40].

4. Summary

It is shown that many different mechanisms of interactions can lead to the formation of strongly correlated (squeezed) state of boson excitations in solids. These correlations are manifested by the quantum number fluctuations which can have a super-Gaussian character at low temperatures. The increase of temperature leads to the increase of thermal number fluctuations exceeding the quantum fluctuations. Therefore, the principal observability of the quantum number fluctuations of bosonic excitations in solids requires the low temperatures.

The quantum number fluctuations of optical-active boson excitations in solids can be observed by the methods of optical spectroscopy, especially by the correlation spectroscopy.

References

1. H. Takahasi, Adv. Commun. Syst. **1**, 227 (1965).
2. D. Robinson, Commun. Math. Phys. **1**, 159 (1965).
3. D. Stoler, Phys. Rev. D **1** 3217 (1970); **4**, 1925 (1971).
4. N.N. Bogoliubov, J. Phys. USSR **11**, 23 (1947).
5. G.J. Milburn, J. Phys. A **17**, 737 (1984).
6. C.M. Caves and B.I. Schumaker, Phys. Rev. D **31** 3068 (1985).

7. R. Loudon and P.L. Knight, J. Mod. Optics **34**, 709 (1987).
8. A.S. Shumovsky, Comm. of JINR E17-92-355 (Dubna, 1992).
9. A.S. Shumovsky, in *Studies in Classical and Quantum Nonlinear Optics*, ed. O. Keller (Nova Sci. Pub., New York, 1995).
10. D.F. Walls and G.J. *Milburn, Quantum Optics* (Springer, Berlin, 1994).
11. O. Madelung, *Introduction to Solid-State Theory* (Springer, Berlin, 1978).
12. H. Haken, *Quantum Field Theory of Solids* (North Holland, Amsterdam, 1983).
13. A.V. Chizhov, B.B. Govorkov, and A.S. Shumovsky, Mod. Phys. Lett. B. **19**, 1233 (1993).
14. A.V. Chizhov, R.G. Nazmitdinov, and A.S. Shumovsky, Quantum Optics **3**, 1 (1991).
15. A.S. Davydov, *Theory of Molecular Excitons* (Plenum, New York, 1971).
16. T.B. Norris, J.K. Rhee, S.Y. Sung, Y. Arikawa, M. Nishioka, and C. Weisbuch, Phys. Rev. D **50**, 14663 (1994).
17. Y. Yamamoto, F. Matinaga, S. Mashida, A. Karlsson, J. Jacobson, G. Björk, and T. Mukai, J. Phys. (France) II, **3**, 39 (1993).
18. S. Pau, G. Björk, J. Jacobson, H. Cao, and Y. Yamamoto, Phys. Rev. B **51**, 7090 (1995).
19. A.S. Shumovsky and Ö.E. Müstecaplioğlu, Phys. Lett. A **209**, 88 (1995).
20. H.I. Yoo and J.H. Eberly, Phys. Rep. **118**, 239 (1985).
21. F.L. Kien and A.S. Shumovsky, Int. J. Mod. Phys. B **5**, 2287 (1991).
22. R. Shore and P.L. Knight, J. Mod. Optics **40**, 1195 (1993).
23. Y. Yamamoto, ed.,*Coherence, Amplification, and Quantum Effects in Semiconductor Lasers* (Wiley, New York, 1991).
24. J. Feldmann, J. Nunnekamp, G. Peter, E. Göbel, J. Kuhl, K. Ploog, P. Dawson, and C.T. Foxon, Phys. Rev. B **42**, 5809 (1992).
25. P.W.M. Blom, C. Smit, J.E.M. Haverkort, and J.H. Wolter, Appl. Phys. Lett. **62**, 2393 (1993).
26. Zheng Hang, Phys. Rev. B **38**, 11865 (1988).
27. T. Hakioğlu, V.A. Ivanov, A.S. Shumovsky and B. Tanatar, Physica C **234**, 167 (1994).
28. T. Hakioğlu, V.A. Ivanov, A.S. Shumovsky, and B. Tanatar, Phys. Rev. B **51**, 15365 (1995).
29. A.M. Jayannavar, Solid State Comm. **71**, 689 (1989).
30. T. Altanhan and B.S. Kandemir, J. Phys.: Condens. Matter **5**, 6729 (1993).
31. Y.R. Shen, *The Principles of Nonlinear Optics* (Wiley, New York, 1984).
32. J. Mostowski and M.G. Raymer, in *Contemporary Nonlinear Optics*, eds. G.P. Agrawal and R.W. Boyd (Academic Press, New York, 1992).
33. A.S. Shumovsky, in *Modern Nonlinear Optics*, eds. M. Evans and S. Kielich (Wiley, New York, 1993).
34. A.S. Shumovsky and B. Tanatar, Phys. Rev. A **48**, 4735 (1993).
35. S. Carusotto, Phys. Rev. A **40**, 1848 (1989).
36. A.S. Shumovsky and B. Tanatar, Phys. Lett. A **182**, 411 (1993).
37. N.N. Bogoliubov Jr, F.L. Kien, and A.S. Shumovsky, Physica A **128**, 82 (1984); **130**, 413 (1985).
38. N.N. Bogolubov Jr, B.I. Sadovnikov, and A.S. Shumovsky, *Mathematical Methods of Statistical Mechanics of Model Systems* (CRC Press, Boca Ration, 1994).
39. L. Knöll and A.S. Shumovsky, Int. J. Mod. Phys. B **4**, 151 (1990).
40. B. Crosignani, P. Di Porto, and M. Bertolotti, *Statistical Properties of Scattered Light* (Academic Press, New York, 1975).

LASER-EXPERIMENTS WITH SINGLE ATOMS IN CAVITIES AND TRAPS

H. WALTHER
Sektion Physik der Universität München
and
Max-Planck-Institut für Quantenoptik
85748 Garching, Fed. Rep. of Germany

1. Introduction

Today frequency tunable lasers enable us to achieve an efficient excitation of atomic resonance lines. In this way it becomes possible to detect single atoms via their fluorescence. An additional advantage is that traps exist for ions[1] as well as for neutral atoms[2] so that the observation of isolated particles is possible. Presently the techniques for trapping atoms are undergoing a rapid development, however, experiments with single isolated atoms using those methods have not yet been realized. Therefore the observation of single trapped particles is still restricted to ions. In the following experiments with single ions will be reviewed; furthermore we will describe the experiments performed with the one-atom maser where single atoms interact with a single mode of a maser field. The atoms in this case are Rydberg atoms with high excitation energies, which can efficiently be detected by ionization in an external electrical field in which the outermost electron is pulled off. We will start this contribution with the review of the one-atom maser.

2. Review of the One-Atom Maser

The most fundamental system to study the generation process of radiation in lasers and masers is to drive a single mode cavity by a single atom. This system, at first glance, seems to be another example of a Gedanken-experiment but such a one-atom maser[3] really exists and can in addition be used to study the basic principles of radiation-atom interaction. The main features of the setup are:

(1) It is the first maser which sustains oscillations with much less than one atom on the average in the cavity.
(2) The maser allows one to study the dynamics of the energy exchange between an atom and a single mode of the cavity field as treated in the Jaynes-Cummings model.[4]
(3) The setup allows one to study in detail the conditions necessary to obtain nonclassical radiation, especially radiation with sub-Poissonian photon statistics in a maser system directly.
(4) It is possible to study a variety of phenomena of a quantum field including nonlocal aspects of the quantum measurement process.

T. Hakioğlu and A. S. Shumovsky (eds.), Quantum Optics and the Spectroscopy of Solids, 69–98.

What are the tools that make this device work? It is the enormous progress in constructing superconducting cavities with high quality factors together with the laser preparation of highly excited atoms – Rydberg atoms – that have made the realisation of such a one-atom maser possible[3]. Rydberg atoms are obtained when one of the outermost electrons of an atom is excited into a level close to the ionization limit. The principle quantum number of the electron is then typically of the order of 60 - 70. Those atoms have quite remarkable properties[5] which make them ideal for the maser experiments: The probability of induced transitions between neighbouring states of a Rydberg atom scales as $n[4]$, where n denotes the principle quantum number. Consequently, a few photons are enough to saturate the transition between adjacent levels. Moreover, the spontaneous lifetime of a highly excited state is very large. We obtain a maser by injecting these Rydberg atoms into a superconducting cavity with a high quality factor. The injection rate is such that on the average there is much less than one atom present inside the resonator. The experimental

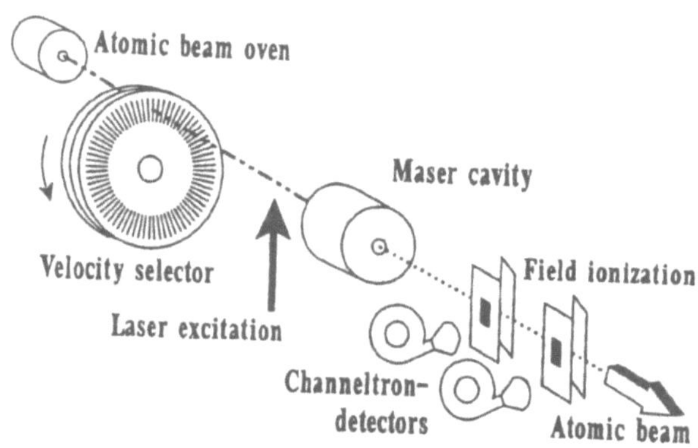

Figure 1. Scheme of the one atom maser. In order to suppress blackbody-induced transitions to neigbouring states, the Rydberg atoms are excited inside a liquid-Helium-cooled environment.

setup of the one-atom maser is shown in Fig. 1. A highly collimated beam of rubidium atoms passes through a Fizeau velocity selector. Before entering the superconducting cavity, the atoms are excited into the upper maser level $63p_{3/2}$ by the frequency-doubled light of a cw ring dye laser. The superconducting niobium maser cavity is cooled down to a temperature of 0.5 K by means of a ^{3}He cryostat. At this temperature the number of thermal photons in the cavity is about 0.15 at a frequency of 21.5 GHz. The quality factor of the cavity can be as high as 3×10^{10} corresponding to a photon storage time of about 0.2 s. Two maser transitions from the $63p_{3/2}$ level to the $61d_{3/2}$ and to the $61d_{5/2}$ level are studied. In a new setup equipped with a dilution refrigerator temperatures in the range of 0.1 K are obtained. Some of the experiments described in this review have been performed with the latter setup. The Rydberg atoms in the upper and lower maser levels are detected by two separate field ionisation detectors. The field strength is adjusted to ensure that in the first detector the atoms in the upper level are ionised, but not those in the lower level; the lower level atoms are then ionised in the second field.

To demonstrate maser operation, the cavity is tuned over the respective transition and the flux of atoms in the excited state is recorded simultaneously. Figure 2 shows the result for $63p_{3/2}$ - $61d_{3/2}$. Transitions from the initially prepared state to the $61d_{3/2}$ level

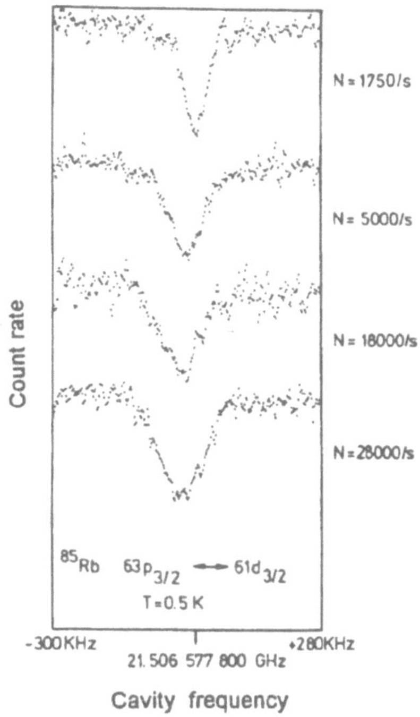

Figure 2. Maser operation of the one-atom maser manifests itself in a decrease of atoms in the excited state when the cavity is tuned through the resonance. The flux of excited atoms N governs the pump intensity. Power broadening of the resonance line demonstrates the multiple exchange of a photon between the cavity field and the atom passing through the resonator.

(21.50658 GHz) are detected by a reduction of the electron count rate. In the case of measurements at a cavity temperature of 0.5 K, shown in Fig. 2, a reduction of the $63p_{3/2}$ signal can be clearly seen for atomic fluxes as small as 1750 atoms/s. An increase in flux causes power broadening and a small shift. This shift is attributed to the ac Stark effect, caused predominantly by virtual transitions to neighbouring Rydberg levels. Over the range from 1750 to 28000 atoms/s the field ionisation signal at resonance is independent of the particle flux which indicates that the transition is saturated. This, and the observed power broadening show that there is a multiple exchange of photons between Rydberg atoms and the cavity field.

For an average transit time of the Rydberg atoms through the cavity of 50 μs and a flux of 1750 atoms/s we estimate that approximately 0.09 Rydberg atoms are in the cavity on the average. According to Poissonian statistics this implies that more than 90% of the events are due to single atoms. This clearly demonstrates that single atoms are able to maintain a continuous oscillation of the cavity with a field corresponding to a mean number of photons between unity and several hundred.

Among the studies performed with the one-atom maser are the measurements of the dynamics of the photon exchange between a single atom and a cavity mode.[6, 7] Before we discuss some experiments with the one-atom maser the theory will be briefly reviewed.

3. Theory of the One-Atom Maser

The simplest form of interaction between a two-level atom and a single quantized mode of the electromagnetic field is described by the Jaynes-Cummings Hamiltonian.[4, 8].

$$
\begin{aligned}
H &= \frac{1}{2}\hbar\omega_o\sigma_z + \hbar\omega a^{\dagger} a + \hbar(g a^{\dagger}\sigma_- + adj.) \\
&= H_o + V,
\end{aligned}
$$

where

$$
\begin{aligned}
H_o &= \frac{1}{2}\hbar\omega_o\sigma_z + \hbar\omega a^{\dagger} a, \\
V &= \hbar(g a^{\dagger}\sigma_- + adj.).
\end{aligned}
$$

Here, ω_o is the atomic transition frequency, ω is the field frequency, a and $a^{\dagger}$ are the photon annihilation and creation operators of the field mode, with $[a, a^{\dagger}] = 1$, σ_z, and σ_+ and σ_- are atomic pseudo-spin operators, with $[\sigma_+, \sigma_-] = \sigma_z$, and

$$
g = \frac{pE_\omega}{2\hbar} sinKZ
$$

is the electric dipole matrix element at the location Z of the atom, where E_ω is the "electric field per photon", $E_\omega = (\hbar\omega/\varepsilon_o V)^{1/2}$.

The Jaynes-Cummings model plays a central role in quantum optics owing to several reasons: it gives the simplest description of Rabi-flopping in a quantised field and the simplest illustration of spontaneous emission; furthermore, it can be solved exactly and thus describes the true quantum dynamics observed with the one-atom maser such as collapse and revivals of the atomic inversion. The model describes the situation realised in the one-atom maser and allows to investigate in detail the complexities of the atom-field dynamics in the simplest of all situations.

In the following a few results following from the Jaynes-Cummings model are reviewed being relevant for the discussions in this paper.

The eigenstates of the Jaynes-Cummings Hamiltonian are

$$
\begin{aligned}
E_{1n} &= \hbar[-\frac{1}{2}\omega_o + (n+1)\omega + \frac{1}{2}(\Omega_n + \delta)] \\
E_{2n} &= \hbar[\ \frac{1}{2}\omega_o + n\omega - \frac{1}{2}(\Omega_n + \delta)]
\end{aligned}
$$

where $\delta = \omega_o - \omega$ is the atom-field frequency detuning. Ω_n is the generalized n-photon Rabi flopping frequency

$$
\Omega_n = \sqrt{4g^2(n+1) + \delta^2}
$$

The corresponding eigenvalues are the one of the dressed atom:

$$
\begin{aligned}
|\,1, n\rangle &= sin\theta_n\,|a, n\rangle + cos\theta_n\,|b, n+1\rangle \\
|\,2, n\rangle &= cos\theta_n\,|a, n\rangle - sin\theta_n\,|b, n+1\rangle,
\end{aligned}
$$

where the states $|a\rangle$ and $|b\rangle$ are the upper and lower atomic states, respectively, and $|n\rangle$ are the number states of the field mode with $a^\dagger a\,|n\rangle = n\,|n\rangle$. The angle θ is defined by means of the relations

$$\begin{aligned} cos\theta_n &= \frac{\Omega_n - \delta}{\sqrt{(\Omega_n - \delta)^2 + 4g^2(n+1)}}, \\ sin\theta_n &= \frac{2g\sqrt{n+1}}{\sqrt{(\Omega_n - \delta)^2 + 4g^2(n+1)}}. \end{aligned}$$

Note, in particular, that

$$cos2\theta_n = -\delta/\Omega_n$$

and

$$sin2\theta_n = 2g\sqrt{n+1}/\Omega_n$$

In the vacuum field n = 0 and on resonance $\omega_o = \omega$ the dressed states are separated by the frequency $\Omega_o = 2g$ generally called vacuum Rabi splitting.

One of the interesting phenomena described by the Jaynes-Cummings model is the dynamical behaviour. When we assume that the atom is initially in the upper state $|a\rangle$ and the field in the number state $|n\rangle$ it follows for the probability of an atom to be in the upper state

$$|\,C_{an}(t)\,|^2 = cos^2(g\sqrt{n+1}\,t).$$

This shows that the upper state population oscillates periodically at the Rabi flopping frequency, similar to the case of classical fields. If the field is initially described by the photon statistics p_n, the above results have to be generalised to:

$$|\,C_a(t)\,|^2 = \sum_n p_n cos^2(g\sqrt{n+1}\,t).$$

It has been shown that in the case where the field mode is initially in a coherent state $|\,C_a(t)|^2$ undergoes a collapse followed by a series of revivals.[6] The collapse is due to the destructive interference of quantum Rabi floppings at different frequencies, a similar phenomenon may also occur with a classical field, however, the revivals are a purely quantum mechanical effect that originates in the discreteness of the quantum field. Collapse and revivals have been observed in a micromaser experiment.[7] In this experiment the interaction time of the atoms with the cavity was varied and the probability was investigated that the atoms leave the cavity in the excited state. As will be shown below the photon statistics in the maser cavity changes when the interaction time is varied, therefore the photon statistics p_n is not a pure distribution. Nevertheless the revival shows up as was also confirmed in a computer simulation of the results on the basis of the Jaynes-Cummings model.[7]

In the following we would like to summarise the results of the quantum-theory of the one-atom maser. Since the atom-field interaction takes place in a closed single mode cavity, there is no spontaneous emission rate into free space modes. Owing to the extremely high quality factors achieved in the superconducting cavities the photon lifetime is extremely long compared to the transit time of the atoms through the resonator. This means that the cavity damping can be practically ignored while an atom interacts with the field. Since the atomic flux is kept small to have at most one atom present inside the cavity at a time, hence the cavity is empty most of the time, therefore cavity damping can be neglected during the rare instances when an atom interacts with the cavity mode.

The one-atom maser theory is therefore based on the following strategy[9]: while an atom is in the cavity, the coupled atom-field system is described by the Jaynes-Cummings Hamiltonian and for the interval between atoms the evolution of the field density matrix is described by a master equation considering damping and in addition the mean number of thermal photons in the cavity. Besides this microscopic theory there is als o a macroscopic theory based on the quantum theory of the laser.[10]

The resulting probability distribution of the photons depends characteristically on the pump rate and on the interaction time t_{int} of the atoms with the cavity field. One obtains the following result for the probability of finding n photons in the maser cavity $P(n)$ in steady state:

$$P(n) = P_o \left[\frac{n_b}{n_b+1}\right]^n \prod_{m=1}^{n} \left[1 + \frac{N}{n_b\gamma}\frac{sin^2(g\sqrt{m}t_{int})}{m}\right],$$

with N being the atomic pump rate, n_b the thermal photon number, and γ the cavity decay rate. P_o is determined by the normalization condition $\Sigma P(n) = 1$. One can now evaluate the mean photon number $\langle n\rangle$ and the field variance in the form of the Q_f parameter[9]:$Q_f = (\langle n^2\rangle - \langle n\rangle^2 - \langle n\rangle)/\langle n\rangle$.

Figure 3 shows the mean photon number as a function of the interaction time of the atoms with the cavity. The photon number is scaled to N_{ex}; here N_{ex} is the average number of atoms that enter the cavity during the cavity decay time $N_{ex} = N/\gamma$. The maser thresholds occurs at $\Theta = 1$ and regions of sub-Poissonian photon statistics are between $\Theta = \pi$ and 2π as well as between 3π and 4π. The sub-Poissonian character leads to a negative Q_f. A large negative value is obtained close before 2π.[8, 9]

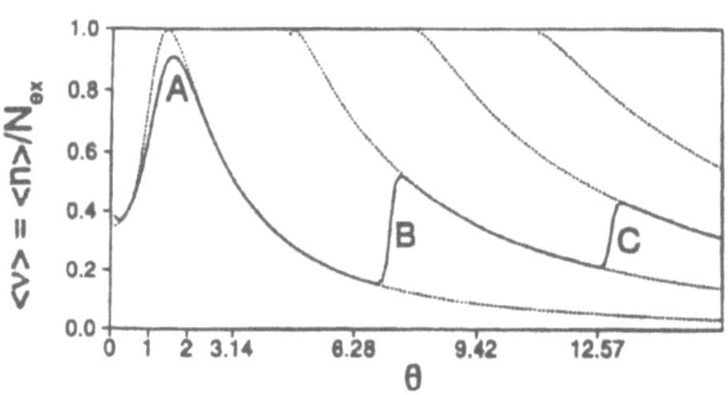

Figure 3. Mean value of the photon number $\nu = n/N_{ex}$ versus the pump parameter Θ, where the value is changed via N_{ex}. The solid line represents the micromaser solution for $\Omega = 36kHz$, $t_{int} = 35\mu s$, and temperature $T = 0.15K$ The dotted lines are semiclascical steady-state solutions corresponding to fixed stable gain=loss equilibrium photon numbers. The crossing points between a line $\Theta = const.$ and the dotted lines correspond to the values where minima in the Fokker-Planck potential $V(\nu)$ occur. For details see text.

To get a more intuitive insight into this effect we recall that the Fizeau velocity selector preselects the velocity of the atoms. Hence, the interaction time is well defined, which leads to conditions usually not achievable in standard masers. This has a very important consequence when the intensity of the maser field grows as more and more atoms transfer their excitation energy to the field: even in the absence of dissipation this increase in photon number is halted when the increasing Rabi frequency leads to a situation where the atoms reabsorb the photon and leave the cavity in the upper state. This situation is close to a quantum-nondemolition case. For any photon number, this situation can be achieved by appropriately adjusting the velocity of the atoms. Then the number distribution of the photons in the cavity is sub-Poissonian.

Unfortunately, the measurement of the nonclassical photon statistics in the cavity is not straightforward. The measurement process of the field involves the coupling to a measuring device, whereby losses lead inevitably to a destruction of the nonclassical properties. The ultimate technique to obtain information about the field employs the Rydberg atoms themselves: for this purpose the population and the statistics of the atoms in the upper and lower maser levels are probed when they leave the cavity. Accordingly, the atoms play a double role: (i) they pump the cavity and (ii) they are used for the diagnostics. These two roles interfere with one another because the detection of the atom in a known final state leads to a quantum mechanical reduction of the photon state inside the resonator. Frequent detection is accompanied by quasipermanent state reduction which can prevent the cavity field from relaxing to the steady state that would be reached if the atoms were left unobserved. Nevertheless, the steady-state properties determine the statistics of the clicks recorded by the atom detectors.

The theoretical treatment of the one-atom maser produces predictions about both the photon field and the emerging atoms. Only the latter can be tested experimentally, and the success of such tests feeds our confidence in the predictions about the quantized radiation field inside the cavity. The pump atoms are statistically independent in the standard one-atom-maser experiments, so that their arrival times are subject to a Poissonian statistics, therefore, the discussion is usually restricted to this standard Poissonian situation.

Inasmuch as the atoms arrive at random, they are recorded at equally random times, and so the only reproducible data are statistical. Consequently, one is led to studying the statistics of the detector clicks. Numerical simulations investigating the effect of repeated atomic measurements on the evolution of the cavity field have been performed by Meystre[11] as well as by Meystre and Wright.[12] The relation between the counting statistics of the detected atoms emerging from the resonator and the photon-number statistics of the field inside the cavity has been studied analytically by Rempe and Walther[13] as well as by Paul and Richter.[14] In Ref.13 the results are also compared with numerical simulations showing good agreement; experimental results are reported by Rempe et al.[15]

In a recent paper a general method for the computation of various statistical properties of the click distribution were presented.[16] This method does not resort to numerical simulations. Naturally, the efficiencies of the detectors - far from the ideal 100 % , unfortunately - are taken into account. A central tool used in those calculations is a nonlinear master equation that governs the dynamics of the photon field in the periods between the detector clicks. The nonlinearity arises from the necessity to distinguish between the notions of observation and detection. When the detectors are active, all emerging atoms are observed but only a fraction is actually detected. Most of the time the experimenter is observing but does not detect - he is listening but does not hear.

In another approach, on the basis of the concept of the counting statistics of light beams, the atomic counting probability, the waiting-time distribution and the "two-atom correlation" function for a Poissonian atomic beam exciting the micro-maser cavity are also calculated. In an analytic treatment it is shown how the waiting-time distribution converges into the atom correlation function for vanishing detection efficiency.[17]

Under steady-state conditions, as mentioned above, the photon number and the photon statistics of the maser field is essentially determined by the dimensionless parameter Θ. The quantity $\langle\nu\rangle = \langle n\rangle/N_{ex}$ shows the following generic behavior (see Fig. 3): It suddenly increases at the threshold value $\Theta = 1$ and reaches a maximum for $\Theta \approx 2$. As Θ further increases, $\langle\nu\rangle$ decreases and reaches a minimum at $\Theta \approx 2\pi$, and then abruptly increases to a second maximum. This general type of behavior recurs roughly at integer multiples of 2π, but becomes less pronounced with increasing Θ. The reason for the periodic maxima of $\langle\nu\rangle$ is that for $\Theta \approx 2\pi$ and multiples thereof the pump atoms perform an almost integer number of full Rabi flopping cycles, and start to flip over at a slightly larger value of Θ, this leading to enhanced photon emission. The maser threshold at $\Theta = 1$ shows the characteristics of a continuous phase transition, whereas the subsequent maxima in $\langle\nu\rangle$ can be interpreted as first-order phase transitions.[8, 9] In the intervals between the phase-transition points, the photon statistics is mostly sub-Poissonian. The field is super-Poissonian for all phase transitions, the large photon number fluctuations above $\Theta = 2\pi$ and multiples thereof being caused by the presence of two maxima in the photon number distribution $P(n)$. They result from the fact that atoms in the upper maser level may or may not tip over to the lower level.[8, 9]

The phenomenon of the two coexisting neighbouring maxima in $P(n)$ was also studied in a semi-heuristic Fokker-Planck (FP) approach.[9] There, the photon number distribution $P(n)$ is replaced by a probability function $P(\nu, \tau)$ with continuous variables $\tau = t/\tau_{cav}$ and $\nu(n) = n/N_{ex}$, the latter replacing the photon number n. The steady-state solution obtained for $P(\nu, \tau), \tau \gg 1$, can be constructed by means of an effective potential $V(\nu)$ showing minima at positions where maxima of $P(\nu, \tau), \tau \gg 1$, are found. Close to $\Theta = 2\pi$ and multiples thereof, the effective potential $V(\nu)$ exhibits two equally attractive minima located at stable gain-loss equilibrium points of maser operation. The mechanism at the phase transitions mentioned is always the same: a minimum of $V(\nu)$ loses its global character when Θ is increased, and is replaced in this role by the next one. This reasoning is a variation of the Landau theory of first-order phase transitions, with $\sqrt{\nu}$ being the order parameter. This analogy actually leads to the notion that in the limit $N_{ex} \to \infty$ the change of the micromaser field around integer multiples of $\Theta = 2\pi$ can be interpreted as first-order phase transitions. In the region of the first-order phase transitions long field evolution time constants τ_{field} are expected. This phenomenon was experimentally demonstrated, as well as related phenomena, such as spontaneous quantum jumps between equally attractive minima of $V(\nu)$, and bistability and hysteresis.[18] Some of those phenomena are also predicted in the two-photon micromaser.[19]

If there are no thermal photons in the cavity - a condition achievable by cooling the resonator to temperatures below 100 mK - very interesting features such as trapping states show up.[20] The investigation of the trapping states is discussed in detail in a recent review.[21]

4. The Photon Statistics of the One-Atom Maser

As discussed above in the experiment on the photon statistics of the one-atom maser field[15] the number N of atoms in the lower maser level is counted for a fixed time interval

T roughly equal to the storage time τ_{cav} of the photons in the cavity. By repeating this measurement many times the probability distribution of finding N atoms in the lower level is obtained. The normalized variance $Q_a = [\langle N^2\rangle - \langle N\rangle^2 - \langle N\rangle]/\langle N\rangle$ is evaluated and is used to characterise the deviation from Poissonian statistics. A negative (positive) Q_a value indicates sub-Poissonian (super-Poissonian) statistics, while Q_a=0 corresponds to a Poisson distribution with $\langle N^2\rangle - \langle N\rangle^2 = \langle N\rangle$. The atomic Q_a can be related to the normalised variance Q_f of the photon number (for details see Refs. 13, 15,16). The experiment[15] shows that the photon field shows a variance $\langle n^2\rangle - \langle n\rangle^2 = 0.3\langle n\rangle$, which is 70 % below the shot noise level; this result agrees with the prediction of the theory.[9, 10]

We emphasise that the reason for the sub-Poissonian atomic statistics is the following: A changing flux of atoms changes the Rabi-frequency via the stored photon number in the cavity. By adjusting the interaction time, the phase of the Rabi-nutation cycle can be chosen so that the probability for the atoms leaving the cavity in the upper maser level increases when the flux and therefore the photon number in the cavity is raised. We observe sub-Poissonian atomic statistics in the case where the number of atoms in the lower state is decreasing with increasing flux and photon number in the cavity. This feedback mechanism is also demonstrated when the anticorrelation of atoms leaving the cavity in the lower state is investigated. Measurements of this 'antibunching' phenomena have also been performed (see Ref. 21 and also Ref. 14 for a detailed review).

The fact that anticorrelation is observed shows that the atoms in the lower state are more equally spaced than expected for a Poissonian distribution. It means when two atoms enter the cavity close to each other the second one performs a transition to the lower state with a reduced probability.

The experimental results presented here clearly show the sub-Poissonian photon statistics of the one-atom maser field. In addition, the maser experiment leads to an atomic beam with atoms in the lower maser level showing number fluctuations which are up to 40% below those of a Poissonian distribution found in usual atomic beams. This is interesting, because atoms in the lower level have emitted a photon to compensate for cavity losses inevitably present. Although this is a purely dissipative process giving rise to fluctuations; nevertheless the atoms still obey sub-Poissonian statistics.

5. Quantum Jumps of the Micromaser Field

The setup used for these measurements is similar to that described above. As before, ^{85}Rb atoms were used to pump the maser. They are excited from the $5S_{1/2}, F = 3$ ground state to $63P_{3/2}, m_J = \pm\frac{1}{2}$ states by linearly polarised light of a frequency-doubled c.w. ring dye laser. The polarisation of the laser light is linear and parallel to the likewise linearly polarized maser field, and therefore only $\Delta m_J = 0$ transitions are excited. Superconducting niobium cavities resonant with the transition to the $61D_{3/2}, m_J = \pm\frac{1}{2}$ states were used. The experiments were performed in a $^3He/^4He$-dilution refrigerator with cavity temperatures $T \approx 0.15K$. The cavity Q-values ranged from $4 \cdot 10^9$ to $8 \cdot 10^9$. The velocity of the Rydberg atoms and thus their interaction time t_{int} with the cavity field were preselected by exciting a particular velocity subgroup with the laser. For this purpose, the laser beam irradiated the atomic beam at an angle of approximately 82°. As a consequence, the UV laser light (linewidth $\approx 2MHz$) is blue-shifted by 50-200 MHz by the Doppler effect, depending on the velocity of the atoms.[18]

As before, information on the maser field and interaction of the atoms in the cavity can be obtained solely by state-selective field ionization of the atoms in the upper or lower maser level after they have passed through the cavity. The field ionization detector was recently modified, so that there is now a detection efficiency of $\eta = (35 \pm 5)\%$. For different t_{int} the atomic inversion has been measured as a function of the pump rate; the coupling constant is about $g = 40\frac{rd}{s}$.

Depending on the parameter range, essentially three regimes of the field evolution time constant τ_{field} can be distinguished.[18] We restrict the discussion here only to the results for intermediate time constants. The maser was operated under steady-state conditions close to the second first-order phase transition (C in Fig. 3). The interaction time was $t_{int} = 47\mu s$ and the cavity decay time $\tau_{cav} = 60ms$. The value of N_{ex} necessary to reach the second first-order phase transition was $N_{ex} \approx 200$. For these parameters, the two maxima in $P(n)$ are manifested in spontaneous jumps of the maser field between the two maxima with a time constant of $\approx 5s$. This fact and the relatively large pump rate lead to the clearly observable field jumps shown in Fig. 4. Due to the large cavity field decay time, the average number of atoms in the cavity was still as low as 0.17. The two discrete values for the counting rates correspond to the metastable operating points of the maser, which correspond to ≈ 70 and ≈ 140 photons. In the Focker-Planck description, the two values correspond to two equally attractive minima in the Focker-Planck potential $V(\nu)$. If one considers, for instance, the counting rate of lower-state atoms (CT2 in Fig. 4), the lower (higher) plateaus correspond to time intervals in the low (high) field metastable operating point. If the actual photon number distribution is averaged over a time interval containing many spontaneous field jumps, the steady-state result $P(n)$ of the micromaser theory is recovered.

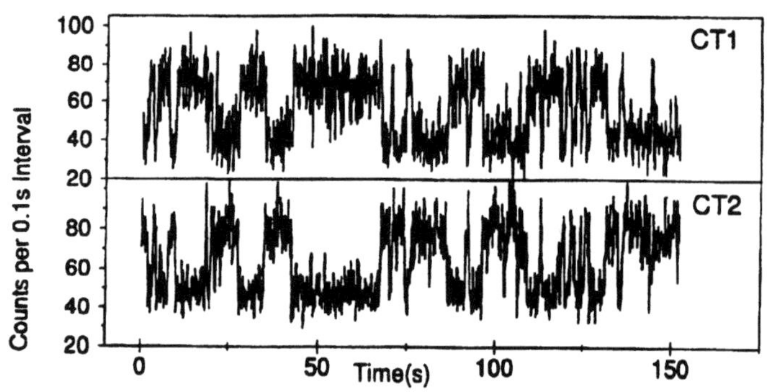

Figure 4. Quantum jumps between two equally stable operation points of the maser field. The channeltron counts are plotted versus time (CT1 = upper state, CT2 = lower state).

In a parameter range where switching occurs much faster than in the case shown in Fig. 4, the individual jumps cannot be resolved any more owing to the reduced N_{ex} necessary in this case. Therefore, a different procedure has to be chosen for the investigation. Those experiments will not be discussed here, details are described in Ref.18.

6. Atomic Interferometry in the Micromaser

In the following we would like to discuss the recent experiment on atomic interferometry in the micromaser.[22] Since a nonclassical field is generated in the maser cavity, we are able to investigate for the first time atomic interference phenomena under the influence of non-classical radiation; owing to the bistable behavior of the maser field the interferences display quantum jumps, thus the quantum nature of the field gets directly visible in the interference fringes. Interferences occur since a coherent superposition of dressed states is produced by mixing the states at the entrance and exit holes of the cavity. Inside the cavity the dressed states develop differently in time, giving rise to Ramsey-type interferences[23] when the maser cavity is tuned through resonance.

The setup used in the experiment is identical to the one described before.[18] However, the flux of atoms through the cavity is by a factor of 5 – 10 higher than in the previous experiments, where the $63P_{3/2} - 61D_{5/2}$ transition was used. For the experiments the $Q-$value of the cavity was 6×10^9 corresponding to a photon decay time of $42ms$.

Fig. 5 shows the standard maser resonance in the uppermost plot which is obtained when the resonator frequency is tuned. This figure corresponds to Fig. 2, however, here the lower state population is plotted instead of the upper state one as shown there.

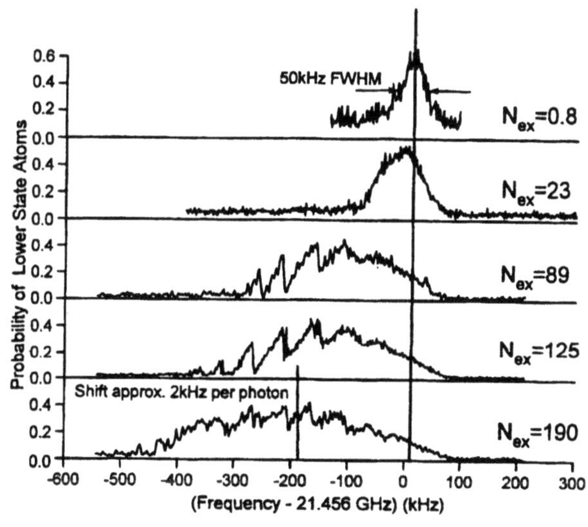

Figure 5. Shift of the maser resonance $63P_{3/2} - 61D_{5/2}$ for fast atoms ($t_{\text{int}} = 35\mu s$). The upper plot shows the maser line for low pump rate ($N_{ex} < 1$). The FWHM linewidth ($50\ kHz$) sets an upper limit of $\approx 5\ \frac{mV}{cm}$ for the residual electric stray fields in th e center of the cavity. The lower resonance lines are taken for the indicated large values of N_{ex}. The plots show that the center of the maser line shifts by about $2\ kHz$ per photon. In addition, there is considerable field-induced line broadening which is approximately proportional to $\sqrt{N_{ex}}$. For $N_{ex} \geq 89$ the lines display periodic structures, which are discussed in the text.

At large values of N_{ex} ($N_{\text{ex}} > 89$) sharp, periodic structures appear. These typically consist of a smooth wing on the low-frequency side, and a vertical step on the high-frequency side. The clarity of the pattern rapidly decreases when N_{ex} increases to 190 or beyond. We will see later that these structures have to be interpreted as interferences. It can be seen that the atom-field resonance frequency is red-shifted with increasing N_{ex}, the shift reaching 200 kHz for $N_{ex} = 190$. Under these conditions there are roughly 100 photons on the average in the cavity. The large red-shift cannot be explained by AC Stark effect, which for 100 photons would amount to about one kHz for the transition used. Therefore it is obvious that other reasons must be responsible for the observed shift.

It is known from previous maser experiments that there are small static electric fields in the entrance and exit holes of the cavity. It is supposed that this field is generated by patch effects at the surface of the niobium metal caused by rubidium deposits caused by the atomic beam or by microcrystallites formed when the cavities are tempered after machining. The tempering process is necessary to achieve high quality factors. The influence of those stray fields is only observable in the cavity holes; in the center of the cavity they are negligible owing to the large atom-wall distances.

Fig. 6 shows the variation of the structure when the interaction time t_{int} between the atoms and the cavity field is changed. The uppermost recording is the same as the second from below in Fig. 5. For large t_{int} no clear substructures can be observed. In the second plot from the top a substructure is still present on the left side, but it is less pronounced than in the uppermost one. The upper two plots in Fig. 6 show that the period of the substructures reduces with increasing interaction time. The substructure disappears for $t_{int} > 47\mu s$. Furthermore, an increasing shift of the whole structure to low frequencies is observed when t_{int} is increased.

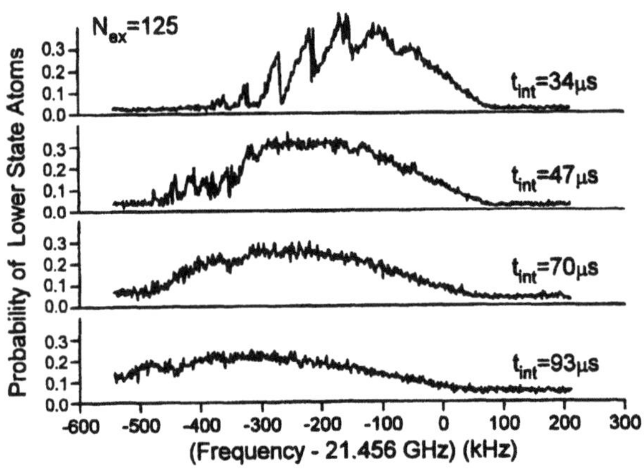

Figure 6. Maser resonance lines for large N_{ex} and the indicated values of t_{int}. The period and visibility of the additional structures reduce when t_{int} is increased. Furthermore the center of the resonance shifts to lower frequencies with increasing t_{int}.

In order to understand the observed structures, we first have to analyze the Jaynes-Cummings dynamics of the atoms in the cavity. This treatment is more involved than that in connection with previous experiments, since the higher maser field requires detailed consideration of the field in the periphery of the cavity, where the additional influence of stray electric fields is more important.

The usual formalism for the description of the coupling of an atom to the radiation field is the dressed atom approach,[24] leading to splitting of the coupled atom-field states, depending on the vacuum Rabi-flopping frequency Ω, the photon number n, and the atom-field detuning δ. We face a special situation at the entrance and exit holes of the cavity. There we have a position-dependent variation of the cavity field, as a consequence of which Ω is position-dependent. An additional variation results from the stray electric fields in the entrance and exit holes. Owing to the Stark-effect these fields lead to a position-dependent atom-field detuning δ.

The Jaynes-Cummings-Hamiltonian only couples pairs of dressed states. Therefore, it is sufficient to consider the dynamics within such a pair. In our case, prior to the atom-field interaction the system is in one of the two dressed states. For parameters corresponding to the periodic substructures in Figs. 5 and 6 the dressed states are mixed only at the beginning of the atom-field interaction and at the end. The mixing at the beginning creates a coherent superposition of the dressed states. Afterwards the system develops adiabatically, whereby the two dressed states accumulate a differential dynamic phase Φ which strongly depends on the cavity frequency. The mixing of the dressed states at the entrance and exit holes of the cavity, in combination with the intermediate adiabatic evolution, generates a situation similar to a Ramsey two-field interaction.

The maximum differential dynamic phase Φ solely resulting from dressed-state coupling by the maser field is roughly 4π under the experimental conditions used here. This is not sufficient to explain the interference pattern of Fig. 5, where we have at least six maxima corresponding to a differential phase of 12π. This means that an additional energy shift differently affecting upper and lower maser states is present. Such a phenomenon can be caused by the above mentioned small static electric fields present in the holes of the cavity. The static field causes a position-dependent detuning δ of the atomic transition from the cavity resonance; as a consequence we get an additional differential dynamic phase Φ. In order to interpret the periodic substructures as a result of the variation of Φ with the cavity frequency, the phase Φ has to be calculated from the atomic dynamics in the maser field.

The quantitative calculation can be performed on the basis of the micromaser theory. A theoretical result of $< n > /N_{ex}$ obtained in this way is shown in Fig. 7. The uppermost plot shows the maser resonance line expected without any static electric field. With increasing DC field strength in the cavity holes the structure changes, the curve for $309 mV/cm$ coming very close to those displayed in Fig. 5 for $N_{ex} = 89$ and 125, and at the top of Fig. 6. We have to stress that the field values indicated in Fig. 7 correspond to the maximum field strength in the cavity holes. The field value in the central part of the cavity is roughly 100 times smaller, and therefore without significance in low-flux maser experiments. Fig. 7 also shows that the qualitative structure of the maser line is the same for all fields larger than about 200 mV/cm.

The calculations also reproduce the experimental finding that the maser line shifts to lower frequencies when N_{ex} is increased. The mechanism for that can be explained as follows: the high-frequency edge of the maser line does not shift with N_{ex} at all, since this part of the resonance is produced in the central region of the cavity, where practically no static electric fields are present. The low-frequency cutoff of the structure is determined by the location where the mixing of the dressed states occurs. With decreasing cavity frequency those points shift closer to the entrance and exit holes, with the difference between the particular cavity frequency and the unperturbed atomic resonance frequency giving a measure of the static electric field at the mixing locations. Closer to the holes the passage behavior of the atoms through the mixing locations becomes non-adiabatic for the following reasons: firstly, the maser field strength reduces towards the holes. This leads to reduced repulsion of the dressed states. Secondly, the stray electric field strongly increases

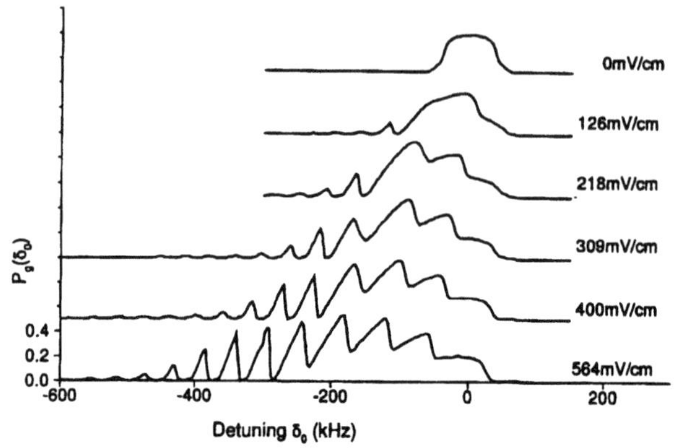

Figure 7. Theoretical maser lines for the indicated values of the static electric field strength in the cavity holes. The theoretical model is explained in the text. In the calculation we use $N_{ex} = 100$ and $\Omega = 45$ *krad/s*. The interaction time is $t_{int} = 35\ \mu s$, and the RMS deviation o f the interaction time is $1\ \mu s$. In order to account for the fluctuations of the dynamic phases induced by the inhomogeneity of the stray electric field s, a Gaussian distribution of the atom-field detuning with a RMS deviation of 5 *kHz* is assumed (see also Ref. 18).

towards the holes. This implies a larger differential slope of the dressed state energies at the mixing locations, and therefore leads to a stronger non-adiabatic passage. At the same time the observed signal extends further to the low frequency spectral region. Since the photon emission probabilities are decreasing towards lower frequencies their behavior finally defines the low-frequency boundary of the maser resonance line. With increasing N_{ex} the photon number n increases. As for larger values of n the photon emission probabilities get larger, also an increasing N_{ex} leads to an extension of the range of the signal to lower frequencies. This theoretical expectation is in agreement with the experimental observation.

In the experiment it is also found that the maser line shifts towards lower frequencies with increasing t_{int} (Fig. 6). This result also follows from the developed model: the red-shift increases with t_{int} since a longer interaction time leads to a more adiabatic behavior in the same way as a larger N_{ex} does.

The calculations reveal that on the vertical steps displayed in the signal the photon number distribution has two distinctly separate maxima similar to those observed at the phase transition points discussed above. Therefore, the maser field should exhibit hysteresis and metastability under the present conditions as well. The hysteresis should show up when the cavity frequency is linearly scanned up and down with a modest scan rate. An experimental result is given in Fig. 8, where each step in the steady-state maser line is, indeed, related to a hysteresis loop.

When the maser is operated in steady-state and the cavity frequency is fixed to the steep side of one of the fringes we also observe spontaneous jumps of the maser field between two metastable field states. The results displayed in Fig. 9 show that the maser field jumps between two values, whereby the lower one corresponds to an almost zero field. The difference between the photon numbers of the two metastable states can be as low as 25.

The calculations also show that on the smooth wings of the more pronounced inter-

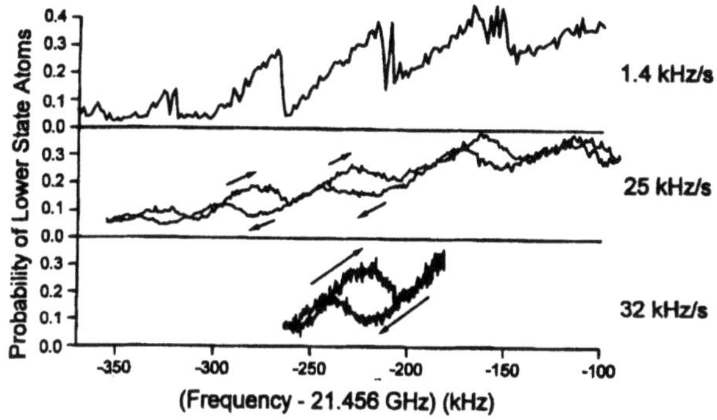

Figure 8. Hysteresis cycle of the maser for $t_{\text{int}} = 35\mu s$ and $N_{\text{ex}} = 84$. Again the atomic inversion is plotted versus the cavity frequency. The upper plot shows a slow scan on the low-frequency wing of the signal. The lower plots show that the vertical steps turn into hysteresis loops when the cavity frequency is rapidly scanned up and down. The values to the right indicate the scan speed.

ference fringes the photon number distribution $P(n)$ of the maser field is strongly sub-Poissonian. This leads us to the conclusion that we observe Ramsey-type interference induced by a non-classical radiation field. The sub-Poissonian character of $P(n)$ results from the fact that on the smooth wings of the fringes the photon gain reduces when the photon number is increased. This feedback mechanism stabilizes the photon number resulting in a sub-Poissonian photon distribution.

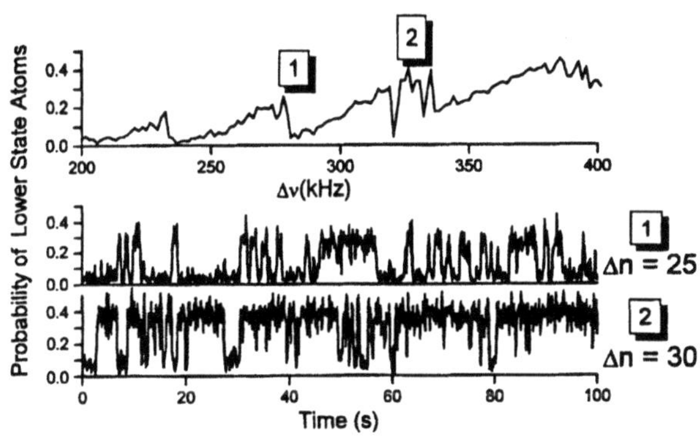

Figure 9. Spontaneous jumps of the maser field for $t_{\text{int}} = 35\mu s$ and $N_{\text{ex}} = 84$. When the maser is operated at locations 1 and 2 of the maser line shown in the upper plot, the atomic inversion as a function of time statistically jumps as displayed in the lower plots. The values Δn denote the differences of the mean photon numbers in the two metastable field states.

The presented maser model explains all the observed experimental facts. The periodic structures in the maser lines are thus interpreted as Ramsey-type interferences. If there was more accurate information on the DC fields in the cavity, a multi-level calculation taking into account the magnetic substructure of the involved fine-structure levels would make sense. Under the present conditions, however, the stray electric field amplitudes close to the cavity holes can only be estimated. The interference structure extends towards lower frequency as far as $\approx 500kHz$ for $N_{ex} \approx 200$, this corresponding to an electric field of $\approx 25mV/cm$ in a distance of roughly a few mm from the cavity holes.

To our knowledge it is the first time that it is demonstrated that non-adiabatic mixing can lead to Ramsey interferences. Besides other phenomena the bistable character of the micromaser field can be observed in jumps of the fringes; the observed interferences thus show a discontinuous behavior owing to the quantum properties of the field - it is the first time that such a discrete quantum behavior is directly observed in interference fringes. One of the applications of the described Ramsey interferometer could be the quantum-non-demolition measurement of the photon number in a cavity along the lines proposed in Ref. 25. For this purpose the atoms in the cavity have to be dispersively coupled to a third level via a second quantum field which could agree with the frequency of another cavity mode.

7. Linewidth and Phase Diffusion of the One-Atom Maser

In the following we would like to discuss another special feature of the one-atom maser: the spectrum. This is determined by the decay of the expectation value of the electric field[21]

$$\langle E(t)\rangle \sim \sum_{n=0}^{\infty}(n+1)^{1/2}\rho_{n,n+1}(t)$$

Hence the micromaser spectrum is different from the other effects discussed so far since it involves the off-diagonal elements $\rho_{n,n+1}$ of the radiation density matrix rather than e.g. the photon statistics, that is, the diagonal elements $\rho_{n,n}$; and it requires their time dependence rather than their steady-state values. It could be shown that the linewidth D of the maser is given by

$$D = 4r sin^2\left[\frac{gt_{int}}{4\sqrt{\langle n\rangle}}\right] + \frac{\gamma(2n_b+1)}{4\langle n\rangle}$$

For small $g\tau/4\sqrt{\langle n\rangle}$ the sine function can be expanded; this leads to the familiar Schawlow-Townes linewidth

$$D = \frac{\alpha + \gamma(2n_b+1)}{4\langle n\rangle}$$

where

$$\alpha = \gamma(\sqrt{N_{ex}}gt_{int}).$$

The complicated pattern of the micromaser linewidth results, in part, from the complicated dependence of $\langle n\rangle$ on the pump parameter, which enters in the denominator. We emphasize that the one-atom maser linewidth goes beyond the standard Schawlow-Townes linewidth. The sine function suggests in the limit of large pumping parameters an oscillatory behaviour of the linewidth confirmed in an exact numerical treatment. (For details see also Refs. 26 and 27).

It was pointed out that the maser linewidth can be measured when two phase coupled microwave fields are used; one before the atoms enter the microwave cavity and one after corresponding, in principle, to a modified Ramsey setup. The first micromaser field creates a superposition of the two maser states being then probed by the second one. The first field is only applied for an initial period in order to "seed" a phase in the micromaser cavity, in the second field the phase diffusion is probed. This technique will provide the basis for future experiments to measure the linewidth. It was tested by computer simulations and analytical calculations which showed that coherent pumping of the maser leads to new and interesting phenomena which cannot be discussed here. The first time coherent injection for the one-atom maser was proposed in the paper by Krause et al.[28] and discussed later in connection with the measurement of the phase diffusion linewidth.[29] Also the phase dynamics of the maser field in steady-state operation was discussed by Wagner et al.[30]; in the latter case only one microwave field is used for probing after the micromaser cavity. The latter scheme was further pursued in connection with the entanglement of states by Wagner et al.[31] The entangled state of the atom-field system occurs since a factorization in field and atom part is not possible. This fact leads to applications of the micromaser for Einstein-Podolsky-Rosen type of experiments.

8. A New Probe of Complementarity in Quantum Mechanics- The One-Atom Maser and Atomic Interferometry

In one of the preceding sections we discussed how a nonclassical field inside the maser cavity can be generated. Such a field is extremely fragile because any attenuation causes a considerable broadening of the photon number distribution. Therefore it is difficult to couple the field out of the cavity while preserving its nonclassical character. But what is the use of such a field? In the present section we want to propose a new set of experiments performed inside the maser cavity to test the 'wave-particle' duality of nature, or better said 'complementarity' in quantum mechanics.

Complementarity lies at the heart of quantum mechanics: Matter sometimes displays wave-like properties manifesting themselves in interference phenomena, and at other times it displays particle-like behaviour thus providing 'which-path' information. No other experiment illustrates this wave-particle duality in a more striking way than the classic Young double-slit experiment. Here we find it impossible to tell which slit light went through while observing an interference pattern. In other words, any attempt to gain 'which-path' information disturbs the light so as to wash out the interference fringes. This point has been emphasised by Bohr in his rebuttal to Einstein's ingenious proposal of using recoiling slits to obtain 'which-path' information while still observing interference. The physical positions of the recoiling slits, Bohr argued, are only known to within the uncertainty principle. This error contributes a random phase shift to the light beams which destroys the interference pattern.

Such random-phase arguments, illustrating in a vivid way how the 'which-path' information destroys the coherent wave-like interference aspects of a given experimental setup, are appealing. Unfortunately, they are incomplete: In principle, and in practice, it is possible to design experiments which provide 'which-path' information via detectors which do not disturb the system in any noticeable way. Such 'Welcher Weg'- (German for 'which-path') detectors have been recently considered within the context of studies involving spin coherence.[32] In this section we describe a quantum optical experiment[33] which shows that the loss of coherence occasioned by 'Welcher Weg' information, that is, by the presence of a 'Welcher Weg'-detector, is due to the establishment of quantum correlations. It is in no way associated with large random-phase factors as in Einstein's recoiling slits.

The details of this application of the micromaser are discussed in Ref. 34. Here only the essential features are given. We consider an atomic interferometer where the two particle beams pass through two maser cavities before they reach the two slits of the Youngs interferometer. The interference pattern observed is then also determined by the state of the maser cavity. The corresponding interference term is given by:

$$\langle \Phi_1^{(f)}, \Phi_2^{(i)} \mid \Phi_1^{(i)}, \Phi_2^{(f)} \rangle,$$

where $| \Phi_i^{(i)} \rangle$ and $| \Phi_j^{(f)} \rangle$ denote the initial and final states of the maser cavity i,j.

Let us prepare, for example, both one-atom masers in coherent states $| \Phi_j^{(i)} \rangle = | \alpha_j \rangle$ of large average photon number $\langle n \rangle = | \alpha_j |^2 \gg 1$. The Poissonian photon number distribution of such a coherent state is very broad, $\Delta m \approx \alpha \gg 1$. Hence the two fields are not changed much by the addition of a single photon associated with the two corresponding transitions. We may therefore write

$$| \Phi_j^{(f)} \rangle \approx | \alpha_j \rangle,$$

which to a very good approximation yields

$$\langle \Phi_1^{(f)}, \Phi_2^{(i)} \mid \Phi_1^{(i)}, \Phi_2^{(f)} \rangle \approx \langle \alpha_1, \alpha_2 \mid \alpha_1, \alpha_2 \rangle = 1.$$

Thus there is a 'radiation' interference cross term different from zero.

When, however, we prepare both maser fields in number states $|n_j\rangle$ the situation is quite different. Since the number states are orthogonal, the interference term disappears whenever a passing atom deposits a photon in one of the cavities therefore no interferences are observed in such a case (For the case that thermal fields are present see Ref. 33).

At first sight this result might seem a bit surprising when we recall that in the case of a coherent state the transitions did not destroy the coherent cross term, i.e. did not affect the interference fringes. However, in the example of number states we can, by simply 'looking' at the one-atom maser state, tell which 'path' the atom took.

The atomic interference experiment in connection with one-atom maser cavities is a rather complicated scheme for a 'Welcher Weg'-detector. There is a much simpler possibility which we will discuss briefly in the following. This is based on the logic of the famous 'Ramsey fringe' experiment. In this experiment two microwave fields are applied to the atoms one after the other. The interference occurs since the transition from an upper state to a lower one may occur either in the first or in the second interaction region. In order to calculate the transition probability, we must sum the two amplitudes and then square, thus leading to an interference term in the same way as the two slit experiment. (For details see Ref. 35).

We conclude this section by emphasising again that this new and potentially experimental example of wave-particle duality and observation in quantum mechanics displays a feature which makes it distinctly different from the Bohr-Einstein recoiling-slit experiment. In the latter the coherence, that is the interference, is lost due to a phase disturbance of the light beams. In the present case, however, the loss of coherence is due to the correlation established between the system and the one-atom maser. Random-phase arguments never enter the discussion. We emphasise that the argument of the number state not having a well-defined phase is not relevant here; the important dynamics are due to the atomic transition. It is the fact that 'which-path' information is made available which washes out the interference cross terms.[34]

9. Experiments with Trapped Ions–The Paul-Trap

In contrast to neutral atoms, ions can easily be influenced by electromagnetic fields because of their charge. Therefore it is possible to isolate a single ion quite easily from its surroundings. In most of the experiments performed so far the Paul trap is used. It consists of a ring electrode and two end caps as shown in Fig. 10. Trapping can be achieved if time varying electric fields[36] are applied between the ring and caps (the two caps are electrically connected). A dc voltage in addition changes the relation of the potential depth along the symmetry axis (vertical direction in Fig. 10) to that in perpendicular direction. The equation of motion of an ion in such a situation is the Mathieu differential equation, well known in classical mechanics, which – depending on the voltages applied to the trap (dc and radio-frequency voltages) – allows stable and unstable solutions. Another way to achieve trapping is the use of a constant magnetic field aligned along the symmetry axis leading to the Penning trap.[1, 37] In this case only a dc voltage has to be applied between ring and cap electrodes.

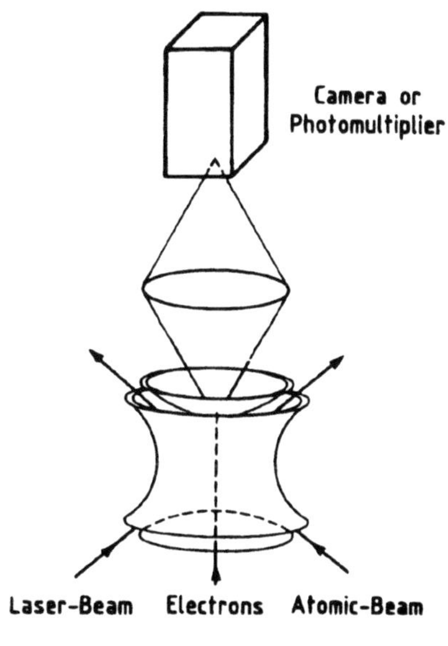

Figure 10. Sketch of the Paul trap. The fluorescence light is observed through a hole in the upper electrode.

In order to produce the ions in the Paul trap a neutral atomic beam is directed through the trap centre and ionised by electrons. Unfortunately, the resulting trapped ions have a lot of kinetic energy rendering them useless for most applications, such as spectroscopy; therefore the ions have to be cooled. This is done by laser light. For this purpose the laser frequency ν is tuned below the resonance frequency, so that the energy of the photon is not sufficient to excite the atom. Crudely, the ion can extract the missing energy from its motion and thus reduce its kinetic energy. In other words, the atomic velocity Doppler-shifts the atom into resonance to bridge the detuning gap Δ between laser and resonance frequency and the atom absorbs the photon of momentum $\hbar k = h\nu/c$. After the absorption process the momentum of the atom in direction opposite to the laser beam is reduced. This leads to a net cooling effect since the reemission of the energy due to atomic fluorescence is isotropic in space. The lowest temperature achievable is determined by the Doppler-limit[38] which is in the milli-Kelvin region. The low temperatures can be obtained within a fraction of a second.

The results discussed in this paper were obtained using a Paul trap with a ring diameter of 5 mm and an end-cap separation of 3.54 mm.[39] This trap is larger than most of the ion traps used in laser experiments.[1, 40] The radio frequency of the field used for dynamic trapping is 11 MHz. The trap is mounted inside a stainless steel, ultrahigh vacuum chamber. We can obtain storage times of hours using a background gas at a pressure of 10^{-10} mbar. The ions are loaded into the trap by means of a thermal beam of neutral atoms (magnesium atoms in our case) which are then ionised close to the centre of the trap by an electron beam entering the trap through a small hole in the lower end-cap (see Fig. 10). In order not to distort the trap potential, the hole is covered by a fine molybdenum mesh. The neutral Mg beam and the laser beam pass through the gaps between the end-cap and the ring electrodes. The laser frequency is shifted by an amount Δ below the $3S_{1/2} \rightarrow 3P_{3/2}$ resonance-transition of $^{24}Mg^+$ at 280 nm to extract kinetic energy from the ions by radiation pressure as discussed above. In this way, a single ion can be cooled to a temperature below 10 mK. The fluorescence from the ions is observed through a hole in the upper end-cap, again covered by a molybdenum mesh. The large size of the trap allows a large solid angle for detecting the fluorescence radiation, either with a photomultiplier, or by means of a photon counting imaging system. To observe the ions, the cathode of the imaging system is placed in the image plane of the microscope objective attached to the trap; in this way images of the ions could be obtained.[41, 42]

10. Phase Transitions of Trapped Ions

The existence of phase transitions in a Paul trap manifests itself by significant jumps in the fluorescence intensity of the ions as a function of the detuning Δ between the laser frequency and the atomic transition frequency.[39] The existence of the two phases - crystal and cloud - can be verified experimentally by direct observation with the help of a highly sensitive imaging system, and theoretically by analysing ion trajectories via Monte Carlo computer simulations.[41, 42]

Fig. 11 shows ion structures as measured with the imaging system. For the measurements only a radio-frequency voltage was applied to the trap electrodes: in this case the potential is a factor of two deeper along the symmetry axis than perpendicular to it, and therefore plane ion structures are observed being perpendicular to the symmetry axis (for details see Refs. 39, 41, 42).

After we have accepted the existence of the ion crystals and the corresponding phase transitions, how do they actually occur? Would the cooling laser not force any cloud immediately to crystallise? A heating mechanism balancing the cooling effect of the laser

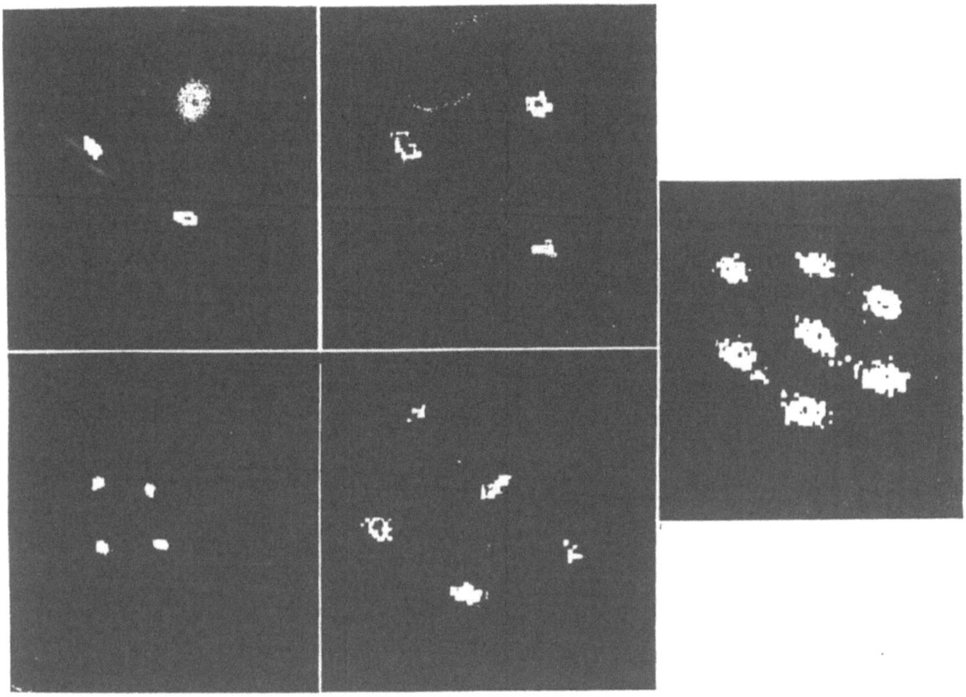

Figure 11. Two, three, four, five and seven trapped ions in the crystallized configuration. The distance between the ions is roughly 20 μm. The scale of the pictures is not completely identical. The trap potent ial is such that the ions arrange themselves practically in two-dimensional structures. There is a slight distortion of the ion structures by a contact potential caused by a coating on the ring electrode with Mg atoms. This distortion is specially visible for five ions.

must be the answer to this puzzle, but what heating mechanism? Since the early days of Paul traps this so-called radio-frequency heating has repeatedly been cited.[1] A deeper understanding however was missing and was provided only recently by a detailed study of the dependence of the cloud→crystal and crystal→cloud phase transitions on the relevant parameters.[41, 42, 43]

11. The Ion Storage Ring

A completely new era of accelerator physics could begin if it were possible to produce, store and accelerate Coulomb crystals in particle accelerators and storage rings. To work with crystals instead of the usual dilute, weakly coupled particle clouds has at least one advantage: the luminosity of accelerators (storage rings) could be greatly enhanced, and (nuclear) reactions whose cross sections are too small to be investigated in currently existing accelerators would become amenable to experimentation.

In the following, we would like to discuss very briefly our recent experiments using a miniature quadrupole storage ring. The storage ring is similar to the one described by Drees and Paul[44] or by Church.[45] We can observe phase transitions of the stored ions and the observed ordered ion structures are quite similar to the ones expected in relativistic storage rings, however, much easier to achieve. The motivation for building this small storage ring came from the fact that micromotion perturbs the ion structures in a Paul trap and only a single trapped ion is free of micromotion.[1] The ring trap used consists of a quadrupole field, leading to harmonic binding of the ions in a plane transversal to the electrodes of the quadrupole and no confinement along the axis (see Fig. 12). Confinement along the axis is achieved, however, by the Coulomb interaction between the ions when the ring is filled; then the total number of ions in the ring determines the average distance between them.

A scheme of the ring trap used for our experiment is shown in Fig. 12.[46, 47] It consists of four electrode rings shown in the insert on the right. The hyperbolic cross section of the electrodes required for an ideal quadrupole field was approximated by a circular one.

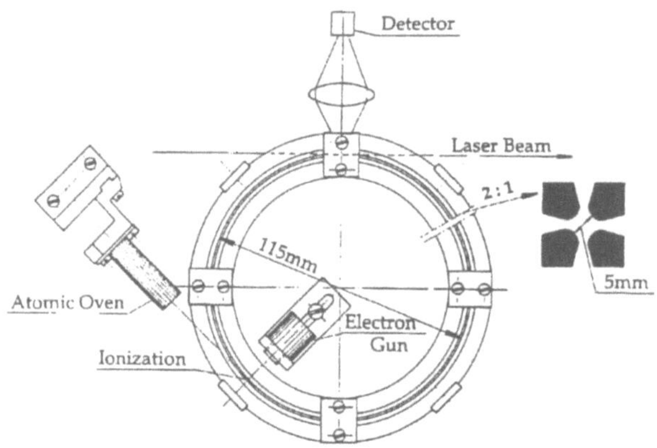

Figure 12. Quadrupole storage ring. The cross section of the electrod e configuration is shown on the insert on the right hand side of the figure. The diameter of the ring is 113 mm and the distance between the electrodes is 5 mm.

The experiments were also performed with Mg^+ ions. The ions were produced between the ring electrodes by ionising the atoms of a weak atomic beam produced in an oven which injected the atoms tangentially into the trap region. The electrons used for the ionisation came from an electron gun the electron beam of which was perpendicular to the direction of the atomic beam. A shutter in front of the atomic beam oven allowed the interruption of the atomic flux. The ultrahigh vacuum chamber was pumped by an ion getter pump. After baking the chamber a vacuum of 10^{-10} mbar could be reached. Under these conditions the number of ions stored in the trap stayed practically constant for several hours.

When laser cooling of the ions is started a sudden change in the fluorescence intensity is observed, resembling very much that seen with stored ions in a Paul trap which indicates a phase transition and the formation of an ordered structure of ions. The ion structure can also be observed using an ultrasensitive imaging system. Pictures of typical ion structures are shown in Fig. 13.

The ions are excited by a frequency tunable laser beam which enters the storage ring tangentially. In the linear configuration the ions are all sitting in the center of the quadrupole field; therefore they do not show micromotion and it is possible to cool them further to temperatures in the micro-Kelvin region.

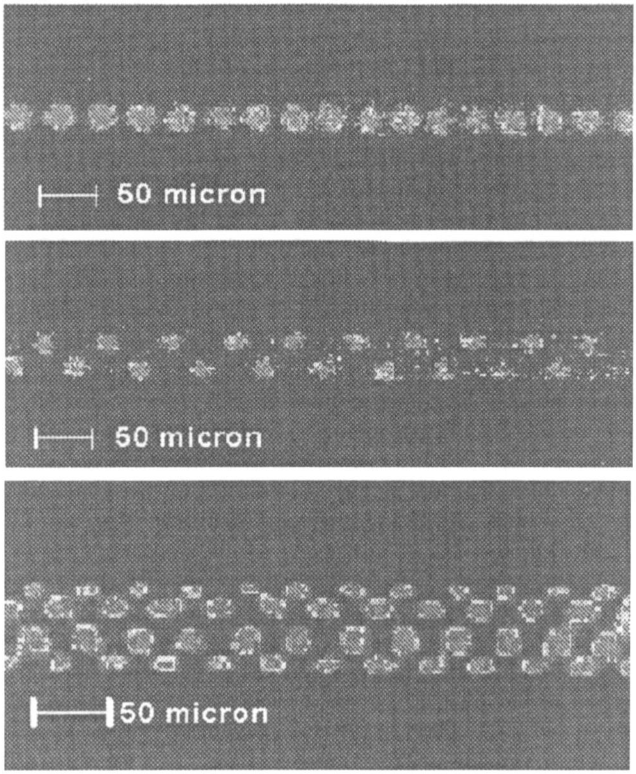

Figure 13. Crystalline structures of laser-cooled $^{24}Mg^+$ ions in the quadrupole storage ring. At a low ion density ($\lambda = 0.29$) the ions form a string along the field axis (a, upper). Increasing the ion density transforms the configuration to a zigzag (b, middle; $\lambda = 0.92$). At still higher ion densities, the ions form ordered helical structures on the surface of a cylinder, e.g., three interwoven helices at $\lambda = 2.6$ (c, lower). As the fluorescent light is projected onto the plane of observation in this case the inner spots are each created by two ions seated on opposite sides of the cylindrical surface, resulting in a single, bright spot.

The new cooling methods proposed by Dalibard et al.[38] can be applied to the Mg^+ ions so that the single-photon recoil limit can be achieved for the cooling process. At this limit the kinetic energy of the ions may reach 'zero-energy' of the harmonically bound ions; the ion structure then reaches its vibrational ground state, i.e. a Mössbauer situation is generated. The observed ion configurations in the quadrupole ring trap are described in a recent paper by Birkl et al.[48] We will review the major results reported in this paper in the next section.

12. Ordered Structures in the Storage Ring

In the molecular dynamics calculations[49], a cylindrically symmetric, static harmonic potential is assumed to describe the confining field. Each particle is subjected to the Coulomb forces of all other particles and to the confining field. The classical equations of motion are integrated for a system of several thousand particles, starting with random positions and velocities, to give the time evolution of the system. Cooling of the stored particles is simulated by scaling down the resulting velocities of the stored particles at defined instants of the integration process. After sufficient cooling, ordered structures such as strings, zigzags, shells, and multiple shells should arise owing to the confining field's harmonic potential. Our experiments are well-suited to checking these predictions. To compare the experimental results with theory, we can use the normalised 'linear particle density' λ which is given by the ion density multiplied by the ratio of Coloumb repulsion and confining force of the trap.[49] Low λ values correspond to a deep potential well or a small number of ions, resulting in an equilibrium structure closely confined to the field axis comprising a string of ions (Fig. 13a). This is the micromotion-free configuration discussed above, and the analogue of the single stored ion in a Paul trap, as in both cases the ions sit in the potential minimum and show no micromotion. For higher values of λ, the structures extend more and more into the off-axis region, giving rise to (in the order of increasing λ) a plane zigzag structure (Fig. 13b) and cylindrical structures with the ions forming helices on cylindrical surfaces. The structure in Fig. 13c consists of three interwoven helices with six ions per pitch. The string and the zigzag have also been observed with laser-cooled Hg^+ ions in a linear trap.[50]

Increasing further the number of ions leads to structures with smaller spacings between the ions where we cannot optically resolve individual ions any more. Images of these structures are presented in Fig. 14. The radial intensity profiles displayed on the right-hand side of the figure provide information about the structures as they give a measure of the radial distribution of the ions.

For increasing λ it becomes energetically more favourable to create a string inside the first ion shell (Fig. 14a) to provide space for more particles. This results in a structure which is a three-dimensional analogue of the plane seven-ion crystal for a Paul trap (Fig. 11). The inner string turns into a second shell at higher densities: a string then develops inside this second shell (Fig. 14b), and so on. Fig. 14c and Fig. 14d show structures consisting of three shells plus string, and four shells, respectively. We have been able to observe all possible structures, from the string up to four shells plus string. The formation of multiple-shell crystalline structures in the quadrupole storage ring contrasts with the observation of shell structures in Penning traps, where the ions do not occupy fixed positions inside the shells.[51] The ion structures of Fig. 13 and Fig. 14 are in very good agreement with the molecular dynamics calculations. For a comparison see Ref. 48.

Our results have important implications for two very different fields. Consider first the physics of low-energy particles: an ion string in a quadrupole ring, being free of micromotion, can be cooled to its vibrational ground state in the confining potential using recently proposed laser cooling techniques.[38] This would place the string in the Lamb-Dicke regime with a vanishing first-order Doppler effect because the spatial amplitude of the motion is smaller than the wavelength of the atomic transition. Furthermore, the second-order Doppler effect, which can only be reduced by further cooling, also disappears making the stored ions very interesting for frequency standards. The large number of ultra-cold ions available in the ring will lead to a high signal-to-noise ratio. Finally,

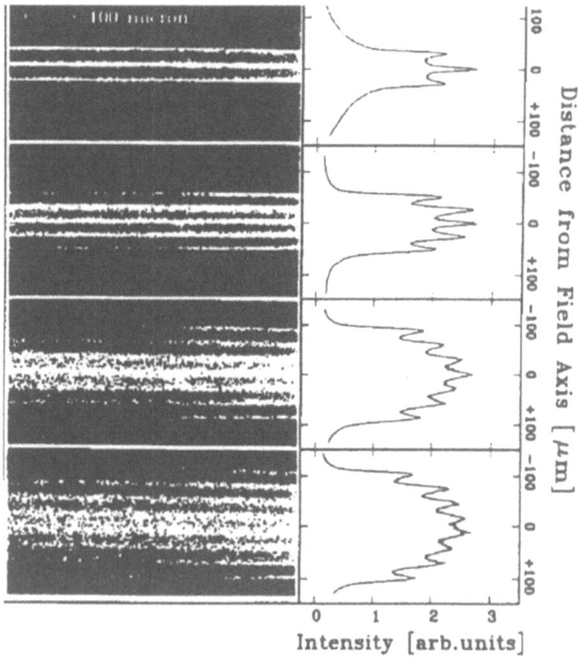

Figure 14. Images and intensity profiles of (from the top): one shell plus string (a; $\lambda \approx 4.3$), two shells plus string (b: $\lambda \approx 12.2$), three shells plus string (c; $\lambda \approx 26$), and four shells (d; $\lambda \approx 34$). There are up to $\approx 8 \times 10^5$ ions stored in the ring for the four-shell structure.

cooled ions in the ring represent a quantum object of macroscopic dimensions - a Wigner crystal. In this limit the photon-recoil will disappear since the ion crystal experiences the Mössbauer effect. First experiments performed with this aim have been published in Ref. 52.

13. Experiments with Single In^+ Ions

In recent years single laser-cooled ions in Paul-traps have been used for many investigations of quantum phenomena and high resolution laser spectroscopy.[53] The trapped ions are also potential candidates for future frequency standards. Several systems have been proposed and are presently persued.[53] In this paper we describe experiments with a single In^+ ion. The relevant energy levels are shown in Fig. 15.

The ${}^1S_0 \rightarrow {}^3P_0$ resonance at 236.5 nm is a hyperfine-induced electric dipole transition and has a natural linewidth of only 1.1 Hz.[54] This line is interesting for an optical frequency standard since it offers a high line $Q = 1.3 \times 10^{15}$ and, in addition, is insensitive against external perturbation. Laser excitation on the ${}^1S_0 \rightarrow {}^3P_1$ intercombination line at a wavelength of 230.6 nm is used to detect the ion via resonance fluorescence. With a natural linewidth of 360 kHz this line is also a relatively narrow optical transition, which

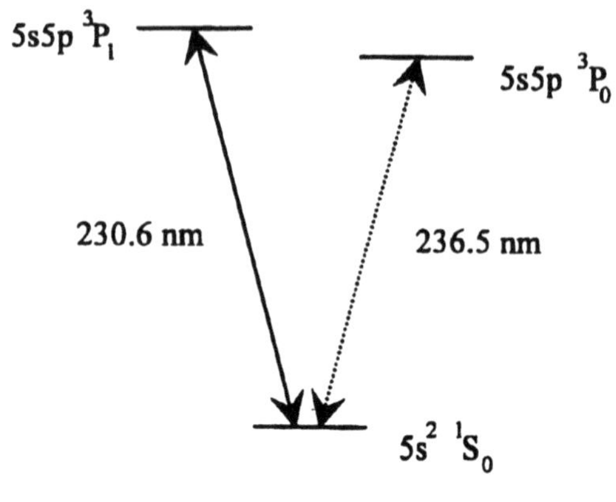

Figure 15. The relevant energy levels of the In^+ ion. The $^1S_0 \rightarrow ^3P_0$ line can be used as an optical frequency standard. The $^1S_0 \rightarrow ^3P_1$ transition is the cooling transition.

gives the advantage that the ion can be simply cooled to the vibrational ground state in the trap potential by sideband cooling.

We recently demonstrated laser cooling of a single indium ion using the $^1S_0 \rightarrow\ ^3P_1$ transition and observed electron shelving in the metastable $^3P_0$ level due to the spontaneous decay on the magnetic dipole transition $^3P_1 \rightarrow\ ^3P_0$.[54] As shown before[55] the $^1S_0 \rightarrow ^3P_0$ clock transition coincides with the center of the tuning range of the frequency quadrupled Nd:YAG $^4F_{3/2} \rightarrow\ ^4I_{9/2}$ laser line. In this paper we describe the first optical-optical double resonance experiment with a single indium ion and excitation of the clock transition with the frequency quadrupled Nd:YAG laser line.[56] Furthermore the application of the trapped indium ion to cavity quantum electrodynamics will be described.

The search for laser excitation of the weak $^1S_0 \rightarrow ^3P_0$ transition requires high output power and a narrow laser linewidth. Diode laser-pumped solid-state lasers provide better inherent frequency stability than other lasers, especially dye lasers, and are therefore technically more suitable as ultrastable lasers. In our previous setup an intracavity frequency doubled standing wave laser cavity was used. The system generated 20μW of tunable UV light at 236.5 nm, which was not sufficient to detect the $^1S_0 \rightarrow ^3P_0$ transition at a high scanning speed being required to search over large spectral regions. The increase of the pump power from 2 W to 4 W failed due to thermal lensing in the Nd:YAG crystal. We now use a new injection locking scheme with an ultrastable diode-pumped master oscillator and a high power diode-pumped slave oscillator, reaching an output power of 0.5 mW at 236.5 nm.[57]

The master oscillator is a V-shaped standing wave cavity. The 3 mm long Nd:YAG crystal is placed between two quarter wave plates to suppress spatial hole burning. A Brewster plate defines the polarization of the laser radiation. The piezo mounted folding mirror is used as the fast tuning element with a dynamic range of 40 kHz. An angle tuned 1 mm etalon and a temperature tuned 5 mm quartz glass etalon lead to stable tunable operation. The single mode output power is 50 mW at 946 nm using a pumping power of 2 W. The master oscillator is frequency stabilized to a zerodur cavity by use of the phase modulation scheme. To get a large tuning range the mirror spacing can be changed with a piezo actuator. The zerodur cavity has a drift rate of about 1 MHz per minute. A deviation of the laser frequency from the zerodur cavity frequency of less than 500 Hz was deduced from the error signal of the feedback loop.[55, 57]

The slave laser has a ring cavity with a Nd:YAG crystal placed in one beam waist and a $KNbO_3$ frequency doubling crystal placed in a second one. The pumping power is 2 W. With a 0.98 reflective input coupler the output power is 5 mW at 473 nm in both directions of the ring resonator. When an optical diode is placed between master and slave oscillator injection locking occurs and the single longitudinal mode and unidirectional oscillation results in an output power of 50 mW at 473 nm. The injection locking range is on the order of 1 MHz and therefore the stabilization of the slave cavity frequency to the master laser frequency using the phase modulation method is required. Phase noise introduced by the injection locking process is estimated to contribute to the laser linewidth on the order of 100 Hz. The second frequency doubling is performed with a Brewster cut BBO crystal placed in an external enhancement cavity.

The laser radiation at 230.6 nm for cooling of the ion is generated by frequency doubling light from a stilbene-3 ring dye laser with a BBO crystal in an external enhancement cavity. Typically 30 μW of UV light is focused to a spot size of about 50 μm at the position of the ion. The linewidth of the laser is about 2 MHz.

Both UV laser beams are initially linearly polarized with the direction orthogonal to each other. They are combined on a polarization beam splitter. Afterwards the co-propagating beams are passed through a quarter-wave plate to provide circular polarized radiation with left-handed and right-handed polarizations. The cooling laser excites the $({}^1S_0, F = 9/2, m_F = 9/2) \rightarrow ({}^3P_1, F = 11/2, m_F = 11/2)$ transition of the isotope ${}^{115}In^+$ (nuclear spin I = 9/2). Circular polarization is used to prepare this two level system by means of optical pumping. The Zeeman component of the clock transition that we are driving is $({}^1S_0, F = 9/2, m_F = 9/2) \rightarrow ({}^3P_0, F = 9/2, m_F = 7/2)$. The magnetic field inside the trap is nulled with a precision of a few milligauss by three orthogonal pairs of solenoids.

The ion is held in a radiofrequency trap of the Paul-Straubel type,[58] consisting of a single ring electrode which is surrounded by additional shield and compensation electrodes for static voltage. In practice the trapping volume is defined by a 1 mm diameter cylindrical hole drilled into a strip of copper-beryllium, 2 mm wide and 0.5 mm thick.[58] We typically apply an AC voltage of 850 V amplitude at 8.7 MHz frequency and 30 V of DC voltage to the strip. For an In^+ ion this yields secular frequencies of 0.8 MHz and 1.0 MHz in the radial and axial direction respectively.

Fluorescence light from the ion is detected using a photomultiplier. With the ion driven below saturation a typical signal count rate is 700 counts per second over a background signal of 50 counts per second, resulting mainly from laser straylight.

The ${}^1S_0 \rightarrow {}^3P_0$ transition of In^+ was observed for the first time in the emission from a strongly hollow cathode discharge by Paschen and Campbell in their extensive study of the In^+ spectrum.[59] In their data two frequency values near 42276 cm^{-1} are given that differ by 0.69 cm^{-1}. Recently a much more precise determination of the transition was performed by Larkins and Hannaford[60] based on the measurement of the frequencies of three allowed transitions, connecting $5s^2\,{}^1S_0$ with $5s5p\,{}^3P_0$ via $5s5p\,{}^3P_1$ and $5s6s\,{}^3S_1$. Their result (42 275.986(7) cm^{-1}) was the starting point for our search for laser excitation of the line.

The excitation spectrum of the ${}^1S_0 \rightarrow {}^3P_0$ line was recorded with a single ion using an optical-optical double resonance method.[61, 62] The detection is based on the fact that an excitation of the metastable ${}^3P_0$ level suppresses the fluorescence of the ${}^1S_0 \rightarrow {}^3P_1$

cooling transition. To prevent light shifts and broadening of the clock transition due to the cooling laser, both lasers were alternately turned on and off for time intervals of 150 ms using a mechanical shutter. The scattered photons from the clock laser contribute to the straylight signal. To avoid this background signal, photons were only counted in the half period when the clock laser was switched off. The integration period of the counting system was set to 50 ms.

Dark periods of the $^1S_0 \rightarrow ^3P_1$ fluorescence may be caused not only by direct laser excitation of the $^3P_0$ level but also by spontaneous decay of $^3P_1$ via $^3P_0$ occurring about once every five minutes in our experiment.[54] To discriminate against the latter events the dark periods in the fluorescence were only registered when the photon count rate remained at the straylight level for at least the first 50 ms after the cooling laser was switched on. The spontaneous decay of $^3P_1$ via $^3P_0$ can be unambiguously recognized when the quantum jump occurs in the on-period of the cooling laser. Those events are consequently not registered.

The Nd:YAG laser was scanned continuously at a rate of 20-30 MHz/min. To facilitate the search for the resonance we intentionally broadened the absorption spectrum of the ion by applying an electric field of 2 V/cm along the axial direction of the Paul-Straubel trap. This field displaces the ion by a few micrometers from the center of the trap and exposes it to the radiofrequency field. The ion is then oscillating at 8.7 MHz, which leads to a frequency modulated optical spectrum, consisting of a set of equally spaced sidebands. The displacement of the ion was chosen to reach a modulation index of about 10, resulting in a total width of 200 MHz. This width was actually observed in the first experiments published in Ref. 56. Presently new measurements with an improved laser system are under way and it is now possible to resolve the sidebands directly.

As a frequency reference Doppler free spectroscopy of $^{130}Te_2$ in the blue spectral range is used. The nearest reference $^{130}Te_2$ line[63] at 21 137.8599(20)cm^{-1} is about 4 GHz away from half the frequency of the indium line. A zerodur cavity with optically contacted mirrors is used as a frequency marker for the infrared radiation. The cavity is suspended in a temperature stabilized high vacuum chamber. Its free spectral range is 559 MHz. From this value we can deduce the absolute frequency of the $^1S_0 \rightarrow ^3P_0$ resonance as 42 275.988(6) cm^{-1}, which is in excellent agreement with the value measured by Larkins and Hannaford.[60]

Finally we would like to mention another application of the described In$^+$ system. The transition $5s5p\ ^3P_1 - 5s5p\ ^3P_0$ is a magnetic dipole transition with a lifetime of 40 s at a wavelength of roughly 9 μm.[54] The long lifetime of this transition allows to perform cavity quantum electrodynamic experiments on the basis of quantum jumps. For this purpose a cavity for the $^3P_1 - ^3P_0$ transition has to be used. The modification of the lifetime by the enhanced vacuum field can then be observed in the dark periods of the fluorescent light of $^3P_1 - ^3S_0$ cooling transition. This is another interesting application of the In$^+$ ion spectroscopy. Experiments in this direction are presently carried out.

References

1. D.J. Wineland, W.M. Itano, J.C. Bergquist, J.J. Bollinger and J.D. Prestage, in *Atomic Physics 9*, eds. R.S. van Dyck, Jr., and E.N. Fortson (World Scientific Publishing, Singapore, 1984).
2. S. Chu, J.E. Bjorkholm, A. Ashkin and A. Cable, in *Atomic Physics* **10**, eds. H. Narumi and S. Shimamura (North-Holland, Amsterdam, 1986) and D.E. Pritchard, K. Helmerson, A.G. Martin, in *Atomic Physics* **11**, eds. S. Haroche, J.C. Gay and G. Grynberg (World Scientific Publishing, Singapore, 1988).
3. D. Meschede, H. Walther and G. Müller, *Phys. Rev. Lett.* **54**, 551 (1985).
4. E.T. Jaynes and F.W. Cummings, *Proc. IEEE* **51**, 89 (1963).
5. For a review, see the articles by S. Haroche and J.M. Raimond, in *Advances in Atomic and Molecular Physics* **20**, eds. D. Bates and B. Bederson, p. 350 (Academic Press, New York, 1985); J.A. Gallas, G. Leuchs, H. Walther and H. Figger, ibid, p. 413.
6. See, for example: J.H. Eberly, N.B. Narozhny and J.J. Sanchez-Mondragon, *Phys. Rev. Lett.* **44**, 1323 (1980), and references therein.
7. G. Rempe, H. Walther and N. Klein, *Phys. Rev. Lett.* **58**, 353 (1987).
8. P. Meystre, in *Progress in Optics* **30**, ed. E. Wolf (North Holland, Amsterdam, 1992)
9. P. Filipowicz, J. Javanainen and P. Meystre, *Opt. Comm.* **58**, 327 (1986); *Phys. Rev. A* **34**, 3077 (1986); *J. Opt. Soc. Am. B* **3**, 906 (1986).
10. L. Lugiato, M.O. Scully and H. Walther, *Phys. Rev. A* **36**, 740 (1987).
11. P. Meystre, *Opt. Lett.* **12**, 669 (1987); P. Meystre, in *Squeezed and Nonclassical Light*, eds. P. Tombesi and E.R. Pike (Plenum, New York and London, 1988), p. 115.
12. P. Meystre, E.M. Wright, *Phys. Rev. A* **37**, 2524 (1988).
13. G. Rempe and H. Walther, *Phys. Rev. A* **42**, 1650 (1990).
14. Paul, T. Richter, *Opt. Comm.* **85**, 508 (1991).
15. G. Rempe, F. Schmidt-Kaler and H. Walther, *Phys. Rev. Lett.* **64**, 2783 (1990).
16. H.J. Briegel, B.G. Englert, N. Sterpi, H. Walther, *Phys. Rev. A* **49**, 2962 (1994).
17. C. Wagner, A. Schenzle, H. Walther, *Opt. Comm.* **107**, 318 (1994).
18. O. Benson, G. Raithel, H. Walther, *Phys. Rev. Lett.* **72**, 3506 (1994).
19. L. Davidovich, J.M. Raimond, M. Brune, S. Haroche, *Phys. Rev. A* **36**, 3771 (1987), M. Brune, J.M. Raimond, P. Goy, L. Davidovich, S. Haroche, *Phys. Rev. Lett.* **59**, 1899 (1987).
20. P. Meystre, G. Rempe and H. Walther, *Opt. Lett.* **13**, 1078 (1988).
21. G. Raithel, Ch. Wagner, H. Walther, L.M. Narducci, M.O. Scully, in *Advances in Atomic, Molecular, and Optical Physics*, Supplement 2, ed. P. Berman (Academic Press, New York, 1994).
22. G. Raithel, O. Benson, H. Walther, Atomic Interferometry with the Micromaser, to be published.
23. N. F. Ramsey, Molecular Beams (Clarendon Press, Oxford, 1956), p. 124.
24. See, for example, C. Cohen-Tannoudji, J. Dupont-Roc, G. Grynberg, Atom-Photon Interactions (John Wiley and Sons, Inc., N.Y., 1992).
25. M. Brune, S. Haroche, V. Lefevre, J. M. Raimond, N. Zagury, *Phys. Rev. Lett.* **65** , 976 (1990).
26. M.O. Scully, H. Walther, G.S. Agarwal, T. Quang, W. Schleich, *Phys. Rev. A* **44**, 5992 (1991).
27. T. Quang, G.S. Agarwal, J. Bergou, M.O. Scully, H. Walther, K. Vogel, W.P. Schleich, *Phys. Rev. A* **48**, 803 (1993); K. Vogel, W.P. Schleich, M.O. Scully, H. Walther, *Phys. Rev. A* **48**, 813 (1993).
28. J. Krause, M.O. Scully and H. Walther, *Phys. Rev. A* **34**, 2032 (1986).
29. R.J. Brecha, A. Peters, C. Wagner, H. Walther, *Phys. Rev. A* **46**, 567 (1992).
30. C. Wagner, R.J. Brecha, A. Schenzle, H. Walther, *Phys. Rev. A* **46**, R5350 (1992).
31. C. Wagner, R.J. Brecha, A. Schenzle, H. Walther, Phys. Rev. A. **47**, 5068 (1993).
32. B.-G. Englert, J. Schwinger and M.O. Scully, *Found. Phys.* **18**, 1045 (1988); J. Schwinger, M.O. Scully and B.-G. Englert, *Z. Phys. D* **10**, 135 (1988); M.O. Scully, B.-G. Englert and J. Schwinger, *Phys. Rev. A* **40**, 1775 (1989).
33. M.O. Scully and H. Walther, *Phys. Rev. A* **39**, 5229 (1989).
34. M.O. Scully, B.-G. Englert and H. Walther *Nature* **351**, 111 (1991).
35. B.-G. Englert, H. Walther and M.O. Scully, *Appl. Phys. B* **54**, 366 (1992).
36. W. Paul, O. Osberghaus and E. Fischer, *Forschungsberichte des Wirtschafts- und Verkehrsministeriums Nordrhein-Westfalen* **415** (1958); E. Fischer, *Z. Phys.* **156**, 1 (1959).
37. H.G. Dehmelt, in *Adv. Atom. Molec. Phys.* **3**, eds. D.R. Bates and I. Estermann (Academic Press, New York, 1967).
38. J. Dalibard, C. Salomon, A. Aspect, E. Arimondo, R. Kaiser, N. Vansteenkiste and C. Cohen-Tannoudji, in *Atomic Physics* **11**, eds. S. Haroche, J.C. Gay and G. Grynberg (World Scientific Publishing, Singapore, 1988).
39. F. Diedrich, E. Peik, J.M. Chen, W. Quint and H. Walther, *Phys. Rev. Lett.* **59**, 2931 (1987).
40. W. Neuhauser, M. Hohenstatt, P. Toschek and H. Dehmelt, *Phys. Rev. Lett.* **41**, 233 (1978); W. Neuhauser, M. Hohenstatt, P. Toschek and H. Dehmelt, *Phys. Rev. A* **22**, 1137 (1980).

41. R. Blümel, J.M. Chen, W. Quint, W. Schleich, Y.R. Shen and H. Walther, *Nature* **334**, 309 (1988).
42. R. Blümel, J.M. Chen, F. Diedrich, E. Peik, W. Quint, W. Schleich, Y.R. Shen and H. Walther, in *Atomic Physics* **11**, eds. by S. Haroche, J.C. Gay and G. Grynberg (World Scientific Publishing, Singapore, 1988).
43. R. Blümel, C. Kappler, W. Quint and H. Walther, *Phys. Rev. A* **40**, 808 (1989).
44. J. Drees and W. Paul, *Z. Phys.* **180**, 340 (1964).
45. D.A. Church, *Journal of Appl. Phys.* **40**, 3127 (1969).
46. H. Walther, in *Proc. of the Workshop on Light Induced Kinetic Effects on Atoms, Ions and Molecules*, eds. L. Moi, S. Gozzini, C. Gabbanini, E. Arimondo and F. Strumia (ETS Editrice, Pisa, 1991).
47. I. Waki, S. Kassner, G. Birkl and H. Walther, *Phys. Rev. Lett.* **68**, 2007 (1992).
48. G. Birkl, S. Kassner and H. Walther, *Nature* **357**, 310 (1992).
49. R.W. Hasse and J.P. Schiffer, *Ann. Physics* **203**, 419 (1990); A. Rahman and J.P. Schiffer, *Phys. Rev. Lett.* **57**, 1133 (1986).
50. M.G. Raizen et al, *J. Mod. Opt.* **39**, 233 (1992).
51. S.L. Gilbert, J.J. Bollinger and D.J. Wineland, *Phys. Rev. Lett.* **60**, 2022 (1988).
52. G. Birkl, J.A. Yeazell, R. Rückerl, *Europhys. Lett.* **27**, 197 (1994).
53. R.C. Thompson, *Advances in Atomic, Molecular and Optical Physics* **31**, 63 (1993).
54. E. Peik, G. Hollemann and H.Walther, *Phys. Rev. A* **49**, 402 (1994).
55. G. Hollemann, E. Peik and H. Walther, *Optics Letters* **19**, 192 (1994).
56. E. Peik, G. Hollemann and H. Walther, Double resonance spectroscopy of a single trapped indium ion, *Physica Scripta*, in print.
57. G. Hollemann, E. Peik, A. Rusch and H. Walther, *Opt. Lett.* **20**, 1 (1995).
58. C.A. Schrama, E. Peik, W.W. Smith and H. Walther, *Opt. Comm.* **101**, 32 (1993).
59. F. Paschen and J.S. Campbell, *Ann. Phys.* **31**, 29 (1938).
60. P.L. Larkins and P. Hannaford, *Z. Phys. D* **27**, 313 (1993).
61. J.C. Bergquist, W.M. Itano and D.J. Wineland, *Phys. Rev. A* **36**, 428 (1987).
62. W. Nagourney, N. Yu and H. Dehmelt, *Opt. Commun.* **79**, 176 (1990).
63. J. Cariou and P. Luc, Atlas du spectre d'absorption de la molécule de tellure, (Laboratoire Aimé-Cotton CNRS II, Orsay, France, 1980).

DRESSED STATES IN ATOMS AND IN EXCITONS

G. COMPAGNO
INFM and Istituto di Fisica dell' Universita, Palermo, Italy

R. PASSANTE
INFM and Istituto per le Applicazioni Interdisciplinari della Fisica, (C.N.R.), Palermo, Italy

AND

F. PERSICO
INFM and Istituto di Fisica dell' Universita, Palermo, Italy
also
INFM and Istituto per le Applicazioni Interdisciplinari della Fisica, (C.N.R.), Palermo, Italy

Abstract.

Due to the coupling with the vacuum fluctuations of the electromagnetic field, a neutral ground-state atom is surrounded by a cloud of virtual photons, leading to the concept of dressed atom. A similar concept has been introduced in solid state physics in connection with the exciton-phonon coupling; this leads to the definition of the polaronic exciton, which can be pictorially represented as an exciton surrounded by a cloud of virtual phonons.

We briefly describe the structure of our model consisting of a two-level dressed atom and the physical nature of its virtual photon cloud. In the second phase of our work, we discuss the similarity of this model with a simplified model of the polaronic exciton in a semiconductor.

We review the quantum theory of measurement in connection with the population of dressed states. We discuss the role of the duration of the measurement with reference to the time scale provided by the vacuum fluctuations which contribute to the virtual cloud.

Finally we introduce the concept of partially dressed states as states whose virtual cloud is initially out of equilibrium and undergoes a regeneration process. We show that appropriate measurements on these states can be performed, in order to detect the time dependence of their population during the reconstruction of the virtual cloud. We discuss the formation of the virtual cloud of a newly created polaritonic exciton and we argue that this process could be experimentally detected. We stress the potentiality of experiments on the partially dressed polaronic exciton, which could provide models for fundamental. and as yet only partially explored, QED process such as positronium creation and annihilation.

T. Hakioğlu and A. S. Shumovsky (eds.), Quantum Optics and the Spectroscopy of Solids, 99–110.

1. Introduction

The aim of this paper is to highlight the importance of cross-fertilization between two of the main traditional branches of physics: Solid State Physics and Quantum Field Theory. This cross-fertilization dates back to the end of the 50's, when a lively exchange of ideas took place between high-energy and condensed matter physics. Some of these ideas were related to symmetry breaking, the renormalization group and elementary excitations. The main interest of this paper is centered on the similarity of the dynamics of some of the elementary excitations in semiconductors and of atoms in free space. The elementary excitation in question is the exciton, and the dynamical process selected for our purposes is caused by the interaction of the exciton with the phonons of the lattice. The analogous process in QED is caused by the interaction of an atom with the photons. As an exciton moves through the crystal, it carries a lattice distortion which changes its physical properties. This lattice distortion can be described as a cloud of virtual phonons surrounding the exciton [1]. These virtual phonons are continuously absorbed and re-emitted by the exciton much in the same way as in the more famous case of polaron [2]. One can then speak of a *dressed exciton* in analogy with the *dressed atom*, which is an atom endowed with a cloud of virtual photons [3]. We will argue that the similarity of the two dressing processes may lead to a deeper understanding of the phenomenon and possibly to devise new experiments in solid state physics, capable of shedding new light on some fundamental problems of QED.

2. Elementary Excitations

As far as we know, the first clear definition of elementary excitation is in terms of the poles of the single-particle propagator and it is due to the Galitskii and Migdal [4] in the context of the problem of electrons in metals, although also other physicists fostered similar ideas [5] and although the idea had been in the air, albeit in qualitative terms, for quite a long time [6]. The point we wish to stress, however, is that the picture of condensed matter which emerged at the end of the 50's was mostly in terms of collective modes and elementary excitations. The line of thought leading to this picture can be simplified as follows. The Hamiltonian of a many-body system is of the form

$$H = H_0 + H_1 \tag{1}$$

where H_0 represents the free particles and H_1 accounts for inter-particle forces. The ground state of H is $|\varphi_0\rangle$ such that

$$H|\varphi_0\rangle = E_0|\varphi_0\rangle. \tag{2}$$

The collective character of the excited states is ensured by the nature of H_1, which distributes the energy among the particles giving rise to the collective modes such as the vibrational modes of a solid [7], or of an electron gas in a metal [8]. From this point of view $|\varphi_0\rangle$ can be considered as the vacuum of collective excitations, whereas the first few excited states can be characterized by the number n_k of quanta populating the collective excitations of type k. Much of the physics of interest in the condensed state is determined by the distribution $\{n_k\}$ of quanta rather than by the detailed nature of the material ground state. This leads to a remarkable simplification of the dynamical problem. For example, the forces binding the atom in a solid can be quite strong, but the interaction between different collective excitations can be weak enough to allow for a perturbative approach [9]. If this is the case, the quanta of the collective excitations are long-lived enough to have a well defined energy, and are called the elementary excitations. This is the contact with quantum field theory, since from this point of view mesons can be considered as elementary excitations of a scalar field, photons as elementary excitations of a two-component vector field and electrons and positrons as the elementary excitations of a two-component spinor field.

From a more quantitative point of view, one can describe the dynamics of an elementary excitations of type k by a complex quantity

$$\overline{E}_k = E_k + i\Gamma_k, \qquad \Gamma_k \ll E_k \tag{3}$$

where E_k is the energy of the elementary excitations and $\Gamma_k = \hbar\tau_k^{-1}$, τ_k is its lifetime which accounts for all the interactions leading to the decay of the elementary excitations. In a second-quantization framework, the approximate eigenstates of H possessing n_k elementary excitations can be obtained by the repeated application of an appropriate creation operator a_k^+ to the vacuum $|\varphi_0\rangle$. Thus, the Hamiltonian of the system can be approximately written as

$$H = \sum_k E_k a_k^+ a_k. \tag{4}$$

The algebra of the a-operators is related to the Bose or Fermi nature of the elementary excitations.

This is the physical basis for a quantum field description of solids. In this way one can describe the whole zoo of phonons, plasmons, excitons, polarons, polaritons, magnons, rotons, ... that populate the nightmares of the overexcited theoretical solid state physicist. A Hamiltonian of the same form is normally used to describe free fields in quantum field theory (photons, electrons, positrons, ...).

3. Exciton versus Atoms

As an example of elementary excitations we will qualitatively describe an exciton. In a pure ideal semiconductor electrons are distributed in bands of different energy. At $T = 0K$ each band is completely full of electrons or empty. The highest full band is called the valence band and the first empty band is the conduction band. These two bands are separated by an energy gap $E_g = E_c - E_v$. Thus, in order to excite an electron from the valence to the conduction band and energy E_g is required. This can be provided e.g. by a photon of appropriate frequency ($\geq E_g/\hbar$) which can be absorbed by the semiconductor. The electron will leave a positively charged hole in the valence band. Neglecting Coulomb attraction, electron and hole can be considered as free particles. The similarity with the QED creation of an electron-positron pair is evident. Taking the Coulomb interaction into account leads to the formation of bound electron-hole states of energy intermediate between E_v and E_c as shown in Fig. 1. These states correspond to elementary excitations which propagate freely throughout the crystal. Their internal structure, when appropriate account is taken of the periodic lattice background and of its dielectric properties, it reminiscent of that of a positronium atom in QED. Indeed, on top of the hydrogen-like discrete set of bound energy levels within the energy gap, they possess a continuum starting at E_c which can be regarded as a kind of ionization limit. It is thus clear that excitons can be regarded as the solid state counterpart of positronium atom.

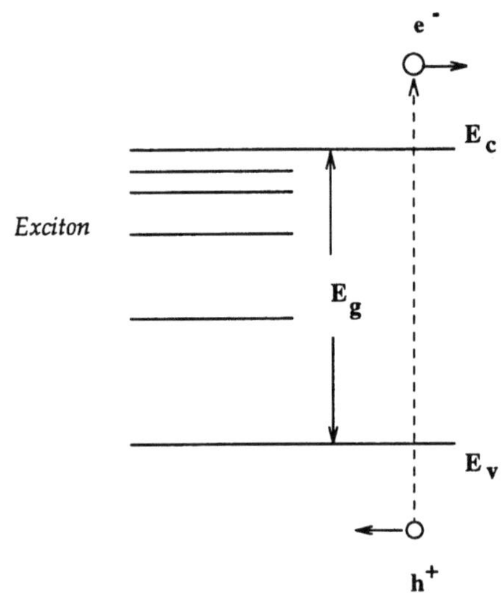

Figure 1.

4. Dressed Atoms and Dressed Excitons

The sources of any field, even in the ground state of the total source-field system, are usually surrounded by a cloud of virtual quanta of the field. Consider for example a two-level atom fixed at the origin of the frame of reference and interacting with the quantized electromagnetic field in the Coulomb gauge and in the electric dipole approximation. This system is described by a Hamiltonian of the form [3]

$$H = H_0 + V_1 + \lambda V_2,$$
$$H_0 = \hbar\omega_0 S_z + \sum_{kj} \hbar\omega_k (a_{kj}^+ a_{kj} + 1/2),$$
$$V_1 = \sum_{kj} (\epsilon_{kj} a_{kj} S_+ + \epsilon_{kj}^* a_{kj}^+ S_-) , \qquad V_2 = \sum_{kj} (\epsilon_{kj} a_{kj}^+ S_+ + \epsilon_{kj}^* a_{kj} S_-) \tag{5}$$

where S_+, S_- and S_z are the usual raising, lowering and population operators for a two-level system of natural frequency ω_0 (see Fig. 2), a_{kj}^+, a_{kj} create and annihilate photons in the normal mode of the field of wavevector k, polarization $j = 1, 2$ and frequency $\omega_k = ck$. For simplicity of notations we omit, here and in the rest of paper, to highlight the vectorial nature of k. ϵ_{kj} is the coupling coefficient between the field mode and the atom, whereas λ is simply a parameter which is kept only for diagnostic purposes throughout the calculations. In a perturbation spirit one obtains the ground state of the Hamiltonian H up to terms

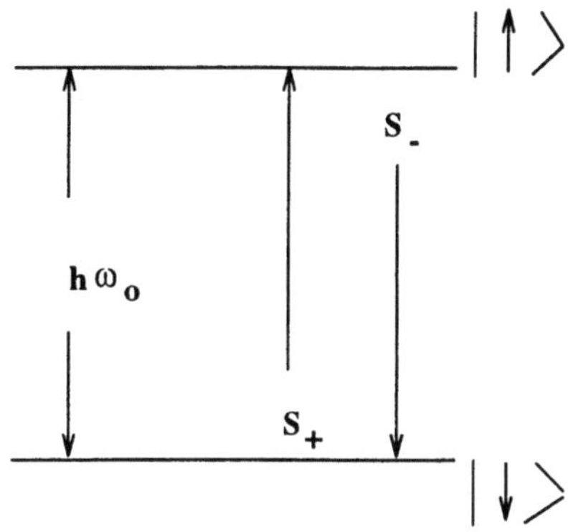

Figure 2.

of order ϵ_{kj}^2 in the form

$$| \{0_{kj}\}, \downarrow\rangle' = [1 + \lambda \frac{1}{E_0 - H_0}(1 - P_0)V_2 + \lambda \frac{1}{E_0 - H_0}(1 - P_0)V_1 \frac{1}{E_0 - H_0}(1 - P_0)V_2 \\ - \frac{\lambda^2}{2}\langle\{0_{kj}\}, \downarrow |V_2 \frac{1}{(E_0 - H_0)^2}(1 - P_0)V_2| \{0_{kj}\}, \downarrow\rangle]| \{0_{kj}\}, \downarrow\rangle \quad (6)$$

and the quantum average number of photons in this ground state as

$$\langle n\rangle = \frac{\lambda^2}{\hbar^2}\sum_{kj}\frac{|\epsilon_{kj}|^2}{(\omega_0 + \omega_k)^2}. \quad (7)$$

Thus, there are photons in the ground state of the system. The presence of λ^2 in (7) shows that these photons originate from the V_2 term in H. This term, acting on the ground state $| \{0_{kj}\}, \downarrow\rangle$ of the atom-field system, creates one photon and excites the atom.Energy is not conserved in this process and for this reason the photons created are called virtual. In a time-dependent picture these photons are reabsorbed after a time τ, and this accounts for the square of λ appearing in (7). The energy unbalance δE is related to τ by the Heisenberg relation

$$\delta E \approx \hbar/\tau. \quad (8)$$

Since for one-photon virtual process $\delta E = \hbar(\omega_0 + \omega_k)$, and in view of (8), the lifetime of the virtual photons

$$\tau \approx (\omega_0 + \omega_k)^{-1} \quad (9)$$

is obtained, as well as the fact that these photons can travel a distance $c\tau \leq c/\omega_0$ from the source before being reabsorbed. Since virtual processes take place continuously, one expects that one-photon fluctuations give rise around the atom to a cloud of virtual photons whose linear dimension are $\propto c/\omega_0$, as shown in Fig. 3. This qualitative picture is confirmed by

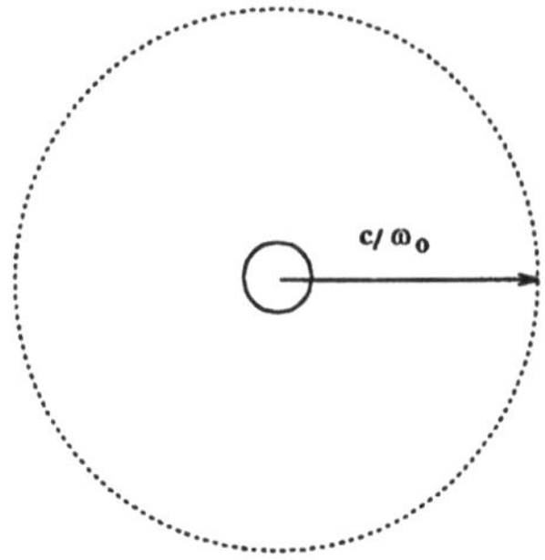

Figure 3.

quantitative evaluation of the energy density around the atom, which leads to a precise definition of what is called the virtual cloud [10]. As mentioned in the Introduction, virtual clouds are of common occurrence in the ground state of most interacting quantum systems. Virtual clouds of photons surround free electrons or holes in semiconductors, yielding a composite elementary excitations called polaron [2], as well as excitons, yielding the excitonic polaron [1]. One is led to suggest that an appropriately modified form of (5), capable of accounting for the translational degrees of freedom, might provide a rough model for a polaronic exciton, at least when the Coulomb electron-hole interaction is stronger than the electron-phonon coupling. In this case the field pertaining to the a and a^+ operators would naturally be the phonon field and the virtual cloud would be a phonon cloud rather than the photon cloud of the atom-photon interaction.

Given that this suggestion is sound, it may be an useful exercise to imagine some kind of measurement to investigate and compare the physical nature of the virtual cloud of photons around an atom and of phonons around an exciton. The population distribution among the energy levels of either object turns out to be sensitive to the presence of virtual cloud, as we will show in the rest of this paper. For the sake of conciseness we will often refer to the atom only, although we assume that our argumentation applies to the exciton as well.

5. Measurement of Bare Population

We adopt a generalization of the textbook quantum theory of measurement in which the finite time τ_m is taken into account. During this time, quantum correlations are established between the macroscopic measuring apparatus (or "pointer" for short) and the object to be measured [11]. If the quantity to be measured is represented by a Hermitian operator A pertaining to an object whose Hamiltonian is H_A, the total Hamiltonian for the measurement is

$$H = H_A + p^2/2M + g(t)Ap \tag{10}$$

where M is the (large) mass of the pointer of momentum p and $g(t)$ is represented in Fig. 4. The pointer is coupled with the object at $t = 0$. The measurement lasts for a time τ_m and is concluded by an observation at $t = \tau_m$ which yields the usual collapse of the

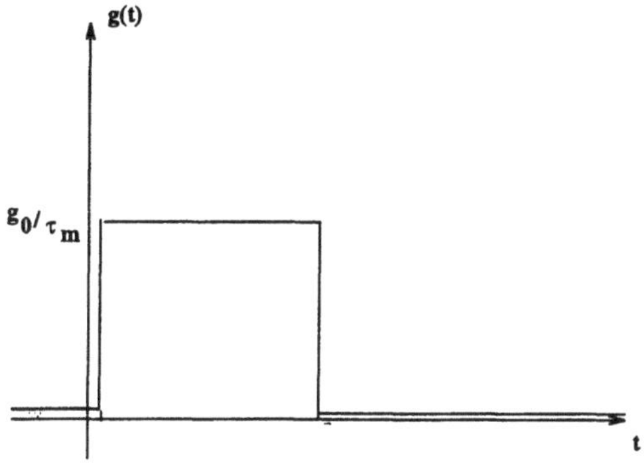

Figure 4.

state of the system. In particular, if we wish to measure the population distribution S_z of a two-level atom (or exciton), we take $A = S_z$ and $H_A = \hbar\omega_0 S_z$. Then the Heisenberg equations of motion yield

$$p(t) = p(0), \quad S_z(t) = S_z(0), \quad q(\tau_m) = g_0 S_z(0) + p(0)\tau_m/M + q(0). \tag{11}$$

Taking quantum averages on a state of the system in which

$$\langle q(0)\rangle = \langle p(0)\rangle = 0$$

and in which the atom is in a superposition $a|\uparrow\rangle + b|\downarrow\rangle$, one has [12]

$$\langle q(\tau_m)\rangle = g_0\langle S_z(0)\rangle, \quad \langle q^2(\tau_m)\rangle - \langle q(\tau_m)\rangle^2 = g_0^2[1/4 - \langle S_z(0)\rangle^2]. \tag{12}$$

This result can be interpreted according to Fig. 5, where $P(q)$ is the probability of finding the pointer at position q. Any observation of q at $t = \tau_m$ yields either of the two possible outcomes $\pm g_0/2$. Moreover $q = \pm g_0/2$ corresponds to $S_z(0) = \pm 1/2$. Thus, the experimental apparatus described by (10) is capable of measuring bare atomic populations, and the result of the measurement is, not unexpectedly, independent of τ_m.

6. Measurement on Dressed Atoms or Dressed Excitons

Suppose now that the atom (or the exciton) is dressed by a single-mode field. Then the measurement Hamiltonian is

$$H = \hbar\omega_0 S_z + \hbar\omega a^+ a + (\epsilon a S_+ + \epsilon^* a^+ S_-) - (\epsilon a^+ S_+ + \epsilon^* a S_-) + \frac{p^2}{2M} + g(t) S_z p. \tag{13}$$

A procedure similar to that used for the bare atom yields

$$\langle q(\tau_m)\rangle = g_0\langle S_z(0)\rangle,$$

$$\langle q^2(\tau_m)\rangle - \langle q(\tau_m)\rangle^2 = \int_0^{\tau_m}\int_0^{\tau_m} [\langle S_z(t')S_z(t'')\rangle - \langle S_z(t')\rangle\langle S_z(t'')\rangle] g(t')g(t'')dt'dt'' \tag{14}$$

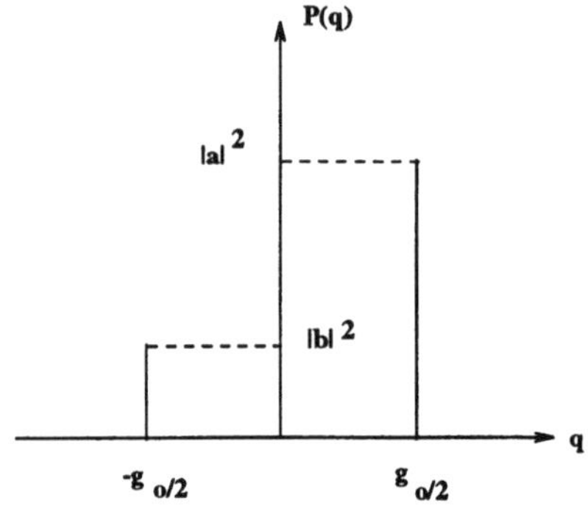

Figure 5.

where the quantum averages are taken on a state where the pointer is at rest at the origin and the atom-field system is in the dressed ground-state $|0,\downarrow\rangle'$ as given by the single-mode version of (6). Evaluating these quantum averages leads to

$$\langle q(\tau_m)\rangle = g_0\langle S_z(0)\rangle,$$
$$\langle q^2(\tau_m)\rangle - \langle q(\tau_m)\rangle^2 = g_0^2\left(\frac{1}{4} - \langle S_z(0)\rangle^2\right)\frac{2[1-\cos(\omega_0+\omega)\tau_m]}{(\omega_0+\omega)^2\tau_m^2}. \quad (15)$$

These expressions reduce to the bare-atom ones for $(\omega_0+\omega)\tau_m \ll 1$. Thus, for s measurement shorter than the duration τ_m of the one-photon quantum fluctuation, the result of the measurement is the same as for a bare atom. For a long measurement, however, $\langle q^2(\tau_m)\rangle - \langle q(\tau_m)\rangle^2$ vanishes so that the probability of finding the pointer at point q is given by Fig. 6. Thus $P(q)$ is single-peaked, although at value of q different from $\pm g_0/2$. This shows that in a long measurement the pointer perceives a new object, which will call the dressed atom. It can also be said that a measurement lasting for a time τ_m can only detect the effects of vacuum fluctuations which last for a time $\tau \ll \tau_m$, whereas slow fluctuations lasting for a time $\tau \gg \tau_m$ cannot be detected.

The single-mode Hamiltonian (13) can be generalized to the multimode case as

$$H = \hbar\omega_0 S_z + \sum_{kj}\hbar\omega_k(a_{kj}^{+}a_{kj}+1/2) + \sum_{kj}[(\epsilon_{kj}a_{kj}S_{+} + \epsilon_{kj}^{*}a_{kj}^{+}S_{-})$$
$$-(\epsilon_{kj}a_{kj}^{+}S_{+} + \epsilon_{kj}^{*}a_{kj}S_{-})] + \frac{1}{2m}p^2 + g(t)S_z p. \quad (16)$$

The relevant quantum average in the dressed ground state of the atom-field system are

$$\langle q(\tau_m)\rangle = g_0\langle S_z(0)\rangle = g_0\left(-\frac{1}{2} + \sum_{kj}\frac{|\epsilon_{kj}|^2}{\hbar^2(\omega_0+\omega_k)^2}\right),$$
$$\langle q^2(\tau_m)\rangle - \langle q(\tau_m)\rangle^2 = g_0^2\sum_{kj}\frac{|\epsilon_{kj}|^2}{\hbar^2(\omega_0+\omega_k)^2}\,\frac{2[1-\cos(\omega_0+\omega_k)\tau_m]}{(\omega_0+\omega_k)^2\tau_m^2}. \quad (17)$$

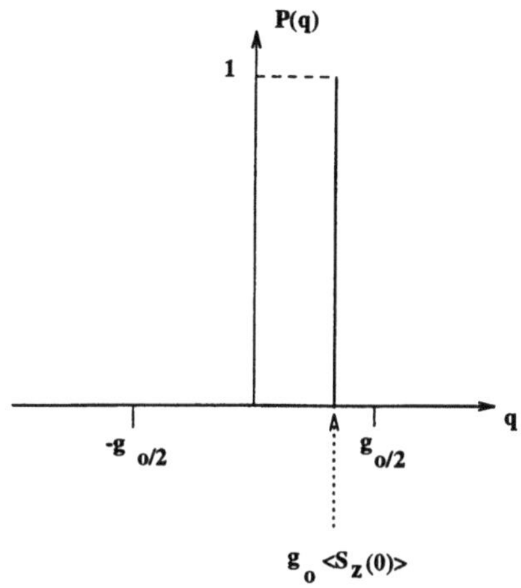

Figure 6.

In the limit of a long measurement the variance in (17) vanishes and the pointer perceives the atom as dressed by all vacuum modes. In the opposite limit of a very short measurement, such that $\tau \ll (\omega_0 + \omega_k)^{-1}$, one has

$$\langle q^2(\tau_m)\rangle - \langle q(\tau_m)\rangle^2 = g_0^2 \sum_{kj} \frac{|\epsilon_{kj}|^2}{\hbar^2(\omega_o + \omega_k)^2} = g_0^2 \left(\frac{1}{4} - \langle S_z(0)\rangle^2\right) \tag{18}$$

which is the result of a bare-atom measurement. In the intermediate case only modes of frequency such that $(\omega_0 + \omega_k) < \tau_m^{-1}$ will contribute to the variance $\langle q^2(\tau_m)\rangle - \langle q(\tau_m)\rangle^2$, whereas modes of higher frequency, such that $(\omega_0 + \omega_k) > \tau_m^{-1}$ will not. Thus the atom will be perceived as dressed only by the high-frequency modes, or as partially dressed.

7. Partially Dressed Atoms and Excitons

The concept of partially dressed atom was proposed by E.L. Feinberg [13], although the fundamental idea is contained in earlier papers by V.L. Ginzburg [14].

Basically a partially dressed atom is a dressed atom which has been deprived of a portion of its equilibrium virtual cloud. Consider for example an atom inside a microscopic cavity as in Fig. 7. The field outside the cavity, which we take here as ideal. vanishes and consequently the virtual cloud cannot extend further out than the cavity radius. If at $t = 0$ the atom is suddenly taken out of the cavity, its virtual cloud will be regenerated gradually and in a causal way, in the sense that reconstruction of its virtual cloud takes place within an outwardly expanding sphere of radius ct [15]

Thus the virtual cloud around a partially dressed atom is out of equilibrium and time-dependent since it tends to assume the equilibrium configuration.

An extreme case of partially dressed atom is for a cavity radius as small as a few atomic radii. Such an extreme case, when the atom is deprived of most of its virtual cloud, will be taken to be a fair realization of a bare atomic state. Strictly speaking, a perfectly bare

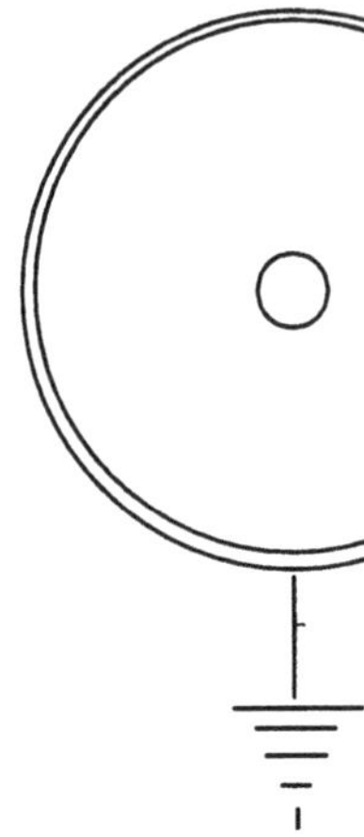

Figure 7.

atom should not possess any virtual cloud at all at $t = 0$, but in relativistic QED this configuration cannot be realized in principle since the complete suppression of the atomic electromagnetic field would also take the atom apart. Thus we will assume and hope that an atom suddenly taken out of a microscopic cavity can approximately represent a bare atom.

A situation, however, where such a stretch of imagination is not necessary is provided by solid state physics. Consider in fact an exciton created by a laser inside the semiconductor gap at $T = 0K$. Since the frequency of the light is in the optical range, i.e $\sim 10^{15}Hz$, the characteristic time for creating the exciton can be taken of the order of $10^{-15}s$. Due to the exciton-phonon interaction, the lattice will tend to readjust around the exciton. Such a readjustment cannot take place in a time shorter than the inverse of the highest frequency of the lattice normal modes, which is the Debye frequency ω_D usually of the order of $\sim 10^{13}Hz$. Thus, although the final result of the readjustment is a polaronic exciton, this final state is attained in a time which is typically about two orders of magnitude larger than the time necessary to create the bare exciton. It is certainly legitimate to regard the exciton as bare at the beginning of this interval, and as getting partially dressed by a virtual phonon cloud in the course of time [16].

Here we will consider a particular aspect of the theory of dressed atoms, namely the time-dependence of the atomic population of a two-level atom described by the Hamiltonian (6). Assume first $g(t) = 0$ and that at $t = 0$ the system is in the bare ground state $|\{0_{kj}\},\downarrow\rangle$. The quantum average of $S_z(t)$ on this state turns out to be, accurate up to terms of order ϵ^2,

$$\langle S_z(t)\rangle = \frac{2}{\hbar^2}\sum_{kj}\frac{|\epsilon_{kj}|^2}{(\omega_0+\omega_k)^2}[1-\cos(\omega_0+\omega_k)t] - \frac{1}{2}. \tag{19}$$

For small t this gives $-1/2$ as it should be. For $\omega_0 t \gg 1$ and using the minimal coupling scheme. expression (19) can be shown to yield

$$\langle S_z(t)\rangle \approx -\frac{1}{2} + \frac{\gamma}{\pi\omega_0}\left[\ln\frac{\omega_M}{\omega_0} - 1 + \frac{\cos\omega_0 t}{(\omega_0 t)^2}\right], \qquad \gamma = \frac{2\pi}{\hbar^2}\sum_{kj}|\epsilon_{kj}|^2\delta(\omega_k - \omega_0) \tag{20}$$

where γ is the spontaneous relaxation rate and ω_M is a cutoff frequency of the order of c/a_0 (a_0 is here the atomic Bohr radius) which is necessary to introduce in view of the electric dipole approximation. In the excitonic case a natural cutoff is provided by the Debye frequency of the crystal. Thus for large t the deviation from the bare-atom value $-1/2$ is proportional to the Lamb shift, apart from oscillations of decreasing amplitude.

Partially dressed states have never been observed directly in QED, where they are very difficult to prepare and detect. Therefore, in view of their conceptual importance, the theoretical calculation and experimental detection of the dressing of an exciton in a semiconductor would be of fundamental nature. Thus the next question we wish to address is: can the time-dependent population of a half-dressed state to be detected in principle?

8. Measurements on Partially Dressed States

The Hamiltonian is the same as (16). We take $g(t)$ to be of the two-parameter form represented in Fig. 8. Thus the measurement starts at $t = t_1$ and ends at $t = t_1 + \tau_m$, at

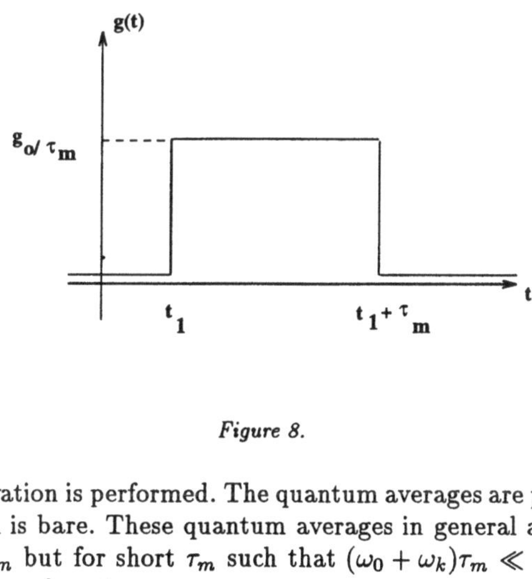

Figure 8.

which time the observation is performed. The quantum averages are performed on an initial state where the atom is bare. These quantum averages in general are rather complicated functions of t_1 and τ_m but for short τ_m such that $(\omega_0 + \omega_k)\tau_m \ll 1$ for any k and up to terms of order ϵ^2 they reduce to

$$\begin{aligned}\langle q(t_1+\tau_m)\rangle &= g_0\langle S_z(t_1)\rangle,\\ \langle q^2(t_1+\tau_m)\rangle - \langle q(t_1+\tau_m)\rangle^2 &= g_0^2\left(\frac{1}{4} - \langle S_z(t_1)\rangle^2\right)\end{aligned} \tag{21}$$

where $\langle S_z(t_1)\rangle$ is given by (19) or (20). The first of these equations shows that for short measurements the pointer is capable of following the time dependence of the atomic population during the dressing process. The second of (21) expresses the fact that in the limit of very short τ_m the variance of the distribution $P(q)$ is the same as for a bare atom possessing a population distribution among bare atomic levels compatible with $\langle S_z(t_1)\rangle$. Thus under appropriate circumstances it is possible to investigate experimentally important aspects of the dressing process by performing local measurements on the atom.

Naturally the question arises whether the same conclusions apply to a half-dressed exciton, at least in the case of small exciton-phonon coupling. Naturally the detailed form of coupling is different from that of the atom-photon case. Nevertheless, the theoretical investigation available on the subject of half-dressed excitons [16] reach conclusions on the excitonic dressing process which are remarkably similar, at least qualitatively, to those of the atomic dressing which we have investigated [17].

9. Conclusions

We have reviewed the applications of the quantum theory of measurement to dressed and partially dressed two-level atoms. In particular, we have concluded that it is possible in principle to follow the time-dependent changes of the atomic population during the process of regeneration of the virtual photon cloud, starting from an effectively bare state. We have argued that these time-dependent changes contain information on the phenomenon of self-dressing in QED. We have suggested that analogous changes should take place in the population of an exciton coupled to the phonons in a semiconductor at low temperature, where these changes might be easier to detect experimentally in view of the different time scales and of the different nature of the forces acting on the exciton. We have pointed out the interest of gaining information on the fundamental problem of self-dressing.

Acknowledgements

The authors acknowledge partial financial support by Comitato Regionale per le Ricerche Nucleari e di Stuttura della Materia and by Ministero della Publica Istituzione e della Ricerca Scientifica e Tecnologica.

References

1. F. Bassani and G. Iadonisi, in *Quantum Field Theory,* ed. F. Mancini (North-Holland, Amsterdam, 1986).
2. See e.g. C.G. Kuper, G.D. Whitfield eds., *Polarons and Excitons* (Oliver and Boyd, 1963).
3. G. Compagno. R. Passante, and F. Persico, *Atom-Photon Interaction and Dressed Atoms* (Cambridge University Press, Cambridge, 1995).
4. V.M. Galitskii and A.B. Migdal, Sov. Phys. JETP **24**, 96 (1958).
5. See e.g. N.M. Hugenholtz and L. Van Hove, Physica **24**, 363 (1958).
6. See e.g. L. Landau and E. Lifshitz, *Statistical Physics* (Pergamon Press, Oxford, 1959).
7. L. Van Hove, MIT Technical Rep. **11**, (Cambridge USA, 1958).
8. D. Bohm and D. Pines, Phys. Rev. **92**, 609, 626 (1953).
9. See e.g. P.W. Anderson, *Concepts in Solids* (Benjamin, New York, 1963).
10. R. Passante, G. Compagno, and F. Persico, Phys. Rev. A **31**, 2827 (1985). R. Passante, E.A. Power, Phys. Rev. A **35**, 188 (1987).
11. A. Peres, W.K. Wootters, Phys. Rev. D **32**, 1968 (1985). A. Peres, Phys. Rev. D **39**, 2943 (1989).
12. G. Compagno, R. Passante, and F. Persico, Europhys. Lett. **12**, 301 (1990).
13. E.L. Feinberg, Sov. Phys. JETP **23**, 132 (1966).
14. V.L. Ginzburg, Dokl. Akad. Nauk SSSR **23**, 775, 899 (1939); **24**, 131 (1939).
15. F. Persico and E.A. Power, Phys. Rev. A **36**, 475 (1987).
16. D.W. Brown, K. Lindenberg, and B.J. West, J. Chem Phys. **84**, 1574 (1986).
17. G. Compagno, R. Passante, and F. Persico, Phys. Rev. A **38**, 600 (1988).

ELECTRON COHERENCE IN QUANTUM WELL INTERSUBBAND TRANSITIONS

A. IMAMOĞLU, H. SCHMIDT, R. J. RAM, K. CAMPMAN AND A. GOSSARD
Department of Electrical and Computer Engineering
University of California, Santa Barbara, CA 93106, USA

1. Introduction and Background

Absorption is a fundamental limitation in coherent light generation in both atomic and semiconductor media. In ordinary lasers, the so-called population inversion requirement is a direct consequence of stimulated absorption that accompanies emission. In nonlinear devices such as sum-frequency generators or parametric amplifiers, absorption determines the maximum possible conversion efficiency or parametric gain. Eliminating the stimulated absorption without affecting the emission properties, would remove these basic restrictions and lead to optical devices with novel properties.

In this work, we analyze electron coherence and interference effects in intersubband transitions and propose a double quantum-well (DQW) structure where absorption is eliminated by a Fano-type destructive quantum interference [1]. The nonreciprocal nature of the quantum interferences in lifetime broadened intersubband transitions allows for efficient generation into the transparent medium and allows us to realize absoption-free coherent optical devices. We concentrate on two new types of optical devices: Semiconductor lasers that operate without a population inversion and resonant nonlinear devices that generate into a transparent medium.

There has been intense theoretical research activity in the field of lasers without population inversion and electromagnetically induced transparency over the last 6 years [2, 3, 4, 5, 6, 7, 8, 10, 11]. Even though the proposed schemes can be divided into two categories, the common feature of all inversionless lasers is a coherence between atomic states which gives rise to quantum interferences that are destructive for absorption and constructive for emission processes. In particular, it has been predicted that if a steady-state coherence between the two upper states of a lifetime broadened three-state system is established, then such a system could exhibit lasing even when the population in the (single) lower-state is higher than the total population in all the excited states combined (i.e. no inversion in any basis-set) [11]. The atomic schemes which depend on coherences between lifetime broadened upper states eliminate the possibility of a hidden inversion, but are unfortunately relatively difficult to realize. This is primarily due to intrinsic atomic properties that cannot be controlled or changed. A second class of inversionless lasers which rely on a coherence betweeen two lower states have proven to be more accessible to experiments; indication of amplification without inversion in the bare-atomic basis have been recently reported by three groups [13, 14, 15]. We note that the search for inversionless lasers in atomic systems has recently resulted in a remarkable experimental demonstration of a

T. Hakioğlu and A. S. Shumovsky (eds.), Quantum Optics and the Spectroscopy of Solids, 111–120.

laser oscillator operating without population inversion [12]. The experiment based on the D_1 transition of a sodium atomic beam, is a realization of the scheme originally proposed in Ref. [11].

Meanwhile, in a significantly different branch of quantum electronics, a very recent experimental result on the first intersubband laser [16] has heralded the beginning of a new field that deals with the realization of coherent sources of radiation in the mid-infrared region. The interest in such lasers were pioneered by the invention of low-dimensional semiconductor heterostructures more than 20 years ago. The earlier attempts in experimental realization have been elusive primarily due to the difficulty in inverting the intersubband transitions. This difficulty is of a fundamental nature, because unlike atomic systems, the excited states (subbands) of a quantum-well structure always have a shorter lifetime than the lower state in the same well, due to optical phonon emission and faster tunneling rates. In their triple quantum-well structure, Faist *et al.* [16] used spatially indirect transitions and resonantly enhanced tunneling decay rates for the lower state to achieve population inversion.

The inversionless laser scheme that we discuss here is a semiconductor analog of the lifetime broadened atomic system described in Refs. [10, 11]. The transparency is achieved by utilizing Fano-type intereferences in intersubband transitions [1]. The coherence between the upper electronic subbands is established by coherent resonant tunneling between two quantum wells. Apart from extending the earlier work in atomic physics to the semiconductor domain, the new scheme promises to be instrumental in the development of new coherent light sources in the (currently unexplored) mid-infrared region of the spectrum by eliminating the inversion requirement.

The transparency resulting from the destructive quantum interference could also be utilized in resonant nonlinear infrared devices [21]. The profile of the nonlinear susceptibility is in general identical to that of emission: Nonlinear conversion is therefore enhanced by a constructive interference. The asymmetric DQW structure that we consider exhibits strong second-order nonlinearities. Elimination of absorption of either the generated or the pump fields by a transparency enables one to use resonantly enhanced susceptibilities for high-efficiency nonlinear conversion via sum-frequency generation or parametric downconversion.

Second harmonic generation in asymmetric quantum wells have attracted considerable interest as an efficient source of coherent infrared radiation [23]. As a result of the success in quasi-phasematching, reabsorption of the generated field has emerged as the fundamental limitation in these devices, restricting the extracted conversion efficiency. In the case of parametric devices, the absorption of the pump field has precluded an experimental demonstration of amplification. It is therefore safe to claim that use of transparent media would allow for major advances in this field, such as the first demonstration of parametric amplification of infrared radiation, which is in turn a first step towards tunable sources of coherent infared radiation.

In Section II, we describe in detail a semiconductor intersubband laser that operates without population inversion in any basis; such a laser will remove a major obstacle in the search for new unipolar lasers covering the infrared region. Nonlinear infrared devices that we discuss in Section III on the other hand, would provide an ideal tunable source of infrared radiation. We discuss the ongoing experiments at UCSB in Section IV and conclude in Section V by discussing possible extensions.

2. Semiconductor Lasers without Inversion

The inversionless DQW laser structure that we propose is outlined in Fig. 1: The two split upper subbands are coupled to the lower subband by a weak probe field at ω_p. The system is lifetime broadened due to incoherent tunneling into the collector region. When the energy splitting (Ω_{32}) is on the order of the lifetime broadening (Γ_3), the two upper subbands are mutually coupled [3, 10]. Both absorption and emission in the DQW structure of Fig. 1 take place via these mutually coupled upper subbands, and are therefore subject to quantum interference effects. The interference is constructive for emission and destructive for absorption. This *nonreciprocity* between the stimulated emission and absorption makes lasing without inversion possible. The DQW structure of Fig. 1 can alternatively

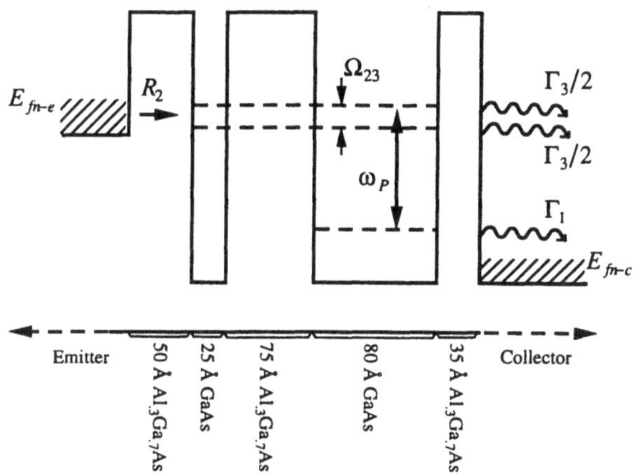

Figure 1. The subband energy-level diagram for the double quantum-well structure

be viewed in the bare-state picture of Fig. 2a: The bare subband ($|2>$) of the narrow well is resonant with the second subband ($|3>$) of the wide-well. These two bare-subbands are strongly coupled by *coherent resonant tunneling* which results in a splitting of the actual (dressed) subbands given by Ω_{23} (Fig. 1) [17]. The atomic scheme that is analogous to the DQW structure of Fig. 2a is shown in Fig. 2b: The bare atomic-states correspond to the bare quantum-well subbands with the same numbering. This atomic scheme has been proposed [10] and analyzed [11] before and was shown to exhibit an electromagnetically induced transparency and lasing without population inversion in any atomic basis. The prediction of an electromagnetically induced transparency was experimentally demonstrated by Boller, Imamoglu, and Harris [18].

In practical systems, dephasing and inhomogeneous broadening are present along with lifetime broadening. It has been previously shown that for steady-state lasing without population inversion [10, 11], three conditions have to be met: 1) The decay rate of the bare-state $|3>$ (Γ_3) has to exceed the dephasing rate (γ_{21}) of the $|2> - |1>$ coherence (ρ_{21}); 2) The coupling strength Ω_{23} of the bare-states $|2>$ and $|3>$ has to exceed $\sqrt{\Gamma_3 \gamma_{21}}$ and the inhomogeneous broadening $\Delta\nu_{inhom}$; 3) The incoherent pumping (R_2) into bare-level $|2>$ should not be accompanied by a rate out into the pumping reservoir (emitter). We shall now discuss these conditions in the context of the laser scheme of Fig. 1 [17].

Coherence dephasing times in undoped semiconductors are limited by electron-electron

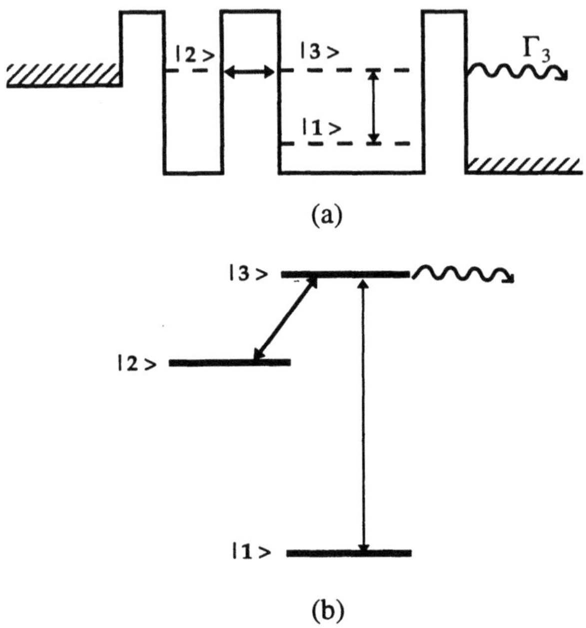

Figure 2. The bare-state energy-level diagrams for (a) the unipolar double quantum-well and (b) the inversionless three-level atomic schemes.

and electron-phonon scattering. The spontaneous optical phonon-emission time [20] from *bare-subband* $|3>$ to subband $|1>$ is about $1\,psec$: This process however, does not dephase the coherence between states $|2>$ and $|1>$; its effect on the laser dynamics is to provide an incoherent rate into state $|1>$ which precludes the possibility of a population inversion. We dominant dephasing mechanism in our structures will be electron-electron scattering, which has a characteristic dephasing time longer than $1\,psec$ at a sheet carrier density $n = 1 \times 10^{12}\,cm^{-2}$ [20]. On the other hand, the relevant tunneling times (Γ_3^{-1} and Ω_{23}^{-1}) can be easily made as short as $100\,fsec$, so that the first condition is satisfied.

The inhomogeneous broadening $\Delta\nu_{inhom}$ in the quantum-well structures is determined by the well-width fluctuations, which in many cases, is the dominant broadening mechanism in intersubband transitions. Recent advances in *growth-interrupt* techniques for binary alloy structures however, have resulted in $\Delta\nu_{inhom} \simeq 1meV$, which is well-below the tunneling rates (Ω_{23} and Γ_3) mentioned above. The second condition is therefore also easily satisfied.

The laser scheme of Fig. 1 is pumped by (incoherent) electron reservoirs of the emitter and collector regions. When we set $E_{fn-e} \geq \hbar\omega_2 + 3kT$, the decay of state $|2>$ into the emitter reservoir is forbidden as there are no avaliable final states due to the Pauli exclusion principle. The third condition is therefore also readily satisfied.

We have carried out calculations on $GaAs/AlGaAs$ DQW structures. We have chosen the widths of the narrow and wide $GaAs$ quantum wells to be 25 Å and 80 Å, respectively. The emitter and the collector regions are doped with Si at a concentration of $5 \times 10^{17}\,\mathrm{cm}^{-3}$

and $6 \times 10^{17}\,\mathrm{cm}^{-3}$, respectively. The quantum wells and the barrier regions are undoped. The emitter region has an Al mole-concentration of $x = 0.15$, which enables us to operate at nearly flatband conditions at an applied bias voltage of 0.135V. The subband energy spacing is about 0.119 eV, which corresponds to an optical transition wavelength of $\lambda_p = 2\pi c/\omega_p = 10.36\,\mu\mathrm{m}$. We have also assumed the Al mole-fraction of the barrier regions to be the same ($y = 0.3$) for further simplification of the growth process. The barrier thickness of $L_{b3} = 35\,Å$ between the wide-well and the collector gives us tunneling decay rates $\Gamma_3 = 4\times10^{12}s^{-1}$ (2.6 meV) and $\Gamma_1 = 5\times10^{11}s^{-1}$ (0.32 meV), for the (bare) subbands $|3>$ and $|1>$, respectively. The coupling strength (or the energy splitting) due to coherent resonant tunneling across the $L_{b2} = 75\,Å$ barrier between the two wells is set so that $\Omega_{23} = 1.15\,\Gamma_3$. Finally, the barrier between the emitter and the narrow well which controls the incoherent electron pumping rate into subband $|2>$ is chosen to be $L_{b1} = 50\,Å$ in order to limit the injected current density to $J = 15\,KA/cm^2$, which in turn corresponds to a sheet carrier density of about $2.7 \times 10^{10}\,cm^{-2}$ in the upper subbands. Within this parameter range, all the requirements for lasing without inversion are satisfied. Also, to maximize the gain, we can *tune* the energy difference between the bare subbands $|2>$ and $|3>$ by changing the bias voltage.

It is important to note that the fundamental properties of the DQW structure are robust against small changes in the parameters chosen above. We carried out calculations of structures with parameters differing from the above values by as much as 20%: The primary effect of these changes is to shift the wavelength at which transparency and lasing without inversion could be observed.

If we neglect the band non-parabolicity effects, the optical properties of the electrons in different **k** states are identical. Therefore the dynamics of a DQW structure with N electrons is determined by the density-matrix equations [19], that are the same as those of the atomic lambda scheme of Fig. 2. Without going through the detailed analysis of the density-matrix equations which are presented elsewhere [11], we evaluate the absorption and gain *cross-sections*, normalized with respect to the resonant single quantum-well absorption cross-section. We remark that the definition of cross-section is somewhat different for a quantum-well intersubband transition: It has the units of length, and therefore an absorption cross-section in this context is really an *absorption length*. Using the solution of the density-matrix equations and assuming that the bare-subbands $|2>$ and $|3>$ have the same energy, we obtain the nonreciprocal absorption and gain lengths

$$\sigma_{abs}(\omega_p) = \sigma_{13}(\omega_{31})\,\frac{4\,\Gamma_3^2\,\Delta\omega_p^2 + \Gamma_3\gamma_{21}\,(\Omega_{23}^2 + \Gamma_3\gamma_{21})}{(\Omega_{23}^2 + \Gamma_3\gamma_{21} - 4\Delta\omega_p^2) + 4\gamma_{21}^2(\Gamma_3 + \gamma_{21})^2\Delta\omega_p^2}\,, \tag{1}$$

and

$$\sigma_{em}(\omega_p) = \sigma_{13}(\omega_{31})\,\frac{\Omega_{23}^2}{\Gamma_3^2 + 2\Omega_{23}^2 + 4\Delta\omega_p^2}\,\frac{\Gamma_3(\Gamma_3 + \gamma_{21})(\Omega_{23}^2 + \Gamma_3\gamma_{21})}{(\Omega_{23}^2 + \Gamma_3\gamma_{21} - 4\Delta\omega_p^2)^2 + 4\gamma_{21}(\Gamma_3 + \gamma_{21})\Delta\omega_p^2} \tag{2}$$

where ω_p denotes the probe-frequency and $\Delta\omega_p = \omega_3 - \omega_1 - \omega_p$. The resonant absorption length $\sigma_{13}(\omega_{31})$ of a single quantum-well in the infinite well approximation satisfies

$$\sigma_{13}(\omega_{31}) \simeq \frac{256e^2}{27\pi^2 n_r m_e^* \epsilon_0 c L_{opt}\Gamma_3}\,. \tag{3}$$

The quantity of interest in a transmission experiment is the net-gain coefficient of the structure, defined by $g(\omega_p) = \sigma_{em}(\omega_p)\,(n_2 + n_3) - \sigma_{abs}(\omega_p)\,n_1$. Here n_1, n_2, and n_3

correspond to the sheet carrier densities in the respective bare-subbands. In such an experiment, one observes the change in the probe-transmission as a function of the probe frequency. Figure 3 shows the normalized transmission of a probe field, which has propagated through a $L = 1\,mm$ long DQW structure detailed in Fig. 1.

Transmission minima occur when the probe field is resonant with either of the two dressed-subband resonances: This is due to the absence of a population inversion ($n_1 > n_2 + n_3$). In between the two minima however, there is a transmission maximum which predicts a net gain of about $e^{0.3}$ for the probe. The magnitude of the net-gain coefficient at the bare-subband resonance can be obtained from Eqs. (1a-b)

$$g(\omega_{31}) = \sigma_{em}(\omega_{31})\,[\frac{\Gamma_3^2}{\Gamma_3^2 + 2\Omega_{23}^2}(n_2 + n_3) - \frac{\gamma_{21}\Gamma_3}{\Omega_{23}^2}n_1] \quad . \tag{4}$$

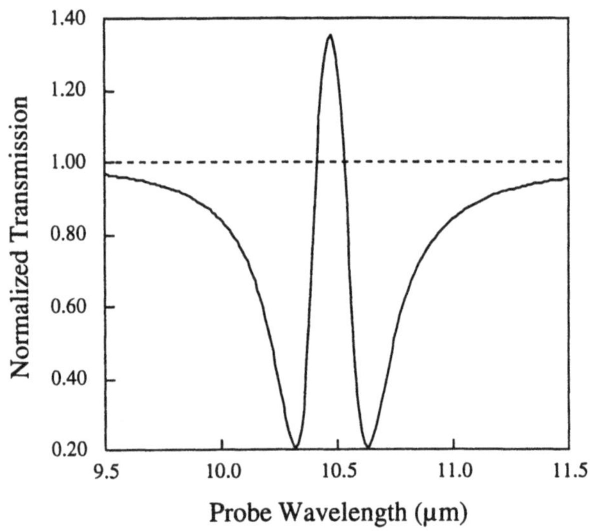

Figure 3. Normalized transmission profile of a probe field ($exp[g(\omega_p)\,L]$) as a function of the probe wavelength. The transmission minima occur at the dressed-subband energies. We assume a $1mm$ long structure, giving a peak gain of $e^{0.3}$. Parameters of the structure are the same as those given in the text.

$g(\omega_{31})$ has a value of $3\,cm^{-1}$ for the structure of Fig. 1, with $n_1 = 1.1(n_2 + n_3) = 3.0 \times 10^{10} cm^{-2}$. We remark that this is a relatively conservative estimate which can easily be increased by a factor of three by increasing the carrier density (without causing significant dephasing). The gain coefficient can also be increased by using a *cascade structure* consisting of 10 or more DQW structures in series: In their recent work, Faist *et al.* [16] used a cascade of 25 triple quantum well structures (with population inversion) to increase their gain coefficient to about $60\,cm^{-1}$, which in turn allowed them to observe laser oscillation. We also note that using semiconductor systems with lower effective mass would increase the gain coefficient due to $1/m_e^*$ dependence (Eq. (3)). In addition, similar transitions energies (ω_{31}) in such systems can be obtained using larger quantum well structures, which implies further reduction in inhomogeneous broadening.

3. Nonlinear Infrared Devices

Nonlinear optical devices provide tunable sources of coherent radiation in the optical and UV regions of the electromagnetic spectrum. In a nonlinear process such as sum frequency generation or parametric amplification, the total conversion efficiency (or gain in the case of parametric amplification) is determined by an interplay of nonlinear and linear susceptibilities. In general, one tries to maximize the generating nonlinear susceptibility, while controlling the linear susceptibility which determines the effective generating length of the medium, either by loss of phase matching (real part) or by reabsorption of the generated field (imaginary part). If one tries to enhance the nonlinearity by tuning to a resonance, the nonlinearity is resonantly enhanced but at the same time, the medium exhibits a rapidly increasing refractive index and becomes opaque due to increased loss. To overcome this fundamental limitation, Harris, Field, and Imamoglu [21] have proposed to use an electromagnetically induced transparency where one obtains resonantly enhanced nonlinear susceptibility with zero refractive index and vanishing loss. The nonreciprocal nature of the nonlinear and linear susceptibilities predicted in Ref. [21] have since been verified experimentally [22].

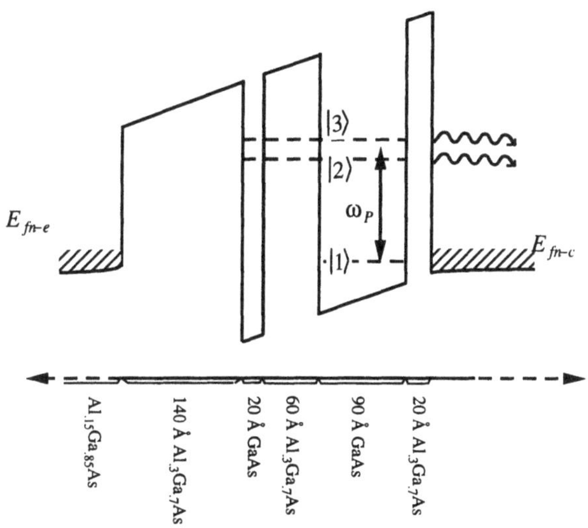

Figure 4. The dressed-state energy-level diagram for the double quantum-well structure that will be used in transparency and nonlinear infrared conversion experiments. This structure is optically pumped and therefore does not require electrical contacts. The dominant broadening mechanism is tunneling into the collector region.

As indicated earlier, the transparency that exists in a DQW structure could also be utilized in *nonlinear infrared sources.* Lack of inversion symmetry in these structures imply that second-order $\chi^{(2)}$ processes would be dominant, in contrast to atomic media where only inherently weaker $\chi^{(3)}$ (and higher order) processes are allowed. The prototypical structure that we consider is outlined in Fig. 4: The structure for nonlinear conversion is actually identical to that used for the transparency experiment, as no pumping into the upper subbands is needed in this case.

In the optically-pumped device of Fig. 4, the fundamental requirement is that the upper-subband splitting dominates over dephasing. The presence of fast tunneling however, is not required for nonlinear devices. For high-quality materials at low carrier densities ($n \leq 1 \times 10^{12}\, cm^{-2}$), the dominant broadening mechanism for excited subbands is spontaneous optical phonon emission (which is a lifetime broadening mechanism). Destructive quantum interferences that we described earlier will also eliminate the loss due to spontaneous optical-phonon decay from (bare) subband $|3>$ to subband $|1>$. An additional advantage of structures which exhibit fast tunneling is relative insensitivity to dephasing processes such as electron-electron scattering.

The resonantly enhanced $\chi^{(2)}$ of the DQW structure could be utilized to generate intense coherent radiation at the transparency frequency via second-harmonic generation. Alternatively, application of a strong pump field at the transparency (i.e. CO_2 laser) can be used for parametric downconversion: Such a process would provide a promising source of tunable coherent infrared radiation. We plan to pursue these experiments in parallel with the experiments aimed at demonstrating inversionless laser oscillation.

4. Quantum Well Experiments

We are carrying out experiments aimed at realizing lasers without inversion and nonlinear conversion using semiconductor intersubband transitions. The first step in our experiments is to demonstrate a Fano-type quantum interference induced transparency in a three-subband DQW structure. The motivation behind a separate measurement aimed at demonstrating the quantum interference induced transparency is twofold: First, Fano-type interference profiles and the associated transparency in field-free quantum-well systems have not been observed. Such an experiment will in essence prove the analogy with the corresponding atomic scheme and constitutes the first step in an experimental realization of nonlinear infrared devices described above. Second, absence of electrical pumping requirement in such an experiment implies that we do not need to make electrical contacts to the DQW structure, since the lower subband of the wide quantum-well will be populated even in the absence of an applied bias. This simplification in the processing stage would allow us to demonstrate that an essential condition (i.e. transparency) for lasing without inversion is indeed satisfied by the DQW structures.

The transparency experiment is currently being carried out in the structure shown in Fig. 4: This structure has resonant excited subbands at $V_{bias} = 0$ and a negligible tunneling decay rate for the bare-subband $|2>$. The specific values of the width of the barrier regions are $L_{b1} = 140$ Å, $L_{b2} = 60$ Å, and $L_{b3} = 20$ Å. The well-widths are chosen to be 20 Å and 85 Å, respectively. We observe the transmission profile of this structure initially using *Fourier transform infrared spectroscopy* (FTIR). The frequency resolution required by the experiment is typically given by the width of the upper subbands: In our structures, these widths are about 2 meV and hence fall well within the FTIR resolution limits. Experiments are carried out at 4 K to minimize the phonon broadening.

The experiment aimed at demonstrating amplification without inversion essentially uses the same experimental technique that we have detailed. The structure in this case is that of Fig. 1. The pumping necessary for stimulated emission is provided by the emitter reservoir under an applied bias of 0.135 $Volts$. An amplifier structure in addition requires layers for optical confinement and ohmic contacts for efficient electrical injection. Optical confinement could be achieved by a lithographically defined ridge waveguide as well as the inclusion of $1\mu m$ thick $AlAs$ layers that straddle the gain region. Since we propose to grow the structure on a semi-insulating substrate (to minimize the free-carrier absorption), we need to introduce highly doped layers above and below the structure. We remark however that the coupling efficiency and the background loss due to free carrier absorption does not affect the amplification profile we intend to observe: For amplification without inversion, all we need to show is a relative gain with respect to the (frequency independent) background loss. Provided that the net-area under the transmission curve is negative, one can prove that the profile shown in Fig. 3 implies amplification without population inversion in any basis set.

Laser oscillation can take place if the net round-trip gain exceeds the background and mirror losses, which are typically around $10\,cm^{-1}$ [16]. As indicated in the previous section, demonstration of laser oscillation would then require a number of DQW structures grown in series: A cascade of 10 DQW structures enhances the gain coefficient so that the total single-pass gain experienced by a probe field propagating through the laser structure is 10 times the single-pass gain. For the parameters of Fig. 1, this means a gain of $30\,cm^{-1}$ per pass, which would allow for laser oscillation. The number of cascaded structures required for laser action could be reduced by coating the cleaved surfaces.

As indicated earlier, the experiments aimed at nonlinear infrared generation into a transparency utilize the DQW structure of Fig. 4. These samples will be processed so as to achieve quasi-phasematching of the pump and generated fields [23]. Initially, we intend to use a tunable CO_2 laser as the pump in a resonant second-harmonic generation experiment. When the length of the DQW structure is smaller than the absorption length (typically $\leq 0.1\,mm$), the absorption is negligible and the generated (weak) second-harmonic signal can be used to determine the $\chi^{(2)}$ profile: This measured profile should indicate the presence of a constructive interference at the transparency frequency. The peak signal in this case however, is obtained when the two-photon pump is exactly resonant with the two (dressed) subbands. For structures longer than the absorption length, the generated second-harmonic signal will provide an evidence for the interplay between the nonlinear (generation) and linear (absorption) susceptibilities. More specifically, we expect to see a sharp peak in second-harmonic signal at the transparency frequency, rather than at the subband resonances.

In a second-harmonic or sum-frequency generation experiment, the transparency is utilized to eliminate the reabsorption of the generated radiation. In the singly-resonant scheme that we are considering, the pump field is off-resonant and does not experience significant loss. As indicated earlier, we can alternatively use the same structure for parametric amplification. The transparency in this case is used to eliminate the absorption of the pump field without sacrificing the parametric gain. The pump source in this experiment is either the second-harmonic of a CO_2 laser generated by another structure, or a CO laser. Another CO_2 laser would be used as the weak signal to be amplified by the parametric device.

5. Conclusion

Since the invention of lasers, population inversion has been thought of as a fundamental requirement for laser action. Removal of the inversion requirement opens up new avenues of research aimed at making new lasers in the previously inaccessible regions of the electromagnetic spectrum.

The inversionless laser that we intend to demonstrate is based on intersubband transitions of a semiconductor heterostructure. The search for such unipolar semiconductor lasers has been hindered by the difficulty in inverting the populations of subbands in quantum wells. We believe that the success of this project will trigger the development of new laser devices covering the infrared region of the spectrum. Even though the initial experiments will be carried out at $4\,K$, the relative temperature insensitivity of both the homogeneous and inhomogeneous broadening mechanisms in unipolar lasers should allow for $77\,K$ operation in the near future. There is great demand for infrared lasers due to applications in areas such as atmospheric spectroscopy, pollution monitoring, and leak detection in manufacturing processes. Some of these applications require tunable sources of coherent infrared devices: The CO_2 laser pumped parametric amplifiers and oscillators could provide ideal tools for such applications.

References

1. U. Fano, Phys. Rev. **124**, 1866 (1961).
2. O. Kocharovskaya and Ya I. Khanin, Pis'ma Zh. Eksp. Teor. Fiz. **48**, 581 (1988) [JETP Lett. **48**, 630 (1988)].
3. S. E. Harris, Phys. Rev. Lett. **62**, 1033-1036 (1989).
4. M. O. Scully, S.-Y. Zhu, and A. Gavrielides, Phys. Rev. Lett. **62**, 2813-2816 (1989).
5. A. Lyras, X. Tang, P. Lambropoulos, and J. Zhang, Phys. Rev. A **40**, 4131-4134 (1989).
6. G. S. Agarwal, S. Ravi, and J. Cooper, Phys. Rev. A **41**, 4721-4726 (1990).
7. O. Kocharovskaya and P. Mandel, Phys. Rev. A **42**, 523-535 (1990).
8. A. Imamoglu, Phys. Rev. A **40**, 2835-2838 (1989).
9. H. M. Doss, L. M. Narducci, M. O. Scully, and J. Y. Gao, Opt. Commun. **95**, 57-63 (1991).
10. A. Imamoglu and S. E. Harris, Opt. Lett. **14**, 1344-1346 (1989).
11. A. Imamoglu, J. E. Field, and S. E. Harris, Phys. Rev. Lett. **66**, 1154-1156 (1991).
12. E. S. Fry, X. Li, D. E. Nikonov, G. G. Padmabandu, G. R. Welch, I. N. Shubin, E. S. Fry, D. E. Nikonov, M. D. Lukin, and M. O. Scully, "Laser Oscillation Without Population Inversion in a Sodium Atomic Beam," (submitted to Phys. Rev. Lett.).
13. A. Nottelmann, C. Peters, and W. Lange, Phys. Rev. Lett. **70**, 1783-1786 (1993).
14. E. S. Fry, X. Li, D. E. Nikonov, G. G. Padmabandu, M. O. Scully, A. V. Smith, F. K. Tittel, C. Wang, S. R. Wilkinson, and S.-Y. Zhu, Phys. Rev. Lett. **70**, 3235-3238 (1993).
15. W. E. van der Veer, R. J. J. van Diest, A. Donszelmann, and H. B. van Linden van den Heuvell, Phys. Rev. Lett. **70**, 3243-3246 (1993).
16. J. Faist, F. Capasso, D. L. Sivco, C. Sirtori, A. L. Hutchinson, and A. Y. Cho, Science **264**, 553-556 (1994).
17. A. Imamoglu and R. J. Ram, Opt. Lett. **19** 1744 (1994).
18. K.-J. Boller, A. Imamoglu and S. E. Harris, Phys. Rev. Lett. **66**, 2593-2596 (1991).
19. H. Haug and S. W. Koch, *Quantum theory of the optical and electronic properties of semiconductors* (World Scientific, Singapore, 1993).
20. J. Shah, *Hot Carriers in Semiconductor Nanostructures* (Academic Press, San Diego. 1992).
21. S. E. Harris, J. E. Field, and A. Imamoglu, Phys. Rev. Lett. **64**, 1107-1110 (1990).
22. G. Z. Zhang, K. Hakuta, and B. Stoicheff, Phys. Rev. Lett. **71**, 3099-3102 (1993).
23. S. J. B. Yoo, M. M. Fejer, R. L. Byer, and J. S. Harris, Jr., Appl. Phys. Lett. **58**, 1724-1726 (1991).

INTERACTION OF TWO-LEVEL ATOMIC SYSTEM WITH A SINGLE-MODE RADIATION FIELD

T. HAKİOĞLU
Physics Department, Bilkent University, Ankara, 06533 Turkey

Abstract. The Dicke model is examined in the limit of large number of atoms and /or large number of excitations. Superfluous instabilities arising from the rotating wave approximation is examined and counterrotating terms are shown to be crucial for the correct dynamical evolution in these limits.

1. Introduction

In this article we will briefly introduce the main results of the problem of interaction of an atomic cluster with "p" atoms with a single mode resonant radiation field in the limit when p and/or the total number of excitations n is large. The principal assumption in this model is that the atom is considered with only two energy levels interacting via photon exchange. The principle reason is that besides the well-observed energy-momentum conservation in the observation time scales, atomic dipole transition between different energy levels are restricted by certain selection rules conserving the total angular momentum and parity due to the vector nature of the interaction. At the zero'th order the atom field interaction is resonant and is described by a two-level transition. The transition between these two levels takes place by the absorption or emission of a single light quantum with an energy exactly matching the energy difference of the two levels in question. The corrections to zero'th order approximation come from the spontaneous radiative processes and Lamb shift, atomic thermal collisions, recoiling effects and Doppler shift. The spontaneous radiative corrections produce a shift $\delta\nu/\nu \sim T/\tau \sim 10^{-3}eV$ where $T \sim 10^{-15}\,s$ is a typical period of the radiation field and $\tau \sim 10^{-12}\,s$ is a typical lifetime for atomic energy scales. These corrections can grow as the number of energy levels in the atom gets larger nevertheless it can still be considered as a perturbative correction[1]. The nonrelativistic Doppler shift in the frequency of radiation for a gas of particles of mass M and temperature T is given in natural units by $\delta\nu/\nu = 7.16\,10^{-7}\,(T/M)^{1/2}$. For a typical example of sodium atom at room temperature one finds $\delta\nu/\nu \simeq 10^{-13}$ which produces a negligible effect. Under more drastic Doppler shifts the Doppler-free spectroscopy can achieve resolutions approaching 1 part in 10^{11} which practically eliminates this effect.

It is clear that the zero'th order approximation is sufficient for most simple atoms except for pure quantum radiative processes, experimental techniques are available to

T. Hakioğlu and A. S. Shumovsky (eds.), Quantum Optics and the Spectroscopy of Solids, 121–138.

suppress higher order effects by shifting them to negligible scales. We will not dwell on the details of the physical justification of two level systems longer and recommend to the reader a good survey by Allen and Eberly[2]. We will now briefly describe our model.

2. The Dicke Model

The simplifications made by the two level atom and a single field mode certainly pays back. In 1954 Dicke introduced a model to study the collective emission-absorption properties of a cluster of two-level atoms[3]. The major additional assumption in this model is that the linear size (i.e. $\mathcal{V}^{1/3}$ where $\mathcal{V}$ describes the volume occupied) of the cluster is much smaller than the wavelenght λ of the emitted and absorbed radiation. Since $\mathcal{V}^{1/3} \ll \lambda$ the cluster-field coupling can be treated as a point-like interaction and all atoms within the cluster interact in phase with the same field strenght. This corresponds to the so called equivalent mode approximation and the atomic cluster is then composed of p two-level indistinguishable dipoles which interact via exchange of quantum of radiation. Each atom is then represented by the complete set of "spin=1/2" dipole operators,

$$\begin{aligned} \hat{L}_j^{(-)} &= |d\rangle_j \; {}_j\langle u| \,, \\ \hat{L}_j^{(+)} &= |u\rangle_j \; {}_j\langle d| \,, \\ \hat{L}_j^{(z)} &= \tfrac{1}{2}\Big(|u\rangle_j \; {}_j\langle u| - |d\rangle_j \; {}_j\langle d|\Big) \end{aligned} \tag{1}$$

Here $|u\rangle_j$ and $|d\rangle_j$ describe the electron states in the excited and ground states of the j'th atom respectively as shown in Fig. (1) below. Eq.'s (1) can be shown easily to satisfy the

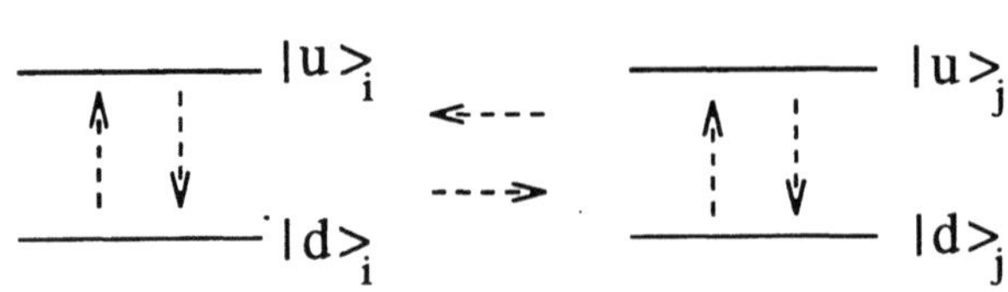

Figure 1. Two atoms in a Dicke cluster. Arrows indicate the exchange of photons.

SU(2) commutation relations,

$$\begin{aligned} [\hat{L}_i^{(-)}, \hat{L}_j^{(+)}] &= \; 2\,\hat{L}_i^{(z)}\,\delta_{i,j} \,, \\ [\hat{L}_i^{(\pm)}, \hat{L}_j^{(z)}] &= \; \mp\,\hat{L}_i^{(\pm)}\,\delta_{i,j} \,. \end{aligned} \tag{2}$$

Since dipoles at each atom are assumed not to couple by direct overlap of the electron wavefunctions localized at each atom[1], the total number of electrons $\hat{N}_j$ at each atom in the cluster is separately conserved which is related to the *total spin* $\hat{L}_j^2$ where,

$$\hat{N}_i = \Big(|u\rangle_{(j)} \; {}_{(j)}\langle u| + |d\rangle_{(j)} \; {}_{(j)}\langle d|\Big) \,, \qquad \text{and} \qquad \hat{L}_i^2 = \frac{\hat{N}_i}{2}\Big(\frac{\hat{N}_i}{2} + 1\Big) \,. \tag{3}$$

[1]Collective effects in a model which inherently has the feature of electron hopping between localized atomic orbitals is an interesting and realistic model in certain cases. This extention will be studied in a separate publication.

In the model we study each atom contains only one transition electron. An arbitrary microscopic state $|\ell\rangle_c$ of the cluster is then represented by,

$$|\ell\rangle_c = |\ell_{z_1}, \ell_{z_2}, \ldots, \ell_{z_n}\rangle \tag{4}$$

where $\ell_{z_j} = \pm 1/2$ $(j = 1, \ldots, n)$ is the eigenvalue of $\hat{L}^z_j$. If the linear size of the cluster is much smaller than the wavelenght of the dipole radiation we can apply the equivalent mode approximation. The whole cluster can then can be considered as a compound dipole with collective dipole operators,

$$\begin{aligned} \hat{\mathcal{L}}_- &= \textstyle\sum_{j=1}^N L_j^{(-)} , \\ \hat{\mathcal{L}}_+ &= \textstyle\sum_{j=1}^N L_j^{(+)} , \\ \hat{\mathcal{L}}_z &= \textstyle\sum_{j=1}^N L_j^{(z)} , \end{aligned} \tag{5}$$

which also respect similar commutation relations à la Eq.'s (2) as,

$$\begin{aligned} [\hat{\mathcal{L}}_-, \hat{\mathcal{L}}_+] &= \ 2\,\hat{\mathcal{L}}_z , \\ [\hat{\mathcal{L}}_\pm, \hat{\mathcal{L}}_z] &= \ \mp \hat{\mathcal{L}}_\pm . \end{aligned} \tag{6}$$

The algebra represented by these commutation relations permits us to find the macroscopic state of the cluster as linear superpositions of the microscopic ones in Eq.'s (4). Here one is tempted to adopt that the conserved quantum number under the action of the collective operators is the total number of atoms (or electrons) in the cluster $\hat{\mathcal{N}}$. The total cluster spin $\hat{\mathcal{L}}$ is then described by,

$$\hat{\mathcal{N}} = \sum_{i=1}^{n} \hat{N}_i , \qquad \text{and} \qquad \hat{\mathcal{L}}^2 = \frac{\hat{\mathcal{N}}}{2}(\frac{\hat{\mathcal{N}}}{2} + 1) \tag{7}$$

Let's adopt Eq.'s (7) temporarily and examine an arbitrary macroscopic state $|\ell\, m\rangle_n$ as a linear superposition of $|j\rangle_n$ as

$$|\ell\, m\rangle_n = \sum_{j=1}^{n} c_j^{(n)}\, |j\rangle_n \qquad \text{with} \qquad \hat{\mathcal{L}}_z\, |\ell\, m\rangle_n = m\, |\ell\, m\rangle_n \quad \text{and} \quad \hat{\mathcal{L}}^2\, |\ell\, m\rangle_n = \ell\,(\ell{+}1)\, |\ell\, m\rangle_n \tag{8}$$

where $|j\rangle_n$ describes a *microscopic* configuration and is nothing but Eq. (4) with j atoms in the *up* (i.e. $+1/2$) and $n-j$ atoms in the *down* (i.e. $-1/2$) spin configuration. A typical microscopic state is then,

$$|j\rangle_n \quad \Longleftrightarrow \quad \mathcal{P}_j \ \, |\underbrace{+,+,\ldots,+}_{j};\overbrace{-,-,\ldots,-}^{n-j}\rangle \tag{9}$$

where $\mathcal{P}_j$ describes any permutation over j up and $n-j$ down spins. From Eq.'s (..) and (..) it is appealing to say that $-p/2 \leq m = j - n/2 \leq p/2$ and $\ell = p/2$. Although everything seems quite straightforward so far there is a subtlety involved such that the natural limit to the total spin ℓ is given by the total number of atoms hence $\ell_{max} = p/2$. However, in the most general case ℓ is a degree of freedom of the system and it is the natural limit for m such that $-\ell \leq m \leq \ell$ and the total number of excited atoms under most general initial conditions may be less than the total number of atoms in the cluster. Another way of approach is that the total cluster spin ℓ is obtained by adding individual spins $\ell_j = 1/2$ and therefore it is allowed to change between $p/2$ and its minimum value, viz. $\ell = 1/2$ for odd total number atoms or $\ell = 0$ for even number of atoms. The value of ℓ is then fixed at the preparation of the initial macroscopic state and is a measure of cooperation between the atoms in their contribution to the radiative properties of the whole. Therefore in general we have $0 \leq \ell \leq p/2$. In this case only those states with $-\ell + p/2 \leq j \leq \ell + p/2$ actively participate in the cooperative effects. Such states can in principle be uniquely determined in the initial state by proper choice of initial number of photons as well as the coefficients c_j in Eq. (8).

Moreover the Dicke Hamiltonian is totally symmetric with respect to the exchange of indices of different atomic dipoles. This implies that the symmetrical or anti-symmetrical initial states never mix in their time evolution. Specific choice for the cooperation number can lead to distinct cooperative quantum effects of radiation. The reader can consult to a vast number of literature in this field of which only a few are listed in the references below[1, 2, 3, 4]. Our specific aim in this section is to briefly investigate the most general case of arbitrary p, ℓ as well as total number of excitations n.

In this general case on has $-\ell \leq m \leq \ell$ with $\ell \leq p/2$. Hence the non-vanishing matrix elements of the collective dipole operators for the arbitrary macroscopic state in Eq. (8) are given as (dropping the index n from the $|\ell\, m\rangle$ states),

$$\begin{aligned} \langle \ell\, m|\hat{\mathcal{L}}_z\,|\ell\, m\rangle &= j - p/2 \\ \langle \ell\, m+1|\hat{\mathcal{L}}_+\,|\ell\, m\rangle &= \sqrt{(\ell - j + p/2)\,(\ell + j - p/2 + 1)} \\ \langle \ell\, m-1|\hat{\mathcal{L}}_-\,|\ell\, m\rangle &= \sqrt{(\ell + j - p/2)\,(\ell - j + p/2 + 1)} \end{aligned} \tag{10}$$

We will use $|\ell\, m\rangle$ as the natural basis in the description of the atomic component of the generalized Dicke state.

The full Hamiltonian for the coupled cluster-single mode radiation state is given by,

$$\hat{\mathcal{H}} = \omega\,(\hat{a}^\dagger\,\hat{a} + \frac{1}{2}) + \xi\,\hat{\mathcal{L}}_z + \gamma\,(\hat{a}^\dagger\,\hat{\mathcal{L}}_- + \hat{a}\,\hat{\mathcal{L}}_+) \tag{11}$$

where ω is the frequency of the radiation, ξ is the difference between the upper and lower atomic energy levels. Usually one considers $\omega = \xi + \Delta$ with $\Delta \neq 0$ describing the detuning from the exact resonance condition (i.e. $\Delta = 0$).

2.1. GENERAL SOLUTION

The exact analytic solution of the Hamiltonian does not exist for an arbitrary number of atoms p and/or arbitrary number of excitations n in a closed form. The solution for arbitrary number of atoms with $n = 1$ has been given by Cummings and Dorri[5] and Seke[6] a more general solution under the same conditions but including the retardation effects of the emitted radiation from each atom was considered earlier by Milonni and Knight[7]. For $p = 1, 2$ and arbitrary n the exact solution is presented by Bužek[8]. The solution of this simplified model is strong geometry dependent results. For arbitrary p number of atoms initially with $n = 1$ the radiation properties of the cluster can be varied from a superradiant (constructive interference) to subradiant (destructive interference).

The case for $3 \leq p, n$ involves collective phenomena which are not present for the simple case when p and/or n are equal to one or two. A general pertubative approach for the $3 \leq p, n$ was formulated by Kozierowsky et al.[9]. Here we will examine the general p and n case primarily focusing on the qualitative aspects of the time evolution arising from the *complexity* of the eigen solutions.

In the Hamiltonian (11) the operator corresponding to the total number of excitations

$$\hat{\mathcal{N}} = \hat{a}^{\dagger}\,\hat{a} + \hat{\mathcal{L}}_z + \frac{p}{2} \tag{12}$$

since $[\hat{\mathcal{N}}, \hat{\mathcal{H}}] = 0$ and the eigenvalue n is an integral of motion. Since $[\hat{\mathcal{L}}^2, \hat{\mathcal{H}}] = 0$ the cooperation number ℓ is also a good quantum number. We can restrict our attention onto a finite dimensional subspace of the Hilbert space corresponding to a given n. A typical state can then be represented at a particular instant by r photons, $n - r$ atoms in the excited energy and $p - n + r$ atoms in the ground state energy level. Since p is fixed and n is determined by the initial conditions, depending on the ℓ values in question there are certain restrictions on possible values that r can take. The conditions $|m| \leq \ell$ with $\ell \leq p/2$ imply $-\ell + n - p/2 \leq r \leq \ell + n - p/2$ hence the total number of states r can take is $2\ell + 1$. A dynamical state is then given in terms of combinations of $|n - r\rangle_c \otimes |r\rangle_f$ satisfying the above restrictions where the subscripts c and f denote the cluster and the field respectively. In order to find the matrix elements of the Hamiltonian, we thus need to replace $m = j - p/2$ in Eq.'s(10) by $m = n - r - p/2$ and,

$$\begin{aligned}
\langle m-1| \otimes \langle r+1|\hat{\mathcal{H}}|r\rangle \otimes |m\rangle &= \gamma\, v_r \\
\langle m+1| \otimes \langle r-1|\hat{\mathcal{H}}|r\rangle \otimes |m\rangle &= \gamma\, v_{r-1} \qquad \text{where} \\
v_r &= \sqrt{r+1} \quad \sqrt{(\ell + n - r - p/2)(\ell - n + r + p/2 + 1)} \\
\text{with} \qquad r_{min} \leq r \leq r_{max} & \qquad \text{and} \qquad \ell \leq p/2\ .
\end{aligned} \tag{13}$$

The equations above represent the generalized case for arbitrary p and n. Here r_{max} and r_{min} determine the maximum allowable range for a given ℓ, n and p. In the most general case,

$$r_{max} = \ell + n - p/2 \qquad \text{and} \qquad r_{min} = \begin{cases} -\ell + n - p/2, & \text{and } \ell \leq n - p/2\ ; \\ 0 & \text{otherwise.} \end{cases} \tag{14}$$

At this point we analyze several distinct situations. Since $r_{min} \geq 0$ and $r_{max} \leq n$ the number of possibilities for certain values of n and p can be classified in $n < p$, $p < n$ and $n = p$ as indicated in Table.1 below.

	$r_{min} = 0$		$r_{min} = 0$ forced bound	$n < p$
a)	and	$\ell = p/2$	unphysical	$p < n$
	$r_{max} = n$		$r_{min} = 0$ natural bound	$n = p$
	$r_{min} = 0$		$\begin{cases} r_{min=0}\ \textit{natural} & \text{if } n>p/2; \\ \text{“}\ \ \textit{forced} & \text{if } n<p/2\,. \end{cases}$	$n < p$
b)	and	$\ell < p/2$	unphysical	$p < n$
	$r_{max} < n$		unphysical	$n = p$
	$r_{min} > 0$		unphysical	$n < p$
c)	and	$\ell = p/2$	allowed	$p < n$
	$r_{max} = n$		unphysical	$n = p$
	$r_{min} > 0$		$\begin{cases} \textit{allowed} & \text{if } n>p/2; \\ \textit{unphysical} & \text{if } n<p/2\,. \end{cases}$	$n < p$
d)	and	$\ell < p/2$	allowed	$p < n$
	$r_{max} < n$		allowed	$n = p$

Table 1.

It is crucial to remember that Dicke's superradiance condition $\ell = p/2$, $m = 0$ can be met initially within all the physically realizable parts (a) and (c) of Table. 1. Dicke's subradiance can be realized within parts (b) and (c) if in the initial state $r_{max} = n - p/2$ which implies the $\ell = 0$, $m = 0$ singlet. The preparation of initial conditions in such a way that the time evolution will be dominated by superradiant or subradiant states is a difficult experimental task. First experiment on superradiant systems was performed in 1973 by Skribanowitz et al.[10] and that for subradiant states has been done in 1985 by Crubellier et al. and Pavolini et al.[11]. A good account on experimental realization of the required symmetry properties for partial (full) observation of superradiance/subradiance has been given in the former reference.

A different approach to subradiance has been suggested by Cummings[5] considering the spatial distribution of the atoms in the "equivalent mode" cluster for arbitrary p as well as number of initial field modes when only one atom is initially excited. The underlying principle behin subradiance, whether it is prepared by a particular initial state or by randomly distributing atoms in the cluster is based on the principle of destructive interference. The very commonly studied case of p atoms and $n = 1$ with field initially in the vacuum state corresponds to the case a in Table. 1 with $n < p$ and $n = p$. In this category, Cummings and Dorri[5] studied the evolution of an asymmetric initial state and for instance Seke[6] examined the symmetric case. The original Jaynes-Cummings model corresponds the case $p = 1$ and arbitrary number of initial photons $r_{in} > p$ and hence $n > p$. The photon number range is therefore given by $n - p \leq r \leq n$. This is contained in case (c) in Table. 1 with $\ell = p/2$ and hence $\ell = 1/2$. Bužek[8] also studied the consequences of the spatial distribution of atoms in the spatially extended cluster (i.e. linear size of the cluster is compatible with the wavelenght of the single resonant field). He considered the case of $p = 2$ with $m = 1$ or $m = 2$ initially with an arbitrary number of photons in the initial field state. Here for the number of initial photons $2 < r_{in}$ hence $2 < n2$ and thus $n - 2 \leq r \leq n$ which can be found within case (c) in Table. 1. Since in Bužek's calculation $\ell = p/2$ and thus only $\ell = 1$ is allowed. He also examined the case where the initial field is in a coherent state again with $p = 2$. For this case each Fock component with a specific number of photons (i.e. r) can be studied independently. Nevertheless, since for each such component it is true that $n-p \leq r \leq n$ we still have the case (c) corresponding to $\ell = p/2$ yielding $\ell = 1$ for $p < n$. For those components with $n < p$ or $n = p$ case (c) is unphysical and thus we must have case a but still with $\ell = 1$.

Recently Kozierowski et al.[9] considered systems with three and more excited atoms in the initial state for arbitrary p. In such systems with $n \geq 3$ super structures modulating the zero'th order collapses and revivals appear arising from the non-equidistant eigenvalue distribution. In their exact results Kozierowski et al. considered the case with $n = 3$ with arbitrary p also using symmetric wavefuntions with $0 \leq r \leq n$. Again, this corresponds to the case (a) in Table. 1 with $\ell = p/2$.

Hence we can see that, for symmetric initial states, or for all cases when $r_{max} = n$ the cooperation number does not need to be mentioned since in this case it is directly implied that $\ell = p/2$. For $\ell = p/2$ Eq.'s (13) can be seen to reduce into a relatively more conventional form via[2]

$$v_r = \sqrt{r+1}\sqrt{(n-r)(p-n+r+1)}\,. \tag{15}$$

We can then write the Hamiltonian in Eq. (11) as

$$\hat{\mathcal{H}} = \omega(\hat{\mathcal{N}} + \frac{1}{2}) - \Delta(\hat{a}^\dagger \hat{a} + 1/2) + \gamma(\hat{a}^\dagger \hat{\mathcal{L}}_- + \hat{a}\hat{\mathcal{L}}_+) \tag{16}$$

We will now rescale the matrix coefficients so that γ is implicitly unity. The resulting matrix to be diagonalized is given by,

$$\begin{pmatrix} -\epsilon - \Delta\, m_{max} & v_{r_{min}} & 0 & \cdots & \cdots & 0 \\ v_{r_{min}} & -\epsilon - \Delta\,(m_{max}-1) & v_{r_{min}+1} & \cdots & 0\ldots & 0 \\ 0 & v_{r_{min}+1} & \cdots & \cdots & \cdots & 0 \\ \vdots & \vdots & \vdots & \vdots & -\epsilon - \Delta\,(m_{min}+1) & v_{r_{max}} \\ 0 & \cdots & \cdots & 0 & v_{r_{max}} & -\epsilon - \Delta\, m_{min} \end{pmatrix} \tag{17}$$

where $m_{max} = n - p/2 - r_{min}$ and $m_{min} = n - p/2 - r_{max}$. Denoting the eigenvalues by $\epsilon^{(s)}$ and eigenvectors by $\psi\rangle^{(s)}$ where $1 \leq s \leq n+1$ is the eigenvalue index, we have for $n = 1$ and arbitrary p,

$$\begin{aligned} \epsilon_1 &= \tfrac{1}{2}\left[(p-1)\Delta + \sqrt{\Delta^2 + 4v_0^2}\right] &\Rightarrow\quad |\psi\rangle_1 &= (\tfrac{4v_0^2}{\Delta^2+8v_0^2})^{1/2}|0\rangle + (\tfrac{\Delta^2+4v_0^2}{\Delta^2+8v_0^2})^{1/2}|1\rangle) \\ \epsilon_2 &= \tfrac{1}{2}\left[(p-1)\Delta - \sqrt{\Delta^2 + 4v_0^2}\right] &\Rightarrow\quad |\psi\rangle_2 &= (\tfrac{\Delta^2+4v_0^2}{\Delta^2+8v_0^2})^{1/2}|0\rangle - (\tfrac{4v_0^2}{\Delta^2+8v_0^2})^{1/2}|1\rangle) \end{aligned} \tag{18}$$

where $v_0 = \sqrt{(\ell+1-p/2)(\ell+p/2)}$. For $p = 1$ and n arbitrary we have the case (c) in Table. (1). The eigenvalues and eigenvectors are

$$\begin{aligned} \epsilon_1 &= \sqrt{\Delta^2/4+n} &\Rightarrow\quad |\psi\rangle_1 &= (\sqrt{1-A^2}\,|n-1\rangle + A\,|n\rangle) \\ \epsilon_2 &= -\epsilon^{(1)} &\Rightarrow\quad |\psi\rangle_2 &= (A\,|n-1\rangle - \sqrt{1-A^2}\,|n\rangle) \end{aligned} \tag{19}$$

[2]Eq. (15) is more appropriate and simple for the case $n \leq p$. For the opposite case its symmetric equivalent with $v_{\tilde{r}} = \sqrt{\tilde{r}+1}\sqrt{(p-\tilde{r})(n-p+\tilde{r}+1)}$ with $0 \leq \tilde{r} \leq p$ and $\tilde{r} = r + (p-n)$ is more appropriate to use. In this case there are $p+1$ eigenvalues as opposed to $n+1$ in the former case.

where $A = (1+\sin\alpha)/\sqrt{2}$ with $\sin\alpha = \frac{\Delta}{2}/\sqrt{\Delta^2/4+n}$. For $p=2$ and n arbitrary we can have $\ell = 1$ or $\ell = 0$. For the former $n-2 \le r \le n$ and we have the case (c) in Table. (1). Whereas for the latter (i.e. $\ell = 0$) we have only one allowed photon number $r = n-1$ and this implies full radiation trapping for the antisymmetric initial state $|\ell = 0\, m = 0\rangle$. For $\Delta \neq 0$ the eigenvalues and eigenvectors are not as simply expressed as in the $\Delta = 0$ case. The eigenvalues and eigenvectors are, for $\ell = 1$,

$$\begin{aligned}
\epsilon_1 &= u + v\,, & \Rightarrow\quad |\psi\rangle_1 &= A_0\,|n-2\rangle + A_1\,|n-1\rangle + A_2\,|n\rangle \\
\epsilon_2 &= u\cos\varphi + v\cos 2\varphi\,, & \Rightarrow\quad |\psi\rangle_1 &= B_0\,|n-2\rangle + B_1\,|n-1\rangle + B_2\,|n\rangle \\
\epsilon_1 &= u\cos 2\varphi + v\cos\varphi\,, & \Rightarrow\quad |\psi\rangle_1 &= C_0\,|n-2\rangle + C_1\,|n-1\rangle + C_2\,|n\rangle
\end{aligned} \tag{20}$$

where u and v are such that,

$$\begin{aligned}
u &= \left\{\Delta + \sqrt{\Delta^2 + (4n-2+\Delta^2)^3}\right\}^{1/3} \\
u &= \left\{\Delta - \sqrt{\Delta^2 + (4n-2+\Delta^2)^3}\right\}^{1/3}
\end{aligned} \tag{21}$$

with $\cos\varphi = -v/(2\,u)$ and

$$\begin{aligned}
A_1 &= \sqrt{\frac{\epsilon_1^2-\Delta^2}{\epsilon_1^2-\Delta^2+\epsilon_1\,(4n-2)+2\,\Delta}}\,, & A_0 &= -\frac{\sqrt{2n-2}}{\epsilon_1-\Delta}\,A_1\,, & A_2 &= -\frac{\sqrt{2n}}{\epsilon_1+\Delta}\,A_1 \\
B_1 &= \sqrt{\frac{(\epsilon_2)^2-\Delta^2}{(\epsilon_2)^2-\Delta^2+\epsilon_2\,(4n-2)+2\Delta}}\,, & B_0 &= -\frac{\sqrt{2n-2}}{\epsilon_2+\Delta}\,B_1\,, & B_2 &= -\frac{\sqrt{2n}}{\epsilon_2-\Delta}\,B_1 \\
C_1 &= -B_1\,, & C_0 &= B_0\,, & C_2 &= B_2
\end{aligned} \tag{22}$$

where $|r\rangle = |n-r\rangle_a \otimes |r\rangle_f$. For $\ell = 0$ we have a zero mode with $|\psi_0\rangle = |\ell = 0\, m = 0\rangle = \frac{1}{2}(|+\,-\rangle - |-\,+\rangle) \otimes |n-1\rangle$. For $n < p$ the photon number is bounded by $0 \le r \le 2$ and thus we have the case (a) with $\ell = p/2$. Whereas for $p < n$ the photon number range is $1 \le r \le 2$ and this again indicates that $\ell = p/2$ which corresponds to the case (c). The corresponding eigenvalues and eigenvectors are obtained from the symmetric case by interchanging n and p in Eq.'s (21) and (22).

We can also briefly mention the results for $n = 3$ and $n = 4$ for arbitrary p. The $n = 3$ case with arbitrary p was solved by Kozierowski[9] using symmetric initial states. For p and n being small and primarily less than two the eigenfrequencies are commensurate and the time dependence of the fluctuating observables are given by regular oscillations. Starting from $3 \le n$ or $3 \le p$ the eigenfrequencies become incommensurate as illustrated above. This is reflected on the time dependence of the atomic population inversion and incommensurate overtones of the eigenfrequencies manifest themselves in the appearance of superimposed modulations on the envelope function.

2.1.1. *Commensurate versus Incommensurate regimes*

If a number of frequencies $\epsilon^{(f)}$ where $f \le (n+1)$ are incommensurate then for a set of integers $s_1, s_2, \ldots, s_n$ the frequency sum

$$s_1\,\epsilon^{(1)} + s_2\,\epsilon^{(2)} + \ldots + s_f\,\epsilon^{(f)} + \ldots + s_n\,\epsilon^{(n)} = 0 \tag{23}$$

can not have a non-trivial solution except $s_1 = s_2 = \ldots = s_n = 0$. This condition can be expressed equivalently by stating that there is at least one frequency $\epsilon^{(f)}$ of which ratio to all other frequencies is an irrational number. The representation of a dynamical observable $G(t)$ in the frequency spectrum

$$\tilde{g}(\omega) = \int_{-\infty}^{\infty} dt\, e^{i\omega t}\, G(t) \tag{24}$$

can display irregular behaviour since

$$\tilde{g}(\omega) = \sum_{s_1, s_2, \ldots, s_n} \Gamma_{s_1, s_2, \ldots, s_n}\, \delta\Big(\omega - \big[s_1\, \epsilon^{(1)} + s_2\, \epsilon^{(2)} + \ldots + s_n\, \epsilon^{(n)}\big]\Big)\,. \tag{25}$$

where in each summation $-\infty < s_i < \infty$ $(i = 1, 2, \ldots, n)$. By investigating the $|\,\tilde{g}(\omega)\,|$ one can determine the nature (i.e. periodic, quasi-periodic, irregular) of its phase space *attractor*. In Fig. (2) below the ratio of the smaller eigenfrequencies to the largest one is plotted for fixed values of n ranging from 3 to 100 with respect to increasing values of p for $n \leq p$ and for $\Delta = 0$. The commensurability of the frequencies is not guaranteed at all parameter values although in the limit $1 \ll p/n$ the ratio of the frequencies approach to rational numbers. The opposite case with $p/n \ll 1$ is symmetric since it can always be mapped into an equivalent $1 \ll p/n$ case by a simple shift in the photon number (see the footnote on page 9). Therefore, the strongest incommensuration in Fig. (2) corresponds to $n = p$. Continuous detuning from resonance is a controlled nevertheless nontrivial way of

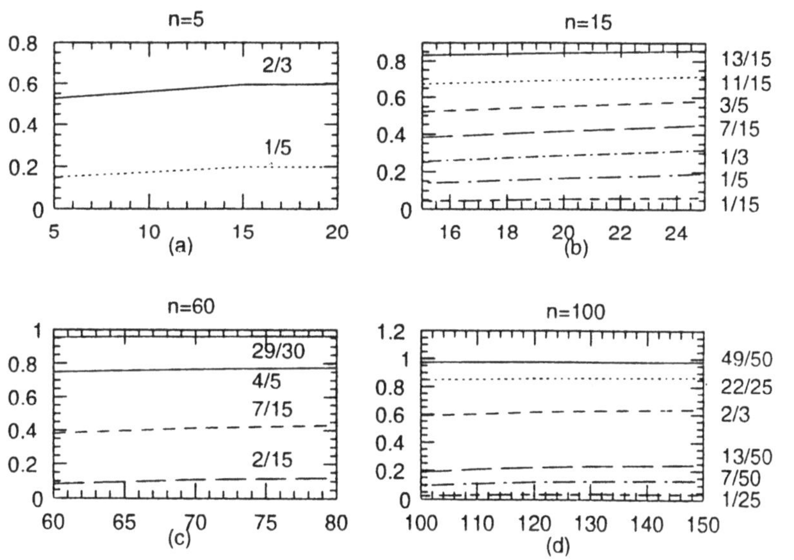

Figure 2. The ratio of the smaller eigenvalues to the largest one (i.e. $\epsilon^{(j)}_{n,p}/\epsilon^{max}_{n,p}$) as $n \leq p$ varies. a) $n = 5$ and from bottom to top $j = 5$ and $j = 3$; b) $n = 15$ and from bottom to top $j = 15, 13, 11, 9, 7, 5, 3$; c) $n = 60$ and from bottom to top $j = 53, 33, 13, 3$; d) $n = 100$ and from bottom to top $j = 97, 87, 75, 35, 13, 3$.

observing incommensuration. Non-zero detuning slightly softens the eigenfrequencies and increases their non-linear dependence on n and p. Strong nonlinearity is observed when

n and p are close to each other. The increase in the allowed range for photon number $r_{max} - r_{min} = 2\ell + 1$ also increases the effect.

Incommensurate frequencies are easily established for a number of parameter values. For instance for $n = 3$ and $n = 4$ at $\Delta = 0$ we list a few cases for which one obtains irrational values for the frequency ratios out of many other possible ones[3]

n	p	$\epsilon^{(1)}/\epsilon^{(3)}$
3	3	$\sqrt{\frac{10-\sqrt{73}}{10+\sqrt{73}}}$
3	4	$\sqrt{\frac{15-\sqrt{153}}{15+\sqrt{153}}}$
4	4	$\sqrt{\frac{25-3\sqrt{33}}{25+3\sqrt{33}}}$
4	5	$\sqrt{\frac{33-\sqrt{553}}{33+\sqrt{553}}}$

n	p	$\epsilon^{(1)}/\epsilon^{(3)}$
4	6	$\sqrt{\frac{30-4\sqrt{89}}{30+4\sqrt{89}}}$
4	7	$\sqrt{\frac{53-\sqrt{1201}}{53+\sqrt{1201}}}$
4	8	$\sqrt{\frac{65-3\sqrt{177}}{65+3\sqrt{177}}}$
4	9	$\sqrt{\frac{75-3\sqrt{233}}{75+3\sqrt{233}}}$

Table. 2

Perhaps a more illustrative quantity to examine is the time evolution of the various moments of the photon number $m^{(k)}(t) = \langle (\hat{a}^\dagger \hat{a})^k \rangle$ as the dynamical observable replacing the role of $\mathcal{G}(t)$ in Eq. (24).

Here there is a competition between various time scales involved. If the eigenfrequencies are labeled as $\epsilon^{n,p}_{[n/2+1]} \le \epsilon^{n,p}_{[n/2]} \le \epsilon^{n,p}_{[n/2-1]} \cdots \le \epsilon^{n,p}_2 \le \epsilon^{n,p}_1$ with $[x]$ representing the integer part of "x", the smallest finite period is given by

$$\begin{aligned} T_1^{n,p} &= 2\pi/(\,\epsilon_1^{n,p} - \epsilon^{n,p}_{[n/2+1]}\,)\,, \qquad \text{if} \quad \text{n is odd} \\ T_1^{n,p} &= 2\pi/(\,\epsilon_1^{n,p} - \epsilon^{n,p}_{n/2}\,)\,, \qquad \text{if} \quad \text{n is even} \end{aligned} \tag{26}$$

with the main difference arising from the fact that for $n = even$ one eigenvalue is always zero yielding an infinite period. On the other the difference between the two smallest eigenfrequencies determine the period of revivals in the atomic population inversion and corresponds to the smallest beat frequency. The revival period is given by,

$$\begin{aligned} T^{(2)}_{n,p} &= 2\pi/(\epsilon^{([n/2])}_{n,p} - \epsilon^{([n/2+1])}_{n,p}\,)\,, \qquad \text{if} \quad \text{n is odd} \\ T^{(2)}_{n,p} &= 2\pi/(\,\epsilon^{(n/2)}_{n,p} - \epsilon^{n/2-1)}_{n,p}\,)\,, \qquad \text{if} \quad \text{n is even} \end{aligned} \tag{27}$$

because of the same reason as above. Comparison of the time scales in (26) and (27) is crucial to understand the dynamical time correlations in $m^{(k)}(t)$. In Fig. (3) below, we examine the ratio $T_1^{n,n}/T_2^{n,n}$ for various n values. As n increases $T_1^{n,n}/T_2^{n,n}$ tends to zero. In the case of arbitrary $n < p$ the same qualitative behaviour is observed.

[3]For the $n = 3$ and $n = 4$ cases the frequency ratios can be represented as $\sqrt{\frac{a+b\sqrt{c}}{a-b\sqrt{c}}}$ where a, b, c are positive integers. It can be shown that if c does not divide a^2 and b^2 the frequency ratio is an irrational number. Such condition is satisfied when c itself or its dividents are prime numbers.

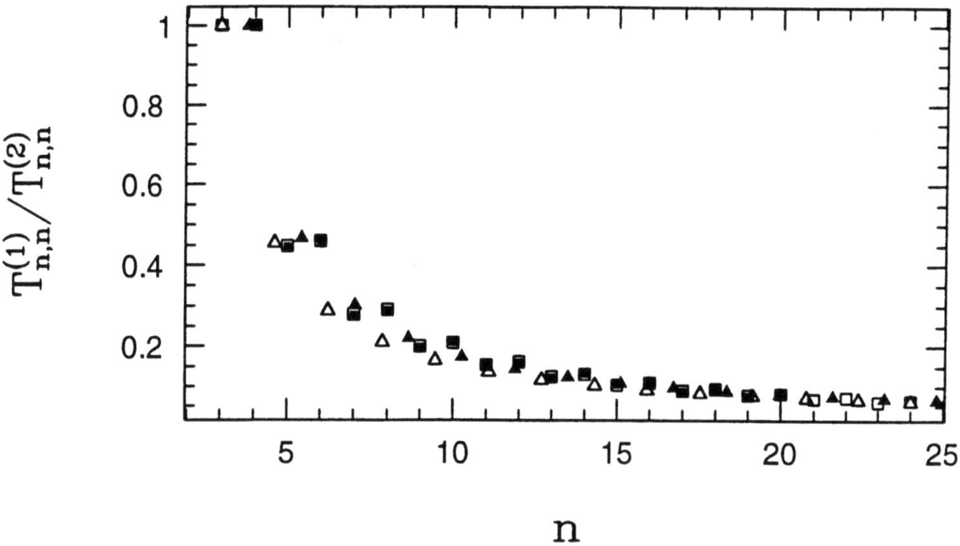

Figure 3. Here the open and solid triangles indicate $n = p = odd$, and $n = p = even$ respectively for $\Delta = 0$. The open and solid rectangles indicate the same for $\Delta = 0.3$.

We can now examine the moments $m^{(k)}(t)$ and their correlations. The time correlations within a single revival (i.e. $T^{(1)}_{n,p} \leq t \leq T^{(2)}_{n,p}$) and those spanning a time interval large enough to intercorrelate more than one revivals (i.e. $T^{(1)}_{n,p} \leq T^{(2)}_{n,p} \leq t$ exhibit qualitatively different behaviour. In the Fig.'s (4) and (5) below $m^{(k)}(t)$ for $k = 1, 2, 4, 6$ are plotted. Fig.'s (4.a,b,c,d) and Fig.'s (4.e,f,g,h) represent $n = p = 9$ and $n = 6$, $p = 9$ respectively. For the former the time dependence of all moments are irregular. The period of the envelope for the latter can be estimated using Kozierowski's perturbative result as[9],

$$T_R = \frac{8\pi}{15}(p - n/2 + 0.5)^{3/2} \simeq 28 \tag{28}$$

which are within ten percent of the numerical calculations (for clarity two full periods are shown in the figure). In Fig.'s (5.a,b,c,d) and Fig.'s (5.e,f,g,h) below $n = p = 40$ and $n = 20$, $p = 40$ are plotted. The same trend continues here and the time dependence of the $n = p$ case is erratic whereas collective collapses and revivals are observed for the $n = p = 40$ case. The revival period can be estimated from Eq. (28) above as $T_R \simeq 156$ which again underestimates the revival period by about ten percent. An interesting observation here is that during the short collapse period of the mean photon number there are strong fluctuations in the number of photons as visible in the higher moments $m^{(k)}(t)$ with $k > 1$ (also note the scale change as k increases). In Fig.'s (6) we numerically confirm for $\Delta = 0.3$ the effect of non-zero detuning driving the system into a strong incommensuration. Fig. (6.a,b,c,d) represents $n = p = 40$, whereas Fig. (6.e,f,g,h) represents $n = 20$, $p = 40$.

3. Physical limitations in the large n limit

The exact solution of the eigen system in Eq.'s (16) and (17) has not been found in a closed analytic form. The eigenvalues of the full Hamiltonian are always symmetrically

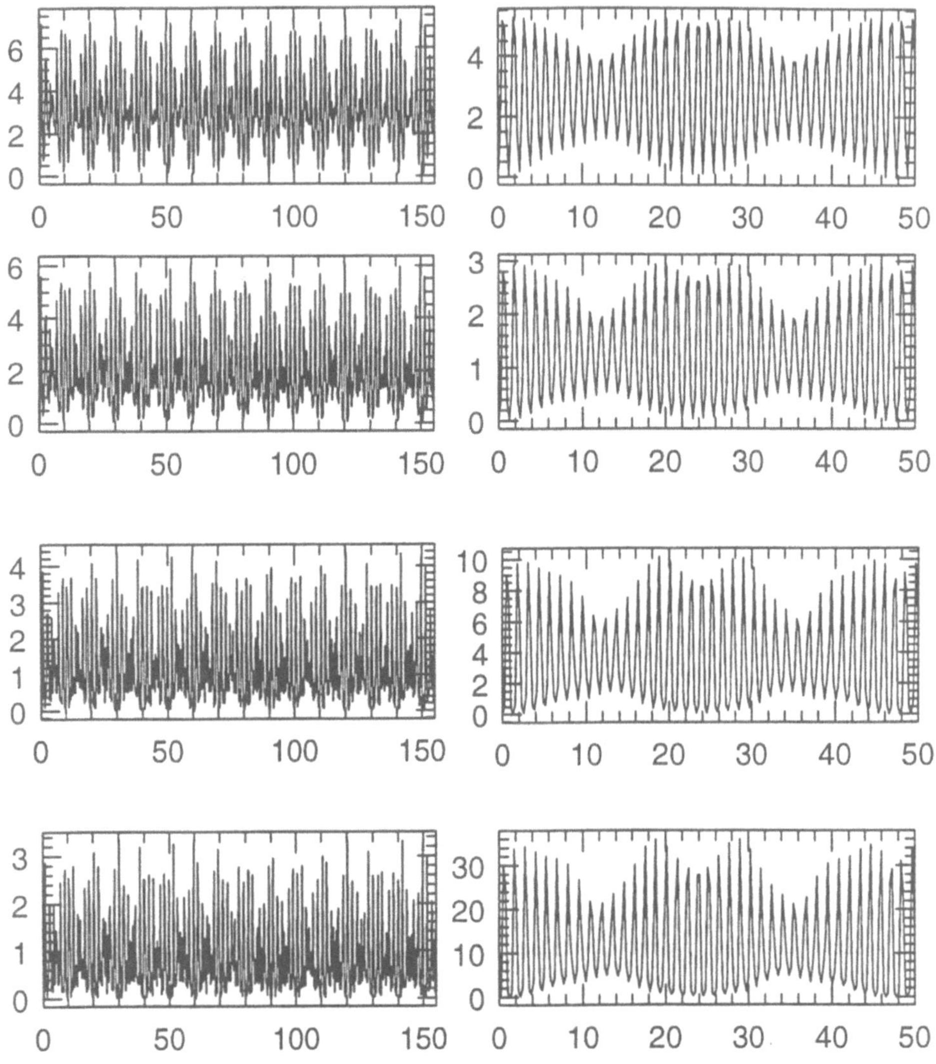

Figure 4. From the top left to the bottom left the plots are (a),(b),(c) and (d) representing $k = 1, 2, 4, 6$ respectively for $n = p = 9$ and for $\Delta = 0$. The vertical axes are scaled for $k = 2$ with $x10$, for $k = 4$ with $x10^3$ and for $k = 6$ with $x10^5$. On the right from top to bottom the plots (c), (d), (e) and (f) represent the same order in k for $n = 6$ $p = 9$ and for $\Delta = 0$. The vertical axes are scaled for $k = 2$ with $x10$, for $k = 4$ with $x10^2$ and for $k = 6$ with $x10^3$.

distributed around the central value ωn as $E_j^{(n)} = \omega n \pm \epsilon_n^{(j)}$. This is a strong signal that one should examine the large n and/or p limit with extra care. The behaviour of the $\epsilon_n^{(j)}$'s as n and/or p increase can be examined using the recursive properties of the determinant. It has been shown previously[12] that the properties of eigensolutions of

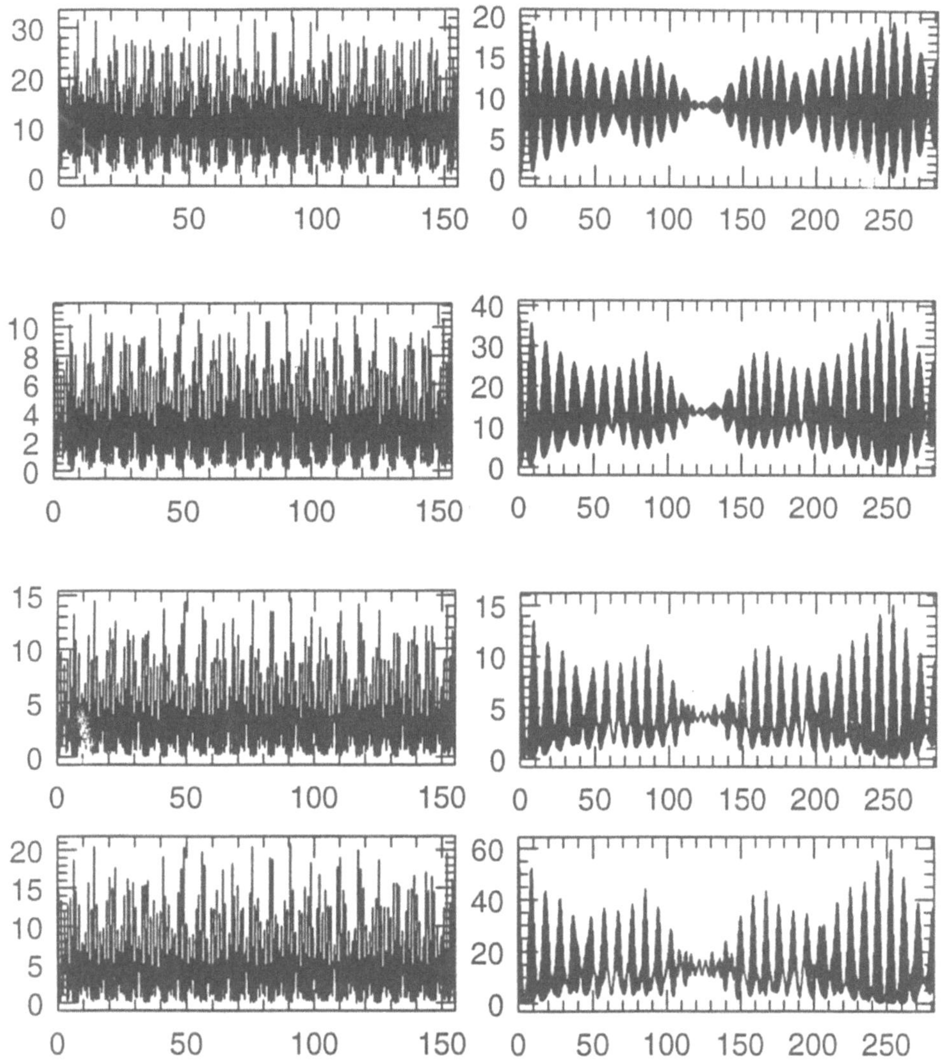

Figure 5. From the left top to the left bottom the plots are (a),(b),(c) and (d) representing $k = 1, 2, 4, 6$ respectively for $n = p = 40$ and for $\Delta = 0$. The vertical axes are scaled for $k = 2$ with $x10^2$, for $k = 4$ with $x10^5$ and for $k = 6$ with $x10^8$. On the right from top to bottom the plots (c), (d), (e) and (f) representing the same order in k for $n = 20$ $p = 40$ and for $\Delta = 0$. The vertical axes are scaled for $k = 2$ with $x10$, for $k = 4$ with $x10^4$ and for $k = 6$ with $x10^6$.

such simple Hamiltonians as (16) can be studied by orthogonal polinomials[13]. The model predicts that $\epsilon_n^{(j)}$ grow much faster than $\omega\, n$ eventually leading to negative energies. This is an artifact of the non-unitarity of the model in (16) arising from the assumption that the counterrotating terms are neligible. For instance the largest eigenvalue can be estimated

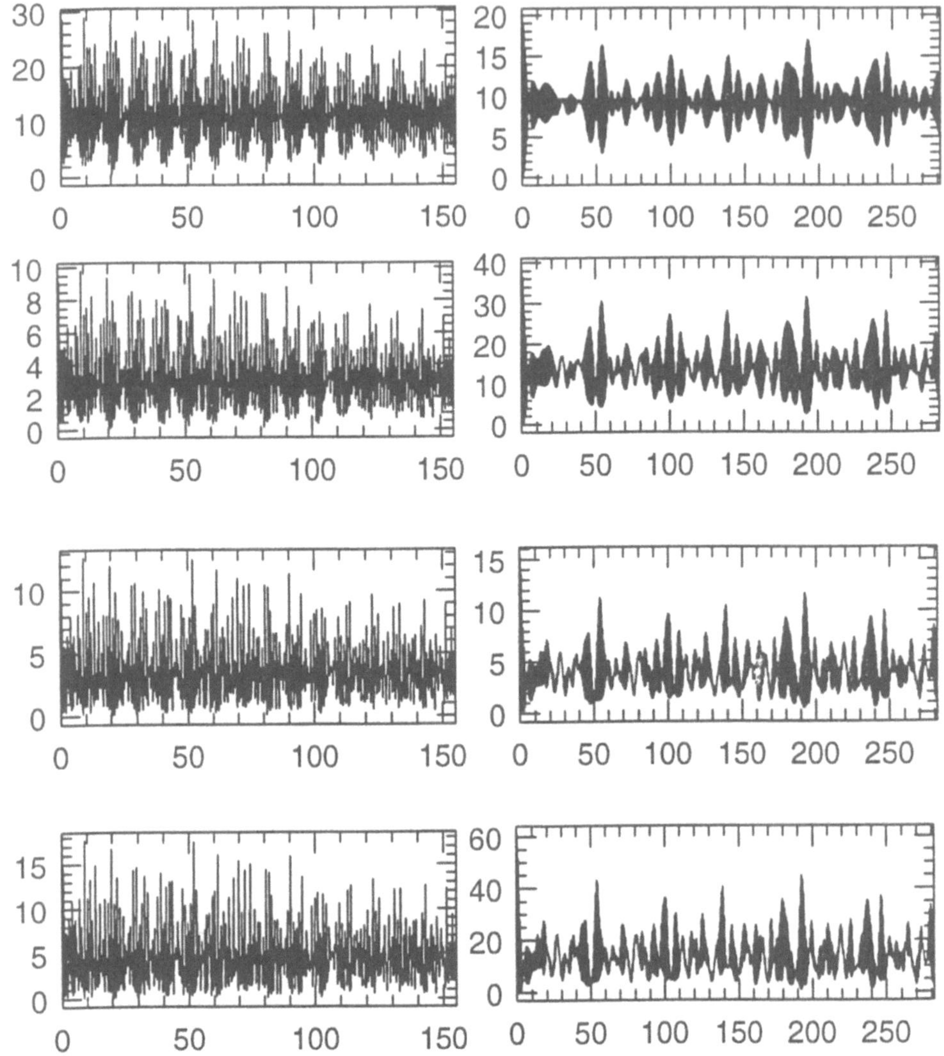

Figure 6. From the left top to the left bottom the plots are (a),(b),(c) and (d) representing $k = 1, 2, 4, 6$ respectively for $n = p = 40$ and for $\Delta = 0.3$. The vertical axes are scaled for $k = 2$ with $x10^2$, for $k = 4$ with $x10^5$ and for $k = 6$ with $x10^8$. On the right from top to bottom the plots (c), (d), (e) and (f) representing the same order in k for $n = 20$ $p = 40$ and for $\Delta = 0.3$. The vertical axes are scaled for $k = 2$ with $x10$, for $k = 4$ with $x10^4$ and for $k = 6$ with $x10^6$.

by the inequality[12, 13, 14],

$$\frac{1}{n}\sum_{r=0}^{n} v_r^2 \le \frac{1}{n}\sum_{r=0}^{n} (\epsilon_r^n)^2 \le \left([\epsilon_j^{(n)}]_{max}\right)^2 \tag{29}$$

the left hand side of the inequality can be computed from Eq. (15) which gives an approximate critical size for the cluster below which the rotating wave approximation can be used. Using (29), this critical size p_c for the Hamiltonian in (16) is given by $p_c = 3\,\omega - 2$ such that $p \ll p_c$ is required to secure the validity of the presented solution based on the rotating wave approximation. The formal solution including the counter rotating terms yields a unitary *S-matrix* but the solution itself can not be given in a simple form[15].

The *counter rotating* terms in the Hamiltonian are represented by $\hat{\mathcal{H}}' = \gamma\,(\hat{a}^\dagger\,\hat{\mathcal{L}}_+ + \hat{a}\,\hat{\mathcal{L}}_-)$. The first order perturbative correction to the energy eigenvalues vanish. The second order correction can be calculated from[4]

$$\epsilon_2^{i,n} = \sum_{i',n'} \frac{|\langle\, i'n'|\mathcal{H}'|i\,n\rangle|^2}{\epsilon_0^{i,n} - \epsilon_0^{i',n'}} \qquad i \le (n+1)\;;\;\; i' \le (n'+1) \tag{30}$$

where only $n' = n \pm 2$ contribute. Here it is important to know the full interaction strenght since the eigenvalues no more globally scale with the first power of the coupling constant γ. Eq. (30) is shown below for $n = p = 20$ and for $\Delta = 0$ and for $\Delta = 0.3$ respectively. In the next table $n = p = 10$ for the same Δ values are given (note that the second order corrections are given in units of γ).

The second order correction $\epsilon_2^{j,n}$ is consistently opposite in sign to the zero'th order eigenvalue $\epsilon_0^{j,n}$. The importance of this result is in the fact that corrections tend to confine the spreading eigenvalues for increasing n and/or p. The net effect is to correct the superfluous instability beyond the critical region $p \sim p_c$ by pushing it to larger values. Additional details about the instability will be presented elsewhere[14].

[4]Since $[\hat{\mathcal{H}}', \hat{\mathcal{N}}] \neq 0$ different n's are mixed. In the Eq. (29) $\epsilon_j^{i,n}$ indicates the j'th order correction to the i'th eigenvalue of the *unperturbed* Hamiltonian (16) with a fixed n.

j	$\epsilon_0^{j,n}$	$\epsilon_2^{j,n}/\gamma$
1	74.60	-268.59
2	65.52	-226.38
3	56.76	-187.74
4	48.34	-152.61
5	40.26	-120.93
6	32.56	-92.64
7	25.24	-67.71
8	18.34	-46.12
9	11.86	-27.85
10	5.79	-12.77
11	0.0	0.0
12	-74.60	268.59
13	-65.52	226.38
14	-56.76	187.74
15	-48.34	152.61
16	-40.26	120.93
17	-32.56	92.64
18	-25.24	67.71
19	-18.34	46.12
20	-11.86	27.85
21	-5.79	12.77

$\Delta = 0$

j	$\epsilon_0^{j,n}$	$\epsilon_2^{j,n}/\gamma$
1	78.66	-243.07
2	69.40	-200.56
3	60.44	-161.82
4	51.79	-126.95
5	43.44	-96.12
6	35.43	-69.35
7	27.79	-46.46
8	20.57	-27.12
9	13.81	-10.85
10	7.52	-3.13
11	1.62	-1.50
12	-4.12	1.22
13	-10.01	42.05
14	-16.21	63.38
15	-22.78	88.69
16	-29.75	116.91
17	-37.14	147.32
18	-44.95	180.06
19	-53.15	215.72
20	-61.72	254.76
21	-70.63	297.35

$\Delta = 0.3$

Table 3. Corrections to the eigenenergies for $n = p = 20$.

j	$\epsilon_0^{j,n}$	$\epsilon_2^{j,n}\gamma$
1	39.72	-71.41
2	31.43	-53.48
3	23.35	-37.82
4	15.46	-24.04
5	7.69	-11.65
6	0.0	0.0
7	-7.69	11.65
8	-15.46	24.04
9	-23.35	37.82
10	-31.43	53.48
11	-39.72	71.41

$\Delta = 0$

j	$\epsilon_0^{j,n}$	$\epsilon_2^{j,n}/\gamma$
1	41.59	-64.53
2	33.12	-45.58
3	24.81	-29.55
4	16.64	-15.97
5	8.64	-1.53
6	0.84	-0.82
7	-6.77	14.19
8	-14.30	28.69
9	-21.91	44.03
10	-29.77	60.10
11	-37.90	77.78

$\Delta = 0.3$

Table 3. Corrections to the eigenenergies for $n = p = 10$

Acknowledgements

The author is gratefull to Dr. A. Miranowicz with whom most parts of this work were discussed during his visit at Bilkent.

References

1. The recent book by L. Mandel and E. Wolf is an excellent source for the treatment of spontaneous processes in two level systems. See Leonard Mandel and Emil Wolf, *Optical Coherence and Quantum Optics* (Cambridge University Press, 1995).
2. L. Allen and J.H. Eberly, *Optical Resonance and Two-Level Atoms*, Dover Publications, (New York 1975).
3. R.H. Dicke, Phys. Rev. **93**, 99 (1954).
4. Nicholas E. Rehler and Joseph H. Eberly, Phys. Rev. **A 3**, 1735 (1970); G.S. Agarwal, Phys. Rev **A 2**, 2038 (1970); R. H. Lehmberg, Phy. Rev. **A 2**, 883 (1970); ibid , 889 (1970).
5. F.W. Cummings and A. Dorri, Phys. Rev. **A 28**, 2282 (1983); F.W. Cummings, Phys. Rev.Lett. **54**, 2329 (1985).
6. J. Seke, Phys. Rev. **A 33**, 739 (1986).
7. W. Milonni and P. Knight, Phys. Rev. **A 10**, 1096 (1974).
8. V. Bužek, Z. Phys. **D 17**, 91 (1990).
9. M. Kozierowski, A.A. Mamedov and S.M. Chumakov, Phys. Rev. **A 42**, 1762 (1990); M. Kozierowski and S.M. Chumakov, Phys. Rev. **A 52**, 4194 (1995).
10. N. Skribanowitz, I.P. Herman, J.C. McGillivray and M.S. Feld, Phys. Rev. Lett. **30**, 309 (1973).

11. A. Crubellier, S. Liberman, D. Pavolini and P. Pillet, J. Phys. **B 18**, 3811 (1985); D. Pavolini, A. Crubellier, P. Pillet, L. Cabaret and S. Liberman, Phys. Rev. Lett. **54**, 1917 (1985).
12. Y. Orlov and V.V. Vedenyapin, Mod. Phys. Lett. **B 9**, 291 (1995).
13. Gabor Szegö, *Orthogonal Polinomials*, (American Mathematical Society, (1939).
14. T. Hakioğlu and A. Miranowicz, unpublished (1996).
15. Ricardo Ramirez and Miguel Orszag, J. Math. Phys. **22**, 1306 (1980).

QUANTUM ESTIMATION THEORY AND OPTICAL DETECTION

G. M. D'ARIANO
Dipartimento di Fisica "A. Volta",
via Bassi 6, I-27100 Pavia, Italy

1. Introduction

In quantum mechanics we call "observable" any physical quantity that can be represented by numbers. An observable is associated in a one-to-one way with a selfadjoint operator $\hat{X}$ acting on the Hilbert space $\mathcal{H}_S$ of the quantum system S, and the spectrum of $\hat{X}$ represents the set of all possible readings from the measurement. Let us consider, for example, an observable with spectrum equal to whole real line $\mathbf{R}$, and with spectral decomposition

$$\hat{X} = \int x\, \mathrm{d}\hat{E}(x) . \tag{1}$$

If the operator $\hat{X}$ is non degenerate, the spectral measure $\mathrm{d}\hat{E}(x)$ is simply the projector on the eigenvector $|x\rangle$ of $\hat{X}$, namely

$$\mathrm{d}\hat{E}(x) = \mathrm{d}x|x\rangle\langle x| , \qquad \hat{X}|x\rangle = x|x\rangle , \qquad \langle x|x'\rangle = \delta(x - x') . \tag{2}$$

Eqs. (1) and (2) supply the physical observable with the minimal mathematical outfit that is needed for stating the basic rule of quantum mechanics—the Born's rule—which at the same time provides the probabilistic interpretation of physical "state". The Born's rule can be enunciated as follows: "If we know in advance that the system is in a (pure) state described by the vector $|\psi\rangle \in \mathcal{H}_S$, we can predict *a priori* the probability $\mathrm{d}P(x)$ that the experimental reading will fall in the range $[x, x + \mathrm{d}x)$ by means of the formula

$$\mathrm{d}P(x) = \langle\psi|\mathrm{d}\hat{E}(x)|\psi\rangle . \tag{3}$$

The *Born's statistical formula* (3) can be further generalized in two ways: i) considering a prior undetermined "mixed" state described by a density operator $\hat{\rho}$; ii) embracing also the description of joint measurements of compatible observables. Compatible observables correspond to commuting operators $\hat{X}_i$ $(i = 1, \ldots, n)$ that share an orthogonal spectral decomposition $\mathrm{d}\hat{E}(\mathbf{x}) \equiv |\mathbf{x}\rangle\langle\mathbf{x}|\mathrm{d}\mathbf{x}$ as follows

$$\hat{X}_i = \int x_i\, \mathrm{d}\hat{E}(\mathbf{x}) , \tag{4}$$

where $\mathbf{x} = (x_1, \ldots, x_n)$ denotes the vector of simultaneous eigenvalues x_i of $\hat{X}_i$ with common eigenvector $|\mathbf{x}\rangle$. Including both generalizations, the statistical formula (3) now

T. Hakioğlu and A. S. Shumovsky (eds.), Quantum Optics and the Spectroscopy of Solids, 139–174.

reads

$$dP(\mathbf{x}) = \langle \mathbf{x}|\hat{\rho}|\mathbf{x}\rangle \equiv \mathrm{Tr}[\hat{\rho}\, d\hat{E}(\mathbf{x})] \,. \tag{5}$$

The Born's rule is very basic: it provides only the interpretation of "observables" and "states" in quantum mechanics. It assumes that one knows in advance what a measuring instrument is and which observables are measured. However, despite any experimental evidence, this assumption cannot be granted from start, because the measuring instrument is a special physical system and, as such, it is itself submitted to the laws of quantum mechanics. In most cases the measuring apparatus is a very complicated system, and some interpretation is already needed to understand what it is and how it works. The Born's rule makes no attempt to provide answers to more "operational" issues as: 1) Given a physical parameter—on the basis of its classical definition, or of the procedure for measuring it—which selfadjoint operator describes the measurement? 2) How to describe the measurement of a physical quantity that apparently does not match any selfadjoint operator? [this is the case of the phase of the electromagnetic field]; 3) How to describe joint measurements of non compatible observables? 4) How to describe instrumental precision/resolution? 5) In which way the state of the system changes after the measurement?

The above issues urge a further generalization of the Born's rule (5) in a way that can be easily recognized at the mathematical level. If a quantum mechanical instrument is to provide information about a physical system S, the probability $dP(\mathbf{x})$ must be governed only by the state of the system, which is represented by a density operator $\hat{\rho}$. However complex the system-apparatus interaction is, quantum mechanics must provide a prevision of the result of the measurement in terms of operators acting on the Hilbert space $\mathcal{H}_S$ of the system S only. Depending on the measurement result $\mathbf{x}$, an operator $d\hat{\Pi}(\mathbf{x})$ will furnish the required probability $dP(\mathbf{x})$ through a rule of the general form

$$dP(\mathbf{x}) = \mathrm{Tr}[\hat{\rho}\, d\hat{\Pi}(\mathbf{x})] \,. \tag{6}$$

In order to have $dP(\mathbf{x})$ as a genuine probability, the operators $d\hat{\Pi}(\mathbf{x})$ in Eq. (6) must be nonnegative (hence selfadjoint)

$$d\hat{\Pi}(\mathbf{x}) \geq 0 \,, \tag{7}$$

as a consequence of positivity of density operators $\hat{\rho}$. Normalization of probability $dP(\mathbf{x})$ is guaranteed by the completeness relation

$$\int d\hat{\Pi}(\mathbf{x}) = \hat{1} \,. \tag{8}$$

The trace rule (6) is intimately connected with the probabilistic interpretation of physical states and their description in terms of density operators. In fact, the linear functional "Tr" guarantees propagation of convex linear combinations from density operators toward probabilities. In mathematical terms, the set of operators $d\hat{\Pi}(\mathbf{x})$ form a mapping that is a *positive operator measure*—more precisely, a *probability-operator measure* (POM)—on $\mathbf{R}^n$. Generally speaking, if $\Delta, \Delta_i \subset \mathbf{R}^n$ denote possible experimental "events", the following map

$$\hat{\Pi}(\Delta) = \int_\Delta d\hat{\Pi}(\mathbf{x}) \tag{9}$$

satisfies the abstract axioms of POM

$$\begin{aligned} \hat{\Pi}(\emptyset) &= 0\,, \qquad \hat{\Pi}(\Delta) \geq 0\,, \qquad \hat{\Pi}(\mathbf{R}^n) = \hat{1}\,, \\ \hat{\Pi}(\cup_i \Delta_i) &= \sum_i \hat{\Pi}(\Delta_i) \qquad \text{for } \cap_i \Delta_i = \emptyset\,. \end{aligned} \tag{10}$$

Axioms (10) can be stated more generally for a probability space Ω in place of $\mathbf{R}^n$, with Ω playing the role of the spectrum of a set of commuting selfadjoint operators in the old Born's rule (5). For simplicity of notation in the following I will consider the case $\Omega \equiv \mathbf{R}^n$, whereas the integration set will be not explicitly written when integrals are extended to the whole space, as in Eq. (8); it is implicit that integrals must be replaced by sums whenever Ω is a discrete set.

Eq. (5) is only a particular case of Eq. (6) with $\mathrm{d}\hat{\Pi}(\mathbf{x}) \equiv \mathrm{d}\hat{E}(\mathbf{x})$ *orthogonal* POM. In the following we will be interested mostly in *nonorthogonal* POM's $\mathrm{d}\hat{\Pi}(\mathbf{x})\mathrm{d}\hat{\Pi}(\mathbf{x}') \neq 0$ for $\mathbf{x} \neq \mathbf{x}'$. How nonorthogonal POM's enters the quantum description of a physical system? As we will see shortly, this happens when a part P of the apparatus—so called "probe"or "ancilla"—itself enters the quantum description of the measurement by its own Hilbert space $\mathcal{H}_P$. Then, if one considers the customary measurement of commuting observables $\hat{X}_i$ now acting on the extended Hilbert space $\mathcal{H}_S \otimes \mathcal{H}_P$ in the uncorrelated joint state $\hat{\rho}_S \otimes \hat{\rho}_P$, the Born's rule reads

$$\mathrm{d}P(\mathbf{x}) = \mathrm{Tr}_{S+P}[\hat{\rho}_S \otimes \hat{\rho}_P |\mathbf{x}\rangle\langle\mathbf{x}|]\mathrm{d}\mathbf{x}\,. \tag{11}$$

The trace in (11) can be evaluated in two successive steps as follows

$$\mathrm{d}P(\mathbf{x}) = \mathrm{Tr}_S\{\hat{\rho}_S \mathrm{Tr}_P[\hat{\rho}_P |\mathbf{x}\rangle\langle\mathbf{x}|]\}\mathrm{d}\mathbf{x}\,. \tag{12}$$

From the point of view of an observer who ignores (deliberately or not) the apparatus P, the Born's rule (12) has to involve operators on the Hilbert space $\mathcal{H}_S$ of the system only, and hence it is written in the form (6) as follows

$$\mathrm{d}P(\mathbf{x}) = \mathrm{Tr}_S[\hat{\rho}_S \mathrm{d}\hat{\Pi}(\mathbf{x})]\,. \tag{13}$$

Comparing Eq. (12) with Eq. (11) leads to the following POM

$$\mathrm{d}\hat{\Pi}(\mathbf{x}) = \mathrm{Tr}_P[\hat{\rho}_P |\mathbf{x}\rangle\langle\mathbf{x}|]\,. \tag{14}$$

The operator $\mathrm{d}\hat{\Pi}(\mathbf{x})$ in Eq. (14) is the partial trace Tr_P over $\mathcal{H}_P$ of an operator acting on $\mathcal{H}_S \otimes \mathcal{H}_P$, an hence it is an operator acting on $\mathcal{H}_S$ only. By definition, $\mathrm{d}\hat{\Pi}(\mathbf{x})$ in Eq. (14) is positive and normalized to identity, i.e. it satisfies the axioms of a POM. It is also clear that for a given probe state $\hat{\rho}_P$ the POM $\mathrm{d}\hat{\Pi}(\mathbf{x})$ in Eq. (14) is generally not orthogonal.

This is the way in which nonorthogonal POM's enter the quantum mechanical description of a measurement, namely through the measuring apparatus. The POM $\mathrm{d}\hat{\Pi}(\mathbf{x})$ depends on the considered experimental setup: for a fixed state $\hat{\rho}_S$ of the system one can have different probability distributions $\mathrm{d}P(\mathbf{x})$ by changing the detector and/or on the detector preparation $\hat{\rho}_P$. As we will see in the following, the correspondence between detectors and POM's is not one-to-one, namely there are many detectors described by the same POM. The notion of POM provides a new concept of physical observable that is more "operational" than the original one, because it is based on the definition of the procedure for performing the measurement. We will examine POM's further in the following sections.

We are now in position to understand what is the meaning of the title of these lectures: "Quantum estimation theory" [1]. Quantum estimation theory analyzes POM's at a purely abstract level. With the purpose of seeking the best strategy for estimating one or more parameters of the system in a fixed state, the theory looks for the pertaining class of POM's, and then seeks the POM that is optimal according to some pre-chosen goodness criterion [for example: maximum likelihood, minimum r.m.s. noise, etc.] In this way the theory allows one to find the best or "ideal" detector for such measurement. Quantum optics is an ideal lab for testing the theory of quantum measurements: in these lectures we will examine some examples of application in this field. In the tool-box of quantum optics we can find simple yet concrete devices for measuring a variety of observables of the electromagnetic field: the homodyne detector, which measures any linear combination of a couple of canonically conjugated observables of the field—the socalled *quadratures*; the heterodyne detector, which jointly measures a couple of conjugated (hence non compatible) quadratures; finally, high-sensitive interferometry, which poses the problem of measuring the phase of the field, a quantity with well defined physical meaning, however with no corresponding selfadjoint operator.

After giving an introductory classification of different types of POM's, in Sect. 2, we will analyze the Naimark's theorem, which assures that every POM can be obtained from conventional observables that involve the measuring apparatus itself. Applications to quantum optics are analyzed in details in Sect. 3, with special emphasis on the heterodyne detector, which achieves the joint measurement of non-commuting observables. Joint measurements are then analyzed in Sect. 4, where general measurements in the phase space are studied, including the measurement of the phase of the field. Quantum estimation theory is reviewed in Sect. 5: here, as a relevant application, a long subsection is devoted to the method for finding the ideal measurement of the phase. The last section 6 analyzes the notion of "instrument", which is more powerful that the concept of POM, as it describes also the back-action on the system after the measurement: the so called "state reduction". Here we will analyze in detail the general scheme for indirect measurements, with two examples—the von Neumann and the Arthurs-Kelly measurements.

2. Probability operator measures (POM)

In this section I briefly analyze the different classes of POM's. The following classification is only for didactic reasons, with the purpose of understanding step by step the new conceptual issues in the generalization from orthogonal spectral resolutions to noncommuting POM's.

2.1. ORTHOGONAL POM'S

Within our general framework, conventional measurements correspond to orthogonal projection-valued measures

$$d\hat{\Pi}(\mathbf{x})d\hat{\Pi}(\mathbf{y}) = d\hat{\Pi}(\mathbf{x})\delta(\mathbf{x}-\mathbf{y})d\mathbf{y}\ , \tag{15}$$

with $\int d\hat{\Pi}(\mathbf{x}) = \hat{1}$. Ideal perfectly *resolved* measurements are non degenerate, namely $d\hat{\Pi}(\mathbf{x}) = d\mathbf{x}|\mathbf{x}\rangle\langle\mathbf{x}|$ is a one-dimensional projector on the Hilbert space. On the other hand, non-ideal *unresolved* measurements carry some degeneration: this is the case, for example, of a set of orthogonal operators $\hat{\Pi}_n \equiv \hat{\Pi}(\Delta_n)$ defined as in Eq. (9), with the subsets Δ_n exhaustive and disjoint.

Example
An obvious example of orthogonal POM is the spectral resolution of a selfadjoint operator $\hat{X}$ with spectrum $\mathbf{R}$. One has

$$\mathrm{d}\hat{\Pi}(x) = \delta(x - \hat{X})dx \equiv |x\rangle\langle x|\mathrm{d}x \ . \tag{16}$$

As a concrete example, in Sect. 3.2 we will analyze the homodyne detector, with $\hat{X}$ representing a quadrature of the electromagnetic field.

2.2. COMMUTING POM'S

A trivial generalization to nonorthogonal POM's is the case of commuting measures, namely

$$[\mathrm{d}\hat{\Pi}(\mathbf{x}), \mathrm{d}\hat{\Pi}(\mathbf{x}')] = 0 \ . \tag{17}$$

As the operators $\mathrm{d}\hat{\Pi}(\mathbf{x})$ are selfadjoint and commute at different $\mathbf{x}$, an orthogonal basis $|\mathbf{z}\rangle\langle\mathbf{z}|$ exists that diagonalizes all of them simultaneously for all $\mathbf{x}$ [here I distinguish between different sets of states only by changing their label, as, for example, $|\mathbf{x}\rangle$ and $|\mathbf{z}\rangle$]. Hence, $\mathrm{d}\hat{\Pi}(\mathbf{x})$ must be of the form

$$\mathrm{d}\hat{\Pi}(\mathbf{x}) = \int \mathrm{d}\mathbf{z}\, m(\mathbf{x}|\mathbf{z})|\mathbf{z}\rangle\langle\mathbf{z}| \ . \tag{18}$$

From the POM axioms it follows that the coefficients $m(\mathbf{x}|\mathbf{z})$ in Eq. (18) are probability densities, namely

$$m(\mathbf{x}|\mathbf{z}) \geq 0 \ , \qquad \int \mathrm{d}\mathbf{x}\, m(\mathbf{x}|\mathbf{z}) = 1 \ . \tag{19}$$

More precisely, $m(\mathbf{x}|\mathbf{z})$ can be interpreted as the conditional probability density of getting outcome $\mathbf{x}$ given that the system is known to be exactly in the state $|\mathbf{z}\rangle\langle\mathbf{z}|$. Therefore, the present kind of measurement describes again a conventional measurement, however with an additional *imprecision* (or extrinsic noise) that makes the outcome $\mathbf{x}$ unpredictable even when it is known a priori that the system is exactly in an eigenstate $|\mathbf{z}\rangle$ of the measured observable.

Example
As an example, consider the following function of the operator $\hat{X}$

$$\mathrm{d}\hat{\Pi}(x) = \frac{1}{\sqrt{2\pi\Delta^2}} \exp\left\{-\frac{(x - \hat{X})^2}{2\Delta^2}\right\} \mathrm{d}x \ , \tag{20}$$

It is easy to see that $\mathrm{d}\hat{\Pi}(x)$ in Eq. (20) is the Gaussian convolution of the orthogonal projector $|x\rangle\langle x|$. A concrete example will be given in Sect. 3.2, where we will analyze the homodyne detector with nonunit quantum efficiency.

2.3. NONCOMMUTING POM'S

The truly nontrivial generalization of the projector spectral decomposition $d\hat{E}(x)$ is the case of a non commuting POM $d\hat{\Pi}(\mathbf{x})$, namely

$$[d\hat{\Pi}(\mathbf{x}), d\hat{\Pi}(\mathbf{x}')] \neq 0 \,. \tag{21}$$

Here, there is no longer an orthogonal basis that diagonalizes all operators $d\hat{\Pi}(\mathbf{x})$ simultaneously: hence, no interpretation is possible in terms of compatible observables, nor the noise can be considered as an additional instrumental imprecision that is added to an ideally sharp measurement. Due to nonorthogonality, the resulting probability distribution is always unsharp for any state $\hat{\rho}_S$ of the system. Therefore, the only possible interpretation of the noise from such measurement is as an "intrinsic unavoidable quantum-mechanical imprecision". As we will see soon, this is the case of the noise arising when one jointly measures two noncommuting observables, or, more generally, when the measuring procedure involves a joint measurement, as in the case of the phase of the electromagnetic field. However, there is no "canonical" measurement of noncommuting observables that corresponds to a given POM, and for this reason the noncommuting POM is usually referred to as *generalized observable.*

It is obvious that, similarly to the classification of commuting POM's, also in the case of noncommuting POM's one could distinguish between: *i)* ideal *resolved* measurements—when $d\hat{\Pi}(\mathbf{x})$ is a 1-dim projector, now ranging over a nonorthogonal (overcomplete) set; *ii) unresolved* measurements—when the POM is degenerate; *iii)* measurements with instrumental imprecision, when the POM is convolved with a conditional probability density. However, in the present case, this classification is mostly academic.

At this point one could notice that POM's provide also new selfadjoint operators available to the theory. Hence, why we do not use them? For example, the following operator is manifestly selfadjoint

$$\hat{X} = \int x \, d\hat{\Pi}(x) \,. \tag{22}$$

Hence, $\hat{X}$ admits also an orthogonal spectral resolution in terms of eigen-vectors $|x\rangle\langle x|$. The operator $\hat{X}$, however, does not describe the same measurement of $d\hat{\Pi}(x)$, apart from giving the correct expectation value $\int x \, dP(x) \equiv \mathrm{Tr}(\hat{\rho}\hat{X})$—and, in fact, the corresponding probability distribution $\langle x|\hat{\rho}|x\rangle$ is different from the experimental one $dP(x) \doteq \mathrm{Tr}[\hat{\rho}\, d\hat{\Pi}(x)]$. Differences are evident already from the second moment, where one has

$$\hat{X}^2 \doteq \left(\int x \, d\hat{\Pi}(x)\right)^2 \neq \int x^2 d\hat{\Pi}(x) \,, \tag{23}$$

as a consequence of non orthogonality of $d\hat{\Pi}(x)$. From the following generalized Schwartz inequality

$$\int x^2 d\hat{\Pi}(x) \geq \left(\int x d\hat{\Pi}(x)\right)^2 \,, \tag{24}$$

one has

$$\overline{\Delta x^2} \geq \langle \Delta \hat{X}^2 \rangle \,, \tag{25}$$

where the over-bar denotes the experimental average $\overline{f} \doteq \int f(\mathbf{x})\mathrm{d}P(\mathbf{x})$, whereas brackets denote the ensemble average $\langle \hat{f} \rangle = \mathrm{Tr}[\hat{f}\hat{\rho}]$. It follows that the "true" variance $\overline{\Delta x^2}$ is larger than the one resulting from the selfadjoint operator $\hat{X}$, despite $\hat{X}$ provides the correct average $\overline{x} \equiv \langle \hat{X} \rangle$ for all states $\hat{\rho}$. In other words, nonorthogonal POM's introduce an additional noise that arises from violations of the operator function calculus, namely

$$\hat{f} = \int f(x)\,\mathrm{d}\hat{\Pi}(x) \neq f(\hat{X}), \qquad \text{where } \hat{X} = \int x\,\mathrm{d}\hat{\Pi}(x). \tag{26}$$

According to Eq. (26), in order to describe the experimental probability distribution $\mathrm{d}P(\mathbf{x}) = \mathrm{Tr}[\hat{\rho}\mathrm{d}\hat{\Pi}(\mathbf{x})]$ for any $\hat{\rho}$, one would need the infinite set of selfadjoint operators

$$\widehat{X^n} = \int x^n \mathrm{d}\hat{\Pi}(x) \neq \hat{X}^n. \tag{27}$$

But, why do we use the POM $\mathrm{d}\hat{\Pi}(x)$ instead of the selfadjoint operator $\hat{X}$, considering that the latter can have sharp probabilities? Simply because $\hat{X}$ does not describe the measurement of the physical parameter that we meant. Moreover, notice that the selfadjoint operators defined in Eq. (22) do not solve the problem of joint measurements, because they generally do not commute, namely

$$\hat{X}_i \doteq \int x_i \mathrm{d}\hat{\Pi}(\mathbf{x}), \quad [\mathrm{d}\hat{\Pi}(\mathbf{x}), \mathrm{d}\hat{\Pi}(\mathbf{x}')] \neq 0 \Longrightarrow [\hat{X}_i, \hat{X}_j] \neq 0. \tag{28}$$

Example

In the following a and $a^\dagger$ will represent the usual annihilation and creation operators of a selected mode of the electromagnetic field, with commutation relation $[a, a^\dagger] = 1$, and with vacuum vector $|0\rangle$, i.e. $a|0\rangle = 0$. It is convenient to adopt the complex notation $f = f(z, \overline{z})$ to denote generic functions of $z \in \mathbf{C}$ (z and $\overline{z}$ are treated as independent variables).

Consider the following POM

$$\mathrm{d}\hat{\Pi}(z, \overline{z}) = \frac{\mathrm{d}^2 z}{\pi} |z\rangle\langle z|, \quad z \in \mathbf{C}. \tag{29}$$

In Eq. (29) $|z\rangle$ denotes the customary coherent state

$$|z\rangle \doteq \hat{D}(z)|0\rangle \doteq \exp(z a^\dagger - \overline{z} a)|0\rangle, \tag{30}$$

which is obtained by displacing the vacuum $|0\rangle$ by the operator $\hat{D}(z)$. It is obvious that $\mathrm{d}\hat{\Pi}(z, \overline{z})$ is not commutative, just because $\langle z|z'\rangle \neq \langle z'|z\rangle$. In Sect. 3.3 we will see that the POM (29) describes the ideal heterodyne detector, which provides the optimal joint measurement of a couple of conjugated quadratures—the optical equivalent of position and momentum of an harmonic oscillator. For the moment, just notice the following identities

$$\hat{X} = \int \mathrm{d}\hat{\Pi}(z, \overline{z})\,\mathrm{Re}z = \frac{1}{2}\left(a + a^\dagger\right), \qquad \hat{Y} = \int \mathrm{d}\hat{\Pi}(z, \overline{z})\,\mathrm{Im}z = \frac{i}{2}\left(a^\dagger - a\right), \tag{31}$$

with $\hat{X}$ and $\hat{Y}$ having commutation $[\hat{X}, \hat{Y}] = i/2$.

2.4. NAIMARK'S THEOREM

In the introductory section we have already seen how nonorthogonal/noncommuting POM's arise in a measurement description that involves the apparatus. In this case, the POM just plays the role of the customary projector in the Born's rule used by an observer who ignores the apparatus. Let us recall in formulas this Born's rule

$$dP(\mathbf{x}) = \mathrm{Tr}_{S+P}[\hat{\rho}_S \otimes \hat{\rho}_P |\mathbf{x}\rangle\langle\mathbf{x}|] = \mathrm{Tr}_S[\hat{\rho}_S d\hat{\Pi}(\mathbf{x})] , \tag{32}$$

with $\hat{\rho}_S$ and $\hat{\rho}_P$ denoting the states of the system S and the probe P respectively, and $\hat{X}_i$ denoting the observables that are measured, with $\hat{X}_i = \int x_i |\mathbf{x}\rangle\langle\mathbf{x}| d\mathbf{x}$ acting on $\mathcal{H}_S \otimes \mathcal{H}_P$. The POM is given by

$$d\hat{\Pi}(\mathbf{x}) = \mathrm{Tr}_P[\hat{1}_S \otimes \hat{\rho}_P |\mathbf{x}\rangle\langle\mathbf{x}|] d\mathbf{x} , \tag{33}$$

namely, the POM is the partial trace over $\mathcal{H}_P$ of the probe preparation $\hat{\rho}_P$ with the projector of the $S+P$ observables. It is very remarkable that every POM can be always represented as in Eq. (33): this is the statement of the Naimark's theorem [2], namely "Given a POM $d\hat{\Pi}(\mathbf{x})$ in the system Hilbert space $\mathcal{H}_S$, there is always an extension $\mathcal{H}_S \otimes \mathcal{H}_P$ of the Hilbert space, a pure state $|\psi_P\rangle$, and an orthogonal POM $|\mathbf{x}\rangle\langle\mathbf{x}| d\mathbf{x}$, such that

$$d\hat{\Pi}(\mathbf{x}) = \mathrm{Tr}_P[\hat{1}_S \otimes |\psi_P\rangle\langle\psi_P|\, |\mathbf{x}\rangle\langle\mathbf{x}|] d\mathbf{x} . \tag{34}$$

As a consequence, using POM's in quantum mechanics is not in conflict with the dictum that "only observables can be measured", because every POM corresponds to a customary observable in a larger Hilbert space. But such an observable is not unique, and may have "unnatural" physical meaning, because it involves the measuring apparatus itself. For the proof of the Naimark theorem the reader is addressed to the original papers [2] or to Ref. [3] (a sketch of the proof is also reported in the Helstrom's book [1] and in the book of A. Peres [4]). In the following, I will illustrate the theorem on the basis of two examples (collected from the same Ref. [1]), which I think can be interesting for applications to quantum optics.

2.4.1. *Example 1: the quantum roulette wheel*

Consider the following (generally non commutative) POM

$$\hat{\Pi}_m = \sum_{i=1}^{M} \zeta_i \hat{E}_m^{(i)} , \qquad m = 1, \ldots, n , \tag{35}$$

where

$$\zeta_i \geq 0 , \quad \sum_{i=1}^{M} \zeta_i = 1 , \qquad \hat{E}_m^{(i)} \hat{E}_n^{(i)} = \delta_{mn} \hat{E}_m^{(i)} , \quad \sum_{m=1}^{n} \hat{E}_m^{(i)} = \hat{1}_S . \tag{36}$$

For fixed i the projectors $\hat{E}_m^{(i)}$ give an orthogonal resolution of the identity. The physical meaning of $\hat{\Pi}_m$ is clear: the POM (35) describes a measuring apparatus where one of M different observables is selected at random at every measurement step, with ζ_i as the probability of the i-th observable.

The Naimark's extension of the POM (35) can be obtained as follows. Consider an M-dimensional Hilbert space $\mathcal{H}_P$, with $\{|\omega_i\rangle\}_{i=1,\ldots,M}$ as an orthonormal basis spanning

$\mathcal{H}_P$. A set of orthogonal projectors $\hat{E}_m\hat{E}_n = \delta_{nm}\hat{E}_m$ in the extended Hilbert space is given by

$$\hat{E}_m = \sum_{i=1}^{M} \hat{E}_m^{(i)} \otimes |\omega_i\rangle\langle\omega_i| \,. \tag{37}$$

The Naimark extension of the POM (35) is given by the projectors in Eq. (37), with the following state preparation of the probe

$$|\psi_P\rangle = \sum_{i=1}^{M} \zeta_i^{1/2}|\omega_i\rangle \,. \tag{38}$$

In fact, one can immediately check that

$$\mathrm{Tr}_P[\hat{1} \otimes |\psi_P\rangle\langle\psi_P|\hat{E}_m] = \sum_{i=1}^{M} \zeta_i \hat{E}_m^{(i)} \equiv \hat{\Pi}_m \,. \tag{39}$$

In this example, the probe P plays the role of a random device corresponding to a sort of "quantum roulette" wheel.

2.4.2. *Example 2: commuting POM's*

Let us consider M elements $\hat{\Pi}_k$ of a finite commuting POM. Upon denoting by $\{|m\rangle\}$ their common orthonormal set of eigenvectors, one has

$$\hat{\Pi}_k = \sum_m \lambda_m^{(k)}|m\rangle\langle m| \,, \qquad k = 1, \ldots, M \,, \tag{40}$$

and, as already mentioned, $\lambda_m^{(k)} \geq 0$ are interpreted as conditional probabilities, with $\sum_{k=1}^{M} \lambda_m^{(k)} = 1$. As in the previous example, let us consider a probe in an M-dimensional Hilbert space $\mathcal{H}_P$ with $\{|\omega_k\rangle\}_{k=1,\ldots,M}$ denoting an orthonormal basis. The following linear combinations

$$|\Lambda_m\rangle = \sum_{k=1}^{M} [\lambda_m^{(k)}]^{1/2}|\omega_k\rangle \,, \tag{41}$$

can be written in terms of unitary transformations of a fixed probe vector $|\psi_P\rangle \equiv |\omega_1\rangle$ as follows

$$|\Lambda_m\rangle = \hat{U}_m|\psi_P\rangle \,, \tag{42}$$

where $\hat{U}_m$ are unitary operators. With the probe preparation $|\psi_P\rangle \equiv |\omega_1\rangle$, the Naimark extension of the POM (40) is given by the orthogonal projectors

$$\hat{E}_k = \sum_m |m\rangle\langle m| \otimes \hat{U}_m^{\dagger}|\omega_k\rangle\langle\omega_k|\hat{U}_m \,. \tag{43}$$

In fact, it is easy to check the following steps

$$\begin{aligned} \mathrm{Tr}_P[|\omega_1\rangle\langle\omega_1|\hat{E}_k] &= \sum_m |m\rangle\langle m|\mathrm{Tr}_P[|\omega_1\rangle\langle\omega_1|\hat{U}_m^{\dagger}|\omega_k\rangle\langle\omega_k|\hat{U}] \\ &= \sum_m |m\rangle\langle m|\mathrm{Tr}_P[\hat{U}_m|\omega_1\rangle\langle\omega_1|\hat{U}_m^{\dagger}|\omega_k\rangle\langle\omega_k|] = \sum_m |m\rangle\langle m||\langle\Lambda_m|\omega_k\rangle|^2 \\ &= \sum_m |m\rangle\langle m|\lambda_m^{(k)} \equiv \hat{\Pi}_k \,. \end{aligned} \tag{44}$$

3. POM's in quantum optics

In this section I illustrate some applications of POM's to quantum optics, where we have detectors for measuring observables of the electromagnetic field, with spectrum either discrete—as for the number of photons—or continuous—as for the quadrature of the field. We will see that by homodyne detection we can measure any linear combination of a couple of canonically conjugated observables—the so called *quadratures* of the field: this is a fortunate situation, which does not occur in the quantum mechanics of massive particles, and that makes possible to detect even the state itself of the field [for this topic see my other set of lectures in this same book [5]]. A long subsection is devoted to the heterodyne detector, which jointly measures two conjugated quadratures of the field. Joint measurements will be analyzed in more detail in Sect. 4.

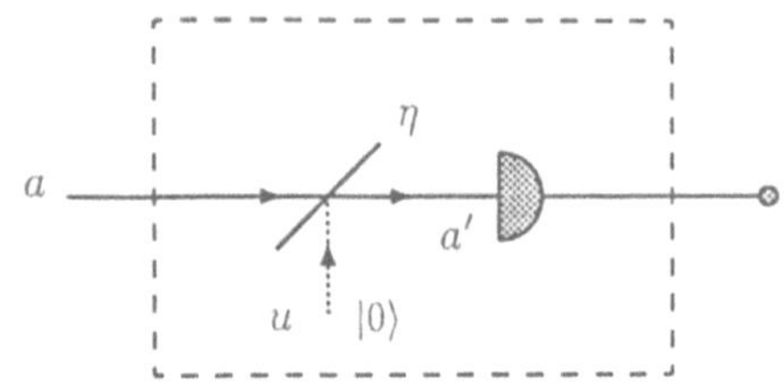

Figure 1. Equivalence of a nonideal ($\eta < 1$) detector with an ideal one preceded by a beam splitter of transmissivity η.

3.1. DIRECT DETECTION

The photon-count distribution for a photodetector (with a photo-tube small with respect to the coherence length of radiation) is given by the Mandel-Kelley-Kleiner formula [6, 7]

$$P_\eta(n) = \left\langle : \frac{(\eta a^\dagger a)^n}{n!} \exp(-\eta a^\dagger a) : \right\rangle , \tag{45}$$

where :: denotes normal ordering, and η is the overall quantum efficiency of the detector ($0 \leq \eta \leq 1$). For simplicity, I consider only monochromatic fields, with a denoting the annihilator of the nonvacuum mode: however, Eq. (45) can be written more generally in the wideband case, where instead of the operator $a^\dagger a$ one has the Pointing flux operator (with time-ordering and integration over the detector time). A simple derivation of Eq. (45) can be found in Ref. [8]. For $\eta = 1$ Eq. (45) gives the POM of ideal photon-number detection. In fact, from the identity

$$|0\rangle\langle 0| = \lim_{\epsilon \to 1} \sum_{l=0}^{\infty} \frac{(-\epsilon)^l}{l!} (a^\dagger)^l a^l \equiv \lim_{\epsilon \to 1} (1-\epsilon)^{a^\dagger a} , \tag{46}$$

and exploiting the recurrence

$$: (a^\dagger a)^n := a^\dagger a (a^\dagger a - 1) \ldots (a^\dagger a - n + 1) , \tag{47}$$

one obtains

$$P_\eta(n) = \sum_{k=n}^{\infty} P(k) \binom{k}{n} \eta^n (1-\eta)^{k-n} , \tag{48}$$

where

$$P(n) \equiv P_1(n) = \langle\, |n\rangle\langle n|\,\rangle . \tag{49}$$

In Eq. (49) $|n\rangle$ denotes the photon-number eigenstate $a^\dagger a|n\rangle = n|n\rangle$. In other words, the POM for $\eta = 1$ is given by

$$\hat{\Pi}_1(n) = |n\rangle\langle n| , \tag{50}$$

whereas, more generally, for $\eta < 1$, using Eqs. (46) and (47) one can see that the probability distribution resulting from Eq. (45) is a Bernoulli convolution of the ideal probability (49), namely the detector POM is given by

$$\hat{\Pi}_\eta(n) = \sum_{k=n}^{\infty} \binom{k}{n} \eta^n (1-\eta)^{k-n} |k\rangle\langle k| . \tag{51}$$

Eq. (51) provides an example of a nonorthogonal commuting POM, with a form similar to Eq. (18) (here for a discrete spectrum) with conditional probability density $m(\mathbf{x}|\mathbf{z})$ given by a Bernoulli distribution.

Now I show that a detector with quantum efficiency $\eta < 1$ is equivalent to an ideal detector preceded by a beam splitter of transmissivity η. Such a "quantum-equivalence" between devices is schematically depicted in Fig. 1, and is relevant for detection theory in quantum optics. We have just to remind that, apart from trivial phase changes, a beam splitter of transmissivity η affects the unitary transformation of fields (see Fig. 2)

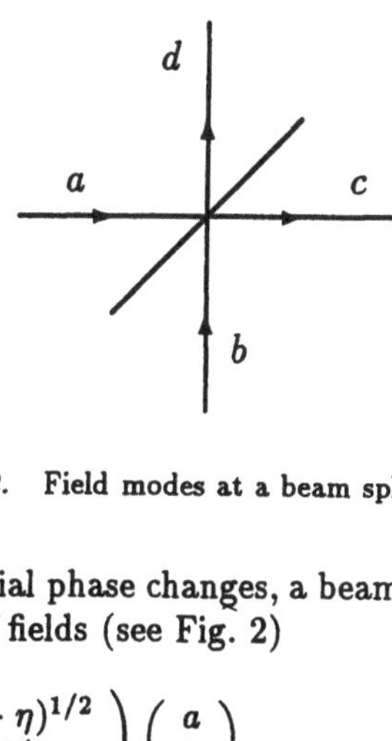

Figure 2. Field modes at a beam splitter.

$$\begin{pmatrix} c \\ d \end{pmatrix} = \hat{U}^\dagger \begin{pmatrix} a \\ b \end{pmatrix} \hat{U} = \begin{pmatrix} \eta^{1/2} & (1-\eta)^{1/2} \\ -(1-\eta)^{1/2} & \eta^{1/2} \end{pmatrix} \begin{pmatrix} a \\ b \end{pmatrix} , \tag{52}$$

where all field modes are considered at the same frequency. Hence, the output mode a' in Fig. 1 is given by the linear combination

$$a' = \eta^{1/2} a + (1-\eta)^{1/2} u . \tag{53}$$

The mode a is entangled with the vacuum mode u, which here plays the role of a "probe" mode. The POM is obtained by partially tracing over the u mode as follows

$$\hat{\Pi}_\eta(n) = \mathrm{Tr}_u[\hat{1}_a \otimes |0\rangle\langle 0|_u |n\rangle\langle n|_{a'}] = {}_u\langle 0| : \exp(-a'^\dagger a')(-a'^\dagger a')^n/n! : |0\rangle_u . \tag{54}$$

Eq. (54) gives $\hat{\Pi}_\eta(n)$ in form of a normal-ordered expectation between coherent (vacuum) states. Here we can use the following identity valid for any function of a linear combination $Ka + Hb$ of two modes a and b

$$ {}_b\langle\beta| : f\left(Ka + Hb, (Ka + Hb)^\dagger\right) : |\beta\rangle_b =: f(Ka + H\beta, \overline{K}a^\dagger + \overline{H\beta}) : , \qquad (55) $$

with $|\beta\rangle_b$ denoting a coherent state for mode b only. Using Eq. (55) we immediately obtain

$$ \hat{\Pi}_\eta(n) =: \exp(-\eta a^\dagger a)\frac{(\eta a^\dagger a)^n}{n!} : , \qquad (56) $$

which gives the probability (45).

3.2. BALANCED HOMODYNE DETECTION

The scheme of a balanced homodyne detector is depicted in Fig. 3. The "signal" mode a is combined by means of a 50-50 beam splitter with a "local oscillator" (LO) mode b operating at the same frequency of a, and prepared in an "intense" coherent state $|z\rangle$. The signal mode a here plays the role of the "system" S, whereas mode b is the "probe" P. The field at the output of the beam splitter is described by a "sum" mode $c = (a + b)/\sqrt{2}$ and a "difference" mode $d = (a - b)/\sqrt{2}$, according to Eqs. (52) for $\eta = 0.5$. These output modes are detected by two identical photodetectors, and finally the difference of photocurrents (at zero frequency) is rescaled by $2|z|$. Thus, the output of the detector is given by the following operator

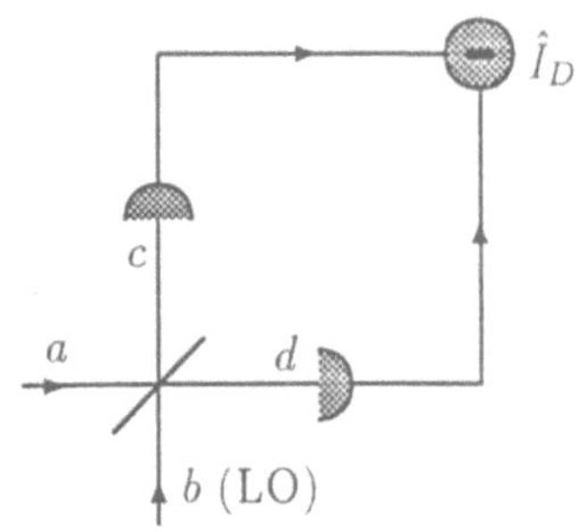

Figure 3. Scheme of the balanced homodyne detector.

$$ \hat{I}_D = \frac{c^\dagger c - d^\dagger d}{2|z|} = \frac{a^\dagger b + b^\dagger a}{2|z|} . \qquad (57) $$

Our intent is to evaluate the POM of the detector, or, in other words, to obtain the probability distribution of the output photocurrent $\hat{I}_D$ for any generic state $\hat{\rho}$ of the signal mode a. It is easier to evaluate the generating function of the moments of $\hat{I}_D$

$$ \chi(\lambda) = \left\langle e^{i\lambda\hat{I}_D}\right\rangle_{ab} , \qquad (58) $$

and then obtain the probability distribution of I_D as the Fourier transform of $\chi(\lambda)$, namely

$$ \mathrm{d}P(I) = \mathrm{d}I \int_{-\infty}^{+\infty} \frac{d\lambda}{2\pi} e^{-i\lambda I} \left\langle e^{i\lambda\hat{I}_D}\right\rangle_{ab} . \qquad (59) $$

Using the Backer-Campbell-Hausdorff (BCH) formula [9] for the $SU(2)$ group, namely

$$ \exp\left(\xi a^\dagger b - \bar{\xi} b^\dagger a\right) = e^{\zeta b^\dagger a}\left(1 + |\zeta|^2\right)^{\frac{1}{2}\left(b^\dagger b - a^\dagger a\right)} e^{-\bar{\zeta} a^\dagger b} , \quad \zeta = \frac{\xi}{|\xi|}\tan|\xi| , \qquad (60) $$

one can normal-order the exponential in Eq. (58) with respect to mode b as follows

$$\chi(\lambda)=\left\langle \exp\left[i\tan\left(\frac{\lambda}{2|z|}\right)b^\dagger a\right]\left[\cos\left(\frac{\lambda}{2|z|}\right)\right]^{a^\dagger a-b^\dagger b}\exp\left[i\tan\left(\frac{\lambda}{2|z|}\right)a^\dagger b\right]\right\rangle_{ab} . \tag{61}$$

The partial trace over b can be evaluated easily as follows

$$\begin{aligned}\chi(\lambda) &= \left\langle \exp\left[i\tan\left(\frac{\lambda}{2|z|}\right)\bar{z}a\right]\left[\cos\left(\frac{\lambda}{2|z|}\right)\right]^{a^\dagger a}\exp\left[i\tan\left(\frac{\lambda}{2|z|}\right)za^\dagger\right]\right\rangle_a \\ &\times \left\langle z\left|\left[\cos\left(\frac{\lambda}{2|z|}\right)\right]^{-b^\dagger b}\right|z\right\rangle ,\end{aligned} \tag{62}$$

with the probe mode b is in the coherent state $|z\rangle$. Using now the customary BCH formula valid for $[\hat{A},[\hat{A},\hat{B}]]=[\hat{B},[\hat{A},\hat{B}]]=0$

$$\exp\hat{A}\exp\hat{B}=\exp\left(\hat{A}+\hat{B}+\frac{1}{2}[\hat{A},\hat{B}]\right) , \tag{63}$$

one can recast Eq. (62) in normal-order with respect to a, namely [10]

$$\chi(\lambda)= \tag{64}$$
$$\left\langle : \exp\left[iz\sin\left(\frac{\lambda}{2|z|}\right)a^\dagger\right]\exp\left[-2\sin^2\left(\frac{\lambda}{4|z|}\right)(a^\dagger a+|z|^2)\right]\exp\left[i\bar{z}\sin\left(\frac{\lambda}{2|z|}\right)a^\dagger\right] :\right\rangle_a .$$

Eq. (64) simplifies greatly in the strong-LO limit $z\to\infty$, where only the lowest order terms in $\lambda/|z|$ are retained, and $a^\dagger a$ is neglected with respect to $|z|^2$. One has

$$\lim_{z\to\infty}\chi(\lambda)=\left\langle : \exp\left[i\frac{\lambda}{2}e^{i\phi}a^\dagger\right]\exp\left[-\frac{\lambda^2}{8}\right]\exp\left[i\frac{\lambda}{2}e^{-i\phi}a^\dagger\right] :\right\rangle_a = \langle\exp[i\lambda\hat{a}_\phi]\rangle_a , \tag{65}$$

with $\phi=\arg z$ and $\hat{a}_\phi=\frac{1}{2}(a^\dagger e^{i\phi}+ae^{-i\phi})$ denoting the so called "quadrature" of the field mode a at phase ϕ with respect to the LO. The generating function in Eq. (65) is equivalent to the POM

$$d\hat{\Pi}(x)=dx\int_{-\infty}^{+\infty}\frac{d\lambda}{2\pi}e^{-i\lambda x}\exp[i\lambda\hat{a}_\phi]=dx\,\delta(x-\hat{a}_\phi) . \tag{66}$$

Hence, in conclusion, the balanced homodyne detector in the strong LO limit achieves the ideal measurement of the quadrature $\hat{a}_\phi$.

It is easy to take into account nonunit quantum efficiency at detectors. According to Eq. (53) one performs the substitutions

$$c \Longrightarrow \eta^{1/2}c+(1-\eta)^{1/2}u , \qquad u,v \text{ vacuum modes} \tag{67}$$
$$d \Longrightarrow \eta^{1/2}d+(1-\eta)^{1/2}v , \tag{68}$$

and now the output current is rescaled by $2|z|\eta$, thus obtaining

$$\hat{I}_D=\hat{a}_\phi+\sqrt{\frac{(1-\eta)}{2\eta}}(\hat{u}_\phi-\hat{v}_\phi)+\mathcal{O}(|z|^{-1}) , \tag{69}$$

with $\mathcal{O}(|z|^{-1})$ denoting terms vanishing as $|z|^{-1}$. Then, by tracing-out the vacuum modes u and v, one obtains

$$\begin{aligned} \mathrm{d}\hat{\Pi}_\eta(x) &= \mathrm{d}x \int \frac{\mathrm{d}\lambda}{2\pi} e^{-i\lambda x} e^{i\lambda \hat{a}_\phi} |\langle 0| e^{i\lambda\sqrt{\frac{1-\eta}{2\eta}} u_\phi} |0\rangle|^2 = \mathrm{d}x \int \frac{\mathrm{d}\lambda}{2\pi} e^{-i\lambda(x-\hat{a}_\phi)} e^{-\lambda^2 \frac{1-\eta}{8\eta}} \\ &= \mathrm{d}x \frac{1}{\sqrt{2\pi\Delta_\eta^2}} \exp\left[-\frac{(x-\hat{a}_\phi)^2}{2\Delta_\eta^2}\right] , \end{aligned} \tag{70}$$

where

$$\Delta_\eta^2 = \frac{1-\eta}{4\eta} . \tag{71}$$

Thus, in the nonideal case the POM is the convolution of the ideal POM with a Gaussian conditional probability: as in the case of photodetection, again nonunit quantum efficiency makes the POM nonorthogonal.

3.3. HETERODYNE DETECTOR

Heterodyne detection provides a method to perform joint measurements of two conjugated quadratures of the field [11]. The detector and the relevant field modes involved in the measurement are outlined in Fig. 4. The input field $\hat{E}_{in}$ impinges into a beam splitter with transmissivity θ, and has nonzero photon number only at frequency $\omega_0 + \omega_{IF}$. The LO works at a different frequency ω_0, and the output photocurrent $\hat{I}_{out}$ is measured at the intermediate frequency ω_{IF}. In the time-domain the measured photocurrent is given by

$$\hat{I}_{out}(t) = \hat{E}^-_{out}(t)\hat{E}^+_{out}(t) , \tag{72}$$

where $\hat{E}^\pm$ denote the usual positive and negative frequency components of the field, containing the annihilation and creation operators, respectively. The output photocurrent analyzed at frequency ω_{IF} is given by

$$\hat{I}_{out}(\omega_{IF}) = \int \mathrm{d}t\, \hat{I}_{out}(t) e^{i\omega_{IF} t} = \int \frac{\mathrm{d}\omega}{2\pi} \hat{E}^-_{out}(\omega+\omega_{IF})\hat{E}^+_{out}(\omega) . \tag{73}$$

The only field modes that are nonvacuum are the signal mode a_s at frequency $\omega + \omega_{IF}$ and the LO b_l at frequency ω_0. The integral in Eq. (73) involves modes at all frequencies: the terms that survive in the strong-LO limit are those linear in b_l or $b_l^\dagger$, namely $b_l^\dagger a_i$ and $b_l a_s^\dagger$, both having frequency difference equal to ω_{IF} (all other nonvacuum modes depicted in Fig. 4 do not involve the LO in Eq. (73)). The detector behaves ideally in the combined limits of strong-LO $z \to \infty$ and perfect transparency $\theta \to 1$, with $\gamma \doteq |z|\sqrt{\theta(1-\theta)}$ kept as constant. In fact, the output rescaled photocurrent $\hat{Z}$

$$\hat{Z} = \lim_{\substack{\theta \to 1, |z| \to \infty \\ \gamma = \mathrm{cost.}}} \gamma^{-1} \hat{I}_{out}(\omega_{IF}) , \tag{74}$$

is given by

$$\hat{Z} = |z|^{-1}(a_s^\dagger b_l + a_i b_l^\dagger) + \mathcal{O}(|z|^{-1}) , \tag{75}$$

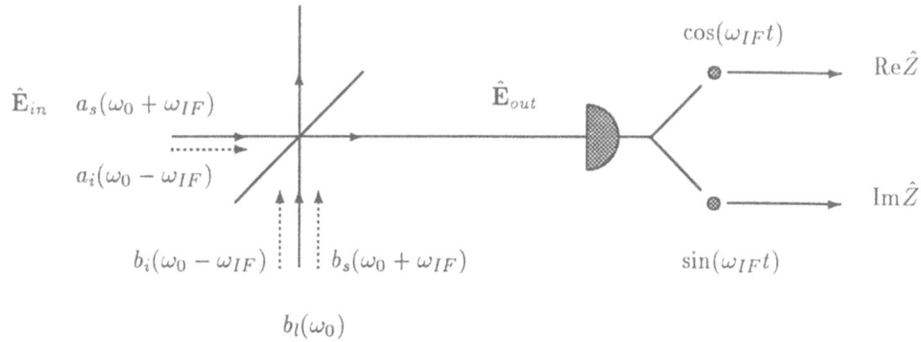

Figure 4. Scheme of the heterodyne detector and relevant field modes involved in the measurement. Dashed lines denote vacuum modes. Signal input modes are denoted by a, LO modes by b. The subindices s, l, and i refer to the frequency of modes: s is for signal band around $\omega_0+\omega_{IF}$, l is for LO band around ω_0, and i is for image band around $\omega_0-\omega_{IF}$. The output photocurrent is detected at the intermediate frequency ω_{IF}.

and after re-phasing the field modes $\hat{Z}$ becomes

$$\hat{Z} = a_s^\dagger + a_i \,. \tag{76}$$

The complex operator $\hat{Z}$ is equivalent to a couple of commuting selfadjoint operators

$$\hat{Z} = \hat{Z}_R + i\hat{Z}_I \,, \qquad [\hat{Z},\hat{Z}^\dagger] = [\hat{Z}_R,\hat{Z}_I] = 0 \,, \tag{77}$$

and is described by a quantum mechanical probability density in the complex plane $p(z_R,z_I)\equiv p(z,\bar{z})$. The probability density is the Fourier transform of the generating function of the moments of $\hat{Z}$, and in complex notation is

$$p(z,\bar{z}) = \int \frac{d^2\lambda}{\pi^2}\langle e^{\lambda\hat{Z}^\dagger-\bar{\lambda}\hat{Z}}\rangle_{si}\, e^{\bar{\lambda}z-\lambda\bar{z}} \,, \tag{78}$$

where $\langle\ldots\rangle_{si}$ denotes the ensemble average on both modes a_s and a_i. In the present case the signal mode a_s represents the "system", whereas the image-band mode a_i is the "probe". The partial trace over the probe is carried out as follows

$$\begin{aligned}\left\langle e^{\lambda\hat{Z}^\dagger-\bar{\lambda}\hat{Z}}\right\rangle_{si} &= \mathrm{Tr}_s\left[\hat{\rho}_s\hat{D}_s(\lambda)\right]\langle 0|\hat{D}_i(-\lambda)|0\rangle_i = \mathrm{Tr}_s\left[\hat{\rho}_s\hat{D}_s(\lambda)\right]e^{-\frac{1}{2}|\lambda|^2}\\ &\equiv \mathrm{Tr}_s\left[\hat{\rho}_s : \hat{D}_s(\lambda) :_A\right] \,,\end{aligned} \tag{79}$$

where $\hat{D}(\alpha)=\exp(\alpha a^\dagger-\bar{\alpha}a)$ denotes the usual displacement operator, and sub-indices s,i pertain signal and image modes, respectively. Anti-normal ordering $::_A$ in Eq. (79) follows from the customary BCH formula (63) applied to the displacement operator. Comparing Eqs. (78) and (79) the POM of the detector is obtained in the form

$$d\hat{\Pi}(z,\bar{z}) = \int \frac{d^2\lambda}{\pi^2} e^{\bar{\lambda}z-\lambda\bar{z}} : \hat{D}_s(\lambda) :_A d^2z \,, \tag{80}$$

and with a little algebra one has[1]

$$\mathrm{d}\hat{\Pi}(z,\bar{z}) = \frac{\mathrm{d}^2 z}{\pi}|z\rangle\langle z| \,. \tag{81}$$

Hence, the POM of the ideal heterodyne detector is the coherent-state projector.

It is not difficult to take into account nonunit quantum efficiency at photodetectors. For simplicity we consider the case of η constant versus ω within the considered frequency band. Similarly to the case of the homodyne detector, we add a vacuum mode to both a_s and a_i

$$a_s \rightarrow \eta^{1/2} a_s + (1-\eta)^{1/2} u \,, \tag{82}$$

$$a_i \rightarrow \eta^{1/2} a_i + (1-\eta)^{1/2} v \,, \tag{83}$$

with u and v denoting the vacuum modes at frequencies $\omega_0 + \omega_{IF}$ and $\omega_0 - \omega_{IF}$ respectively. Upon rescaling the output photocurrent by an additional factor $\eta^{1/2}$, we obtain

$$\hat{Z} = a_s + \sqrt{\frac{1-\eta}{\eta}} u + a_i^\dagger + \sqrt{\frac{1-\eta}{\eta}} v^\dagger \,. \tag{84}$$

The two modes u and v enter the definition of the new enlarged "probe" of the detector, and must be traced out. In this way one obtains the POM

$$\mathrm{d}\hat{\Pi}_\eta(\alpha,\bar{\alpha}) = \mathrm{d}^2\alpha \int \frac{\mathrm{d}^2\lambda}{\pi^2} e^{\alpha\bar{\lambda}-\bar{\alpha}\lambda} : \hat{D}_s(\lambda) :_A {}_{uv}\langle 0|\hat{D}_u(\lambda_\eta)\hat{D}_v(-\lambda_\eta)|0\rangle_{uv} \,, \tag{85}$$

where $\lambda_\eta = \sqrt{\frac{1-\eta}{\eta}}\lambda$. The POM (85) is the Gaussian convolution of the ideal POM

$$\mathrm{d}\hat{\Pi}_\eta(\alpha,\bar{\alpha}) = \mathrm{d}^2\alpha \int \frac{\mathrm{d}^2\lambda}{\pi^2} e^{\alpha\bar{\lambda}-\bar{\alpha}\lambda} : \hat{D}_s(\lambda) :_A e^{-|\lambda_\eta|^2} = \int \frac{\mathrm{d}^2 z}{\pi\Delta_\eta^2} e^{-\frac{|z-\alpha|^2}{\Delta_\eta^2}} \mathrm{d}\hat{\Pi}(z,\bar{z}) \,, \tag{86}$$

where $\Delta_\eta^2 = \frac{1-\eta}{\eta}$.

Before continuing further, it is instructive to see an alternative derivation of the heterodyne POM (81). We have seen that the heterodyne detector measures the complex

[1] Here is the proof:

$$\int \frac{\mathrm{d}^2\lambda}{\pi} : \hat{D}(\lambda) :_A = \int \frac{\mathrm{d}^2\lambda}{\pi} e^{-|\lambda|^2} e^{\lambda a^\dagger} e^{-\bar{\lambda} a} = \int \frac{\mathrm{d}|\lambda|^2}{2} e^{-|\lambda|^2} \sum_{n=0}^{\infty} \frac{(-)^n}{(n!)^2} |\lambda|^{2n} (a^\dagger)^n a^n = \sum_{n=0}^{\infty} \frac{(-)^n}{n!} (a^\dagger)^n a^n \,,$$

and using the identity (46) one has

$$\int \frac{\mathrm{d}^2\lambda}{\pi} : \hat{D}(\lambda) :_A = |0\rangle\langle 0| \,.$$

Using the last equation we obtain

$$\int \frac{\mathrm{d}^2\lambda}{\pi} e^{\alpha\bar{\lambda}-\bar{\alpha}\lambda} : \hat{D}(\lambda) :_A = \int \frac{\mathrm{d}^2\lambda}{\pi} e^{-|\lambda|^2} e^{\lambda(a^\dagger - \bar{\alpha})} e^{-\bar{\lambda}(a-\alpha)} = \hat{D}^\dagger(-\alpha)|0\rangle\langle 0|\hat{D}(-\alpha) = |\alpha\rangle\langle\alpha| \,.$$

photocurrent $\hat{Z} = a_s + a_i^\dagger$, with a_i playing the role of the probe. This assertion is translated into mathematical terms as follow

$$d\hat{\Pi}(z,\bar{z}) = d^2z\, {}_i\langle 0|\delta_2(z-\hat{Z})|0\rangle_i\,, \tag{87}$$

where δ_2 is the Dirac delta in the complex plane

$$\delta_2(\alpha-\beta) = \int \frac{d^2\lambda}{\pi^2} e^{(\alpha-\beta)\bar{\lambda}-(\bar{\alpha}-\bar{\beta})\lambda}\,. \tag{88}$$

The form of Eq. (87) does not depend on the operator ordering, because $[\hat{Z},\hat{Z}^\dagger]=0$. In particular, we can use normal ordering with respect to a_i, corresponding to anti-normal ordering with respect to a_s. Then, the vacuum expectation is evaluated just upon setting a_i to zero, namely

$$d\hat{\Pi}(z,\bar{z}) = d^2z : \delta_2(z-a_s) :_A\,. \tag{89}$$

By definition, one has

$$:\delta_2(z-a_s) := \int \frac{d^2\lambda}{\pi^2} e^{(z-a_s)\bar{\lambda}-(\bar{z}-a_s^\dagger)\lambda} :_A = \int \frac{d^2\lambda}{\pi^2} : e^{z\bar{\lambda}-\bar{z}\lambda} : \hat{D}_s(\lambda) :_A\,, \tag{90}$$

namely Eq. (81).

4. Joint measurements

From the derivation of the POM of the heterodyne detector we can understand the basis of a joint measurement of two non commuting observables. The heterodyne detector performs a joint measurement of any couple of conjugated quadratures, say for example

$$\hat{X} = \frac{1}{2}(a_s + a_s^\dagger)\,, \qquad \hat{Y} = \frac{i}{2}(a_s^\dagger - a_s)\,, \tag{91}$$

with

$$[\hat{X},\hat{Y}] = \frac{i}{2}\,. \tag{92}$$

The method for jointly measuring $\hat{X}$ and $\hat{Y}$ consists of making a conventional measurement of two commuting currents $\hat{Z}_R$ and $\hat{Z}_I$ that have the same expectation values of $\hat{X}$ and $\hat{Y}$, namely

$$\langle\hat{X}\rangle = \langle\hat{Z}_R\rangle\,, \qquad \langle\hat{Y}\rangle = \langle\hat{Z}_I\rangle\,, \tag{93}$$

or, in complex notation

$$\langle a_s\rangle = \langle\hat{Z}\rangle\,. \tag{94}$$

Eq. (94) emphasizes the fact that measuring $\hat{X}$ and $\hat{Y}$ jointly is equivalent to "measuring the complex operator" a_s. Now we will see that the price to pay for jointly measuring non commuting observables is an additional noise. In fact, let us evaluate the r.m.s. fluctuations for $\hat{X}$, $\hat{Y}$, and for $\hat{Z}_R$, $\hat{Z}_I$, and then compare the respective results. One has

$$\langle\Delta\hat{Z}_R^2\rangle = \langle\Delta\hat{X}^2\rangle + \langle\Delta\hat{X}_i^2\rangle = \langle\Delta\hat{X}^2\rangle + \frac{1}{4}\,, \tag{95}$$

and, similarly

$$\langle\Delta\hat{Z}_I^2\rangle = \langle\Delta\hat{Y}^2\rangle + \langle\Delta\hat{Y}_i^2\rangle = \langle\Delta\hat{Y}^2\rangle + \frac{1}{4}\,, \tag{96}$$

where $\hat{X}_i$ and $\hat{Y}_i$ are the same quadratures as in Eq. (91), but for the image-band mode a_i. It follows that the experimental probability distribution of the photocurrent $\hat{Z}$ has the same average of the complex field a_s, but with an additional noise. From Eqs. (95) and (96) we deduce the "experimental" Heisenberg relation [12, 13]

$$\langle\Delta\hat{Z}_R^2\rangle\langle\Delta\hat{Z}_I^2\rangle \geq \frac{1}{4}\,, \tag{97}$$

which should be compared with the customary inequality

$$\langle\Delta\hat{X}^2\rangle\langle\Delta\hat{Y}^2\rangle \geq \frac{1}{4}|\langle[\hat{X},\hat{Y}]\rangle|^2 \geq \frac{1}{16}\,. \tag{98}$$

Notice that the usual Heisenberg relation pertains the intrinsic uncertainties of a couple of conjugated observables, and thus can be used only to analyze conventional measurements of one of the two observables at a time (the uncertainty of the other observable refers to a "preparation" before the measurement). On the other hand, the case of joint measurements is described by the new Heisenberg inequality (98): here, the "experimental" noise is double than the "theoretical" one, and their relative factor 2 is usually referred to as "the additional 3 decibels (3dB) noise due to the joint measure". Such noise is of quantum origin, and is unavoidable. This can be easily understood with the aid of the following argument. The 3dB noise originates from the vacuum fluctuations of the image-band mode, which is needed in order to have a commuting current $\hat{Z}$. For this purpose one needs to add $a_i^\dagger$—not a_i—to the signal annihilator a_s, and this produces the anti-normal ordering for the POM, which corresponds to the 3dB Gaussian convolution

$$\begin{aligned} \mathrm{d}\hat{\Pi}(z,\bar{z}) &= \mathrm{d}^2 z\,{}_i\langle 0|\delta_2(z-\hat{Z})|0\rangle_i = \mathrm{d}^2 z : \delta_2(z-a_s) :_A \\ &= \int \frac{\mathrm{d}^2\lambda}{\pi^2} e^{z\bar{\lambda}-\bar{z}\lambda} e^{-\frac{1}{2}|\lambda|^2} \hat{D}_s(\lambda)\,. \end{aligned} \tag{99}$$

In the following we will see that the 3dB additional noise is equivalent to measuring each quadrature with effective quantum efficiency $\eta = \frac{1}{2}$.

4.1. MARGINAL JOINT MEASUREMENTS

It is clear that once a method for measuring the complex field a is given, then any function of the field can be measured. Such measuring scheme resembles a "classical" measurement in the phase space, where one jointly measures the canonical pair and then evaluates functions of it. Operatively, the measurement works similarly to the classical case, namely, after detecting the complex photocurrent $\hat{Z}$ and obtaining the reading $z \in \mathbf{C}$, one evaluates the function $w = f(z,\bar{z})$. What is the POM that gives the probability distribution for w? The answer is simple: the probability density for w is just the marginal probability of $p(z,\bar{z})$, namely

$$p(w) = \int \mathrm{d}^2 z\, p(z,\bar{z})\, \delta(w - f(z,\bar{z}))\,. \tag{100}$$

Hence, the POM is the marginal POM of $\mathrm{d}\hat{\Pi}(z,\bar{z})$

$$\mathrm{d}\hat{\Pi}(w) = \mathrm{d}w \int \mathrm{d}\hat{\Pi}(z,\bar{z})\, \delta(w - f(z,\bar{z}))\,. \tag{101}$$

The Dirac delta function in Eq. (101) must be defined carefully on the complex plane, depending on the particular analytic form of the function f. Using Eq. (87) one obtains

$$\mathrm{d}\hat{\Pi}(w) = \mathrm{d}s\, \mathrm{d}w\, {}_i\langle 0|\delta(w - f(\hat{Z},\hat{Z}^\dagger))|0\rangle_i = \mathrm{d}w : \delta(w - f(a,a^\dagger)) :_A \,. \tag{102}$$

In the following we will examine some relevant choices for the function f.

4.1.1. *Field quadrature*

Field quadrature corresponds to the function $f(z,\bar{z}) = \mathrm{Re}(ze^{-i\phi})$ of the field. In this way, from a joint measurement of any couple of conjugated quadratures, one obtains a marginal probability distribution for any desired single quadrature $\hat{a}_\phi$. In fact, from Eq. (102) one has

$$\begin{aligned} \mathrm{d}\hat{\Pi}(x) &= \mathrm{d}x : \delta(x - \hat{a}_\phi) :_A = \mathrm{d}x \int_{-\infty}^{+\infty} \frac{\mathrm{d}\lambda}{2\pi} e^{i\lambda x} e^{-\frac{1}{8}|\lambda|^2} e^{-i\lambda \hat{a}_\phi} \\ &= \mathrm{d}x \frac{2}{\sqrt{\pi}} \exp\left[-2(x - \hat{a}_\phi)\right]\,, \end{aligned} \tag{103}$$

which is a Gaussian with variance $\Delta^2 = \frac{1}{4}$, as expected from Eq. (95): this is just the 3dB noise due to the joint measure. Comparing Eq. (103) with Eq. (71) we immediately recognize that this noise corresponds to an effective quantum efficiency $\eta = \frac{1}{2}$.

4.1.2. *Field intensity*

This case corresponds to the function $f(z,\bar{z}) = |z|^2$. One has

$$\begin{aligned} \mathrm{d}\hat{\Pi}(w) &= \mathrm{d}w : \delta(w - a^\dagger a) :_A = \mathrm{d}w \int \frac{\mathrm{d}\lambda}{2\pi} e^{-i\lambda w} : e^{i\lambda a^\dagger a} :_A \\ &= \mathrm{d}w \int \frac{\mathrm{d}\lambda}{2\pi} e^{-i\lambda w} (1 - i\lambda)^{-a^\dagger a - 1} = \frac{w^{a^\dagger a}}{(a^\dagger a)!} e^{-w} \mathrm{d}w\,. \end{aligned} \tag{104}$$

The POM (104) is an unsharp version of the ideal POM (50). Notice that the function in Eq. (104) is not a Poisson, because here it is regarded as a function of w—not versus $a^\dagger a$.

4.1.3. *Phase of the field*

This case is particularly interesting, as in practice it is the only way to define a quantum mechanical measurement of the phase of the field, namely through the measurement of the polar angle of a complex photocurrent. It is instructive to analyze briefly the experimental procedure for obtaining the marginal phase distribution. This is illustrated in Fig. 5, where, as an example, a computer simulation of the experimental procedure is illustrated for a squeezed state. Each experimental event consists of a reading of the complex heterodyne photocurrent, which is represented by a point plotted in the complex plane of the field amplitude. The phase value inferred from the event is the polar angle of the point itself. The experimental histogram of the phase distributions is obtained upon dividing the plane into "infinitesimal" angular bins of equal width $\delta\phi$, from $-\pi$ to π, then counting the number of points which fall into each bin. In Fig. 5 the simulated experimental histogram (10^4 events) is compared with the theoretical probability as obtained from the marginal phase POM of the heterodyne detector. Formally, the marginal phase POM is given by

$$\mathrm{d}\hat{\Pi}(\phi) = \mathrm{d}\phi : \delta(\phi - \arg a) :_A , \tag{105}$$

where the meaning of the δ function is the marginal integral of the δ_2 distribution over the polar modulus on the complex plane, namely

$$\delta(\phi - \arg z) = \int_0^\infty \rho\, \mathrm{d}\rho\, \delta_2(\rho e^{i\phi} - z) = \int_0^\infty \mathrm{d}r \int_{\mathbf{C}} \frac{\mathrm{d}^2\lambda}{2\pi^2} \exp\left[\sqrt{r}(e^{i\phi} - z)\bar{\lambda} - \mathrm{c.c.}\right] . \tag{106}$$

One has

$$\begin{aligned} \mathrm{d}\hat{\Pi}(\phi) &= \frac{\mathrm{d}\phi}{2\pi} \int_0^\infty \mathrm{d}r \int_{\mathbf{C}} \frac{\mathrm{d}^2\lambda}{\pi} e^{\sqrt{r}(e^{i\phi}\bar{\lambda} - e^{-i\phi}\lambda)} e^{-\bar{\lambda}a} e^{\lambda a^\dagger} \\ &= \frac{\mathrm{d}\phi}{2\pi} \int_{\mathbf{C}} \frac{\mathrm{d}^2\lambda}{\pi} e^{-\frac{1}{2}|\lambda|^2} \frac{\hat{D}(\lambda)}{[\mathrm{Im}(\lambda e^{-i\phi}) - i0^+]^2} . \end{aligned} \tag{107}$$

Equivalently, evaluating the marginal POM of the coherent-state projector, one has

$$\mathrm{d}\hat{\Pi}(\phi) = \frac{\mathrm{d}\phi}{2\pi} \int_0^\infty \mathrm{d}r |\sqrt{r}e^{i\phi}\rangle\langle\sqrt{r}e^{i\phi}| = \frac{\mathrm{d}\phi}{2\pi} \sum_{nm=0}^{\infty} e^{i(n-m)\phi} \frac{\Gamma[\frac{1}{2}(n+m)+1]}{\sqrt{n!m!}} |n\rangle\langle m| . \tag{108}$$

We will discuss this POM later and compare it with the ideal one coming from quantum estimation theory.

5. Quantum estimation theory

Quantum estimation theory analyzes POM's at a purely abstract level, with the purpose of seeking the best strategy for estimating one or more parameters of a quantum system. The theory looks for the general class of POM's that describe the specific measurement, then optimizes the POM according to some prefixed goodness criterion. In general, one can say that the problem resorts to seeking the best strategy for estimating a set of parameters $\theta = \{\theta_1, \theta_2, \ldots, \theta_m\}$ of the density operator $\hat{\rho}(\theta)$ of the system (for example the position and momentum of a particle, the amplitude of a field mode, etc). The observational strategy for estimating θ is expressed by a POM that pertains a generic apparatus along with its "data processing rule", i.e. the evaluation of a function of the experimental result. Let us denote by $\mathrm{d}\hat{\Pi}(\theta)$ the generic POM. Generally, the result of the measurement—i.e. the estimated values—are different from the "true" ones, and we will denote the true values by θ and the estimated values by θ_*. Then, the joint conditional probability density $p(\theta_*|\theta)$ of estimating θ_* for true values θ, is given by

$$p(\theta_*|\theta)\mathrm{d}^m\theta = \mathrm{Tr}[\hat{\rho}(\theta)\mathrm{d}\hat{\Pi}(\theta_*)] . \tag{109}$$

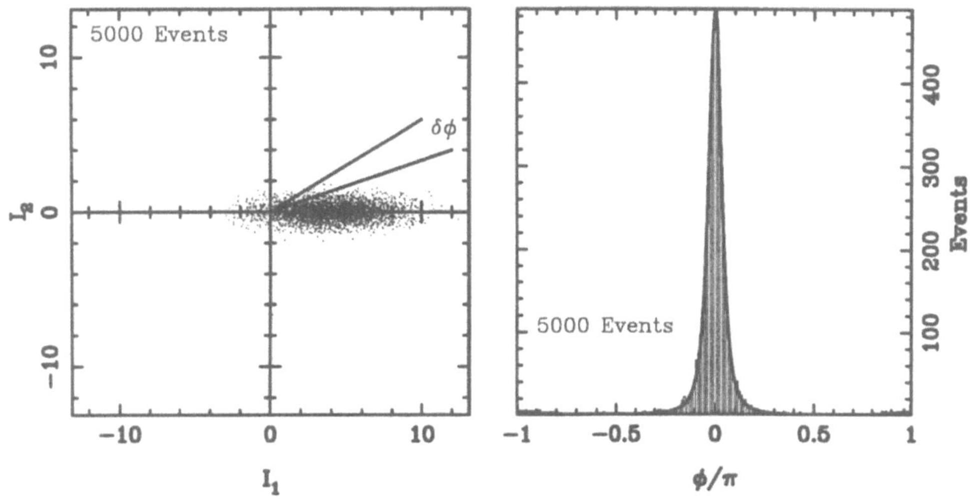

Figure 5. Computer simulation of a heterodyne phase detection experiment for a squeezed state with 4.53 squeezing photons and 20 photons in total. The histogram for 5000 events is compared with the theoretical result from the POM (108).

The goodness of the POM is considered on the basis of a cost function $C(\theta_*, \theta)$, which assesses the cost of errors in the estimates. Examples of cost functions are the delta-function cost

$$C(\theta_*, \theta) = -\prod_{k=1}^{m} \delta(\theta_{k*} - \theta_k) , \tag{110}$$

and the quadratic cost

$$C(\theta_*, \theta) = -\sum_{k=1}^{m} (\theta_{k*} - \theta_k)^2 . \tag{111}$$

More generally, in Eqs. (110) one could consider different weights for every component, or one can introduce a positive cost matrix in the quadratic case (111). We must provide also an *a priori* probability density $z(\theta)$ for the estimanda. Then, the average cost incurred when the strategy represented by $d\hat{\Pi}(\theta)$ is employed, is given by

$$\overline{C} = \int d^m\theta \, z(\theta) C(\theta_*, \theta) p(\theta_* | \theta) = \mathrm{Tr} \int \hat{\mathcal{W}}(\theta_*) \, d\hat{\Pi}(\theta_*) , \tag{112}$$

where $\hat{\mathcal{W}}(\theta)$ denotes the selfadjoint risk operator

$$\hat{\mathcal{W}}(\theta_*) = \int d^m\theta \, z(\theta) C(\theta_*, \theta) \, \hat{\rho}(\theta) . \tag{113}$$

We want now to select the POM that minimizes the average cost $\overline{C}$ under the constraints

$$d\hat{\Pi}(\theta) \geq 0 , \qquad \int d\hat{\Pi}(\theta) = \hat{1} . \tag{114}$$

This can be done as follows. Both operators $d\hat{\Pi}(\theta)$ and $\hat{\mathcal{W}}(\theta)$ are limited from below: then the minimum $\overline{C}$ is achieved when their product $d\hat{\Pi}(\theta)\hat{\mathcal{W}}(\theta)$ under trace and integration in Eq. (112) is minimum. It follows that the equations for the optimal POM can be written as follows

$$[\hat{\mathcal{W}}(\theta) - \hat{\mathcal{Y}}]d\hat{\Pi}_0(\theta) = 0\,, \qquad \hat{\mathcal{W}}(\theta) - \hat{\mathcal{Y}} \geq 0\,, \tag{115}$$

where $\hat{\mathcal{Y}}$ denotes the minimum risk operator (also called Lagrange operator)

$$\hat{\mathcal{Y}} = \min_{\theta}\{\hat{\mathcal{W}}(\theta)\}\,. \tag{116}$$

It is clear that the solution of Eqs. (115) minimizes the operator under trace in Eq. (112). In fact, upon integrating the first of Eqs. (115), from the definition of $\hat{\mathcal{Y}}$ it follows that

$$\int \hat{\mathcal{W}}(\theta)d\hat{\Pi}(\theta) \geq \hat{\mathcal{Y}} \int d\hat{\Pi}_0(\theta) = \hat{\mathcal{Y}}\,. \tag{117}$$

Hence, $\hat{\mathcal{Y}}$ is also the minimum integral under trace in Eq. (112), i.e.

$$\hat{\mathcal{Y}} = \int \hat{\mathcal{W}}(\theta)d\hat{\Pi}_0(\theta)\,, \qquad \overline{C}_{min} = \mathrm{Tr}\hat{\mathcal{Y}}\,. \tag{118}$$

Notice that the Lagrange operator is selfadjoint by definition, and hence

$$\hat{\mathcal{Y}} = \int \hat{\mathcal{W}}(\theta)d\hat{\Pi}_0(\theta) = \int d\hat{\Pi}_0(\theta)\hat{\mathcal{W}}(\theta)\,, \tag{119}$$

namely the optimal POM also satisfies the hermitian conjugated of equation (115).

In general, solving Eqs. (115) and (118) is a difficult task. In the following subsection we will analyze in some detail a relevant example: the canonical measurement of the phase.

5.1. CANONICAL MEASUREMENT OF THE PHASE

The estimation problem is the following: to estimate the phase-shift ϕ of a fixed density matrix $\hat{\rho}_0$ undergoing the unitary transformation

$$\hat{\rho}(\phi) = e^{-ia^\dagger a\phi}\hat{\rho}_0 e^{ia^\dagger a\phi}\,. \tag{120}$$

First, we observe that ϕ is defined on a circle (a 2π-window), because $a^\dagger a$ is an integer operator. Then, we notice that the family of states $\{\hat{\rho}(\phi)\}$ is "covariant", namely it is of the form

$$\hat{\rho}(\phi) = \hat{U}_\phi \hat{\rho}_0 \hat{U}_\phi^\dagger\,, \tag{121}$$

where $\hat{U}_\phi$ are unitary operators representing a group—in the present case, the abelian group $U(1)$ of rotation along one axis. For a covariant estimation problem, the optimal POM must be itself covariant. This should be true at least if we want a likelihood $p(\phi|\phi)$ which is independent on ϕ [for a general study of covariant estimation problems see Ref. [14]] i.e.

$$p(\phi|\phi) = \mathrm{Tr}[\hat{U}_\phi \hat{\rho}_0 \hat{U}_\phi^\dagger d\hat{\Pi}(\phi)] = \mathrm{const.}\,, \tag{122}$$

which, due to invariance of trace under cyclic permutations leads to

$$\mathrm{d}\hat{\Pi}(\phi) = \hat{U}_\phi \mathrm{d}\hat{\Pi}(0)\hat{U}_\phi^\dagger \,. \tag{123}$$

Hence, the problem is restricted to find only the operator $\mathrm{d}\hat{\Pi}(0) \doteq \mathrm{d}\phi\,\hat{\xi}_0/(2\pi)$. Consistently with the assumption of likelihood constant vs ϕ, we consider the case of *a priori* unknown parameter ϕ, with uniform probability density $z(\phi) = \frac{1}{2\pi}$. For the moment, we address only the max-likelihood estimation problem, corresponding to the cost function

$$C(\phi_*, \phi) = -\delta_{2\pi}(\phi_* - \phi) \,, \tag{124}$$

where $\delta_{2\pi}$ denotes the 2π-periodic delta function. With the above choices, the risk operator is given by

$$\hat{\mathcal{W}}(\phi_*) = \int \frac{\mathrm{d}\phi}{2\pi} C(\phi, \phi_*)\, \hat{U}_\phi \hat{\rho}_0 \hat{U}_\phi^\dagger = -\frac{1}{2\pi} \hat{U}_{\phi_*} \hat{\rho}_0 \hat{U}_{-\phi_*} \,, \tag{125}$$

and the Lagrange operator becomes diagonal with $a^\dagger a$, namely

$$\hat{\mathcal{Y}} = -\int \frac{\mathrm{d}\phi}{(2\pi)^2} \hat{U}_\phi \hat{\xi}_0 \hat{\rho}_0 \hat{U}_\phi^\dagger \,, \tag{126}$$

$$\langle k|\hat{\mathcal{Y}}|l\rangle = -\int \frac{\mathrm{d}\phi}{(2\pi)^2} e^{-i(k-l)\phi} \langle k|\hat{\xi}_0 \hat{\rho}_0|l\rangle = -\delta_{kl} \frac{1}{2\pi} \langle k|\hat{\xi}_0 \hat{\rho}_0|k\rangle \,. \tag{127}$$

Thanks to covariance, the estimation problem resorts to seeking the solution $\mathrm{d}\hat{\Pi}(0)$ of the following equations only

$$[\hat{\mathcal{W}}(0) - \hat{\mathcal{Y}}]\mathrm{d}\hat{\Pi}(0) = 0 \,, \qquad \hat{\mathcal{W}}(0) - \hat{\mathcal{Y}} \geq 0 \,. \tag{128}$$

Notice that from Eq. (125) one has $\hat{\mathcal{W}}(0) = -\frac{1}{2\pi}\hat{\rho}_0$: hence, Eqs. (128) can be written as follows

$$[\hat{\rho}_0 + 2\pi\hat{\mathcal{Y}}]\xi_0 = 0 \,, \qquad \hat{\rho}_0 + 2\pi\hat{\mathcal{Y}} \leq 0 \,. \tag{129}$$

The problem is still too difficult, and we restrict attention to the case of pure states $\hat{\rho}_0 = |\psi\rangle\langle\psi|$. We seek solutions of Eqs. (129) in the form

$$\mathrm{d}\hat{\Pi}(0) = \frac{\mathrm{d}\phi}{2\pi} |\gamma\rangle\langle\gamma| \,, \tag{130}$$

where $|\gamma\rangle$ is a (generally non normalizable) vector in the Hilbert space. The Lagrange operator has the following nonvanishing matrix elements

$$\langle k|\hat{\mathcal{Y}}|k\rangle = -\frac{1}{2\pi} \langle k|\gamma\rangle\langle\gamma|\psi\rangle\langle\psi|k\rangle \equiv -\frac{1}{2\pi} \gamma_k \psi_k^* \sum_{n=0}^{\infty} \gamma_n^* \psi_n \,, \tag{131}$$

where $\gamma_k \doteq \langle k|\gamma\rangle$ and $\psi_k \doteq \langle k|\psi\rangle$. Completeness of $\mathrm{d}\hat{\Pi}(\phi)$ implies that

$$\delta_{nm} = \langle n| \int \mathrm{d}\hat{\Pi}(\phi)|m\rangle = \int \frac{\mathrm{d}\phi}{2\pi} e^{-i\phi(n-m)} \langle n|\gamma\rangle\langle\gamma|m\rangle = \delta_{nm} |\langle n|\gamma\rangle|^2 \,, \tag{132}$$

which requires $|\gamma_k| = 1$. Moreover, reality of $\hat{\mathcal{Y}}$ needs $\arg(\gamma_k) = \arg(\psi_k)$ for $\psi_k \neq 0$. Hence, we write

$$\gamma_k = \begin{cases} \psi_k/|\psi_k| \doteq e^{i\chi_k} , & \psi_k \neq 0 \\ 1 , & \psi_k = 0 \end{cases} \tag{133}$$

leading to

$$\langle k|\hat{\mathcal{Y}}|k\rangle = -\frac{1}{2\pi}|\psi_k|\sum_{n=0}^{\infty}|\psi_n| \,. \tag{134}$$

Now, we only need to check Eqs. (129). The second equation means that for any vector $|v\rangle$ in the Hilbert space, one has

$$0 \geq \langle v|\hat{\rho}_0 + 2\pi\hat{\mathcal{Y}}|v\rangle = \Big|\sum_{n=0}^{\infty} v_n^*\psi_n\Big|^2 - \sum_{n=0}^{\infty}|v_n|^2|\psi_n|\sum_{k=0}^{\infty}|\psi_k| \,. \tag{135}$$

This bound is satisfied according to the Schwartz inequality

$$\begin{aligned} \Big|\sum_{n=0}^{\infty} v_n^*\psi_n\Big|^2 &= \Big|\sum_{n=0}^{\infty} v_n^*|\psi_n|\gamma_n\Big|^2 = \Big|\sum_{n=0}^{\infty} v_n^*|\psi_n|^{1/2}\gamma_n|\psi_n|^{1/2}\Big|^2 \\ &\leq \sum_{n=0}^{\infty}|v_n|^2|\psi_n|\sum_{k=0}^{\infty}|\psi_k| \,. \end{aligned} \tag{136}$$

It remains to show that the first one of Eqs. (129) is also satisfied. One has

$$\begin{aligned} \langle k|[\hat{\rho}_0 + 2\pi\hat{\mathcal{Y}}]\xi_0|m\rangle &= \langle k|\psi\rangle\langle\psi|\gamma\rangle\langle\gamma|m\rangle + 2\pi\langle k|\hat{\mathcal{Y}}|k\rangle\langle k|\gamma\rangle\langle\gamma|m\rangle \\ &= [\langle k|\psi\rangle\langle\psi|\gamma\rangle - \langle k|\psi\rangle\langle\psi|\gamma\rangle\langle\gamma|k\rangle\langle k|\gamma\rangle]\langle\gamma|m\rangle \\ &= \langle k|\psi\rangle\langle\psi|\gamma\rangle[1 - |\gamma_k|^2]\langle\gamma|m\rangle = 0 \,, \end{aligned} \tag{137}$$

where we have considered that due to Eq. (119) also the following identity holds true

$$\hat{\mathcal{Y}} = -\frac{1}{2\pi}\hat{\xi}_0\hat{\rho}_0 \,. \tag{138}$$

In summary, we have proved the following assertion: the POM for estimating a phase shift of a pure state $|\psi\rangle\langle\psi|$ with max-likelihood cost-function is given by

$$\mathrm{d}\hat{\Pi}(\phi) = \frac{\mathrm{d}\phi}{2\pi}e^{-ia^\dagger a\phi}|\gamma\rangle\langle\gamma|e^{ia^\dagger a\phi} \equiv \frac{\mathrm{d}\phi}{2\pi}\sum_{nm=0}^{\infty} e^{-i(n-m)\phi+i(\chi_n-\chi_m)}|n\rangle\langle m| \,, \tag{139}$$

where the phases χ_n depend on the state as follows

$$\chi_n = \begin{cases} \arg(\langle n|\psi\rangle) , & \langle n|\psi\rangle \neq 0 \\ 0 , & \langle n|\psi\rangle = 0 \end{cases} \tag{140}$$

In practice, it is not too restrictive to consider states with a well defined phase—i.e. with $\chi_n = n\nu$—which are just rotated by an angle ν of real positive states having all $\chi_n = 0$. In this case the optimal POM takes the canonical form

$$\mathrm{d}\hat{\Pi}(\phi) = \frac{\mathrm{d}\phi}{2\pi}\sum_{nm=0}^{\infty} e^{-i(n-m)\phi}|n\rangle\langle m| \,. \tag{141}$$

We have find the optimal POM according to the max-likelihood criterion that corresponds to a δ-like cost function. However, the same POM is optimal for any cost-function of the form $C(\phi_*, \phi) = C(\phi - \phi_*)$ where $C(\phi)$ is an even 2π-periodic function on $\mathbf{R}$ satisfying

$$\int_0^\infty C(\phi) \cos k\phi \, \mathrm{d}\phi \leq 0 \,, \quad k = 1, 2 \ldots \tag{142}$$

In fact, any function $C(\phi)$ satisfying Eq. (142) has Fourier series of the form

$$C(\phi) = c_0 - \sum_{k=0}^{\infty} c_k \cos k\phi \,, \qquad c_k \geq 0 \,. \tag{143}$$

Then, consider a general covariant POM

$$\mathrm{d}\hat{\Pi}(\phi) = \frac{\mathrm{d}\phi}{2\pi} \sum_{nm=0}^{\infty} e^{-i(n-m)\phi} \xi_{nm} |n\rangle\langle m| \,, \tag{144}$$

with $\xi_{nm} = \langle n|\hat{\xi}_0|m\rangle$ and $\hat{\xi}_0$ a generic selfadjoint operator. The average cost is given by

$$2\pi\overline{C} = c_0 - \sum_{k=1}^{\infty} c_k \int \frac{\mathrm{d}\phi}{2\pi} \cos k\phi \sum_{nm=0}^{\infty} e^{-i(n-m)\phi} \xi_{nm} \langle m|\psi\rangle\langle\psi|n\rangle \tag{145}$$

$$= c_0 - \sum_{k=1}^{\infty} c_k \frac{1}{2} \sum_{|n-m|=k} \langle\psi|n\rangle \xi_{nm} \langle m|\psi\rangle \,. \tag{146}$$

Positivity of $\mathrm{d}\hat{\Pi}(\phi)$ implies that $|\xi_{nm}| \leq \sqrt{\xi_{nn}\xi_{mm}} = 1$, hence

$$\sum_{|n-m|=k} \langle\psi|n\rangle \xi_{nm} \langle m|\psi\rangle \leq \sum_{|n-m|=k} |\langle\psi|n\rangle||\langle m|\psi\rangle| \,, \tag{147}$$

and the equality is achieved only if $\xi_{nm} = \gamma_n^* \gamma_m$, with $|\gamma_n| = 1$. It follows that the minimum cost is

$$2\pi\overline{C}_{min} = c_0 - \frac{1}{2} \sum_{k=1}^{\infty} c_k \sum_{nm=0}^{\infty} |\langle\psi|n\rangle||\langle m|\psi\rangle| \tag{148}$$

and this is attained by the POM (139).[2]

Notice that in the category (142) one also has the following cost functions

$$4 \sin^2 \frac{\phi}{2} = 2 - 2\cos\phi \,, \tag{149}$$

$$\min\{\phi, 2\pi - \phi\} = \frac{\pi}{2} - \frac{4}{\pi} \sum_{k=0}^{\infty} \sum_{k=0}^{\infty} \frac{\cos(2k+1)\phi}{(2k+1)^2} \,, \tag{150}$$

$$|\sin \frac{\phi}{2}| = \frac{2}{\pi} - \frac{4}{\pi} \sum_{k=1}^{\infty} \frac{\cos k\phi}{4k^2 - 1} \,, \tag{151}$$

$$-\delta_{2\pi}(\phi) = -\frac{1}{2\pi} - \frac{1}{\pi} \sum_{k=0}^{\infty} \cos k\phi \,. \tag{152}$$

[2] This is another derivation of the optimal POM for the quantum estimation problem of phase shift of pure states, but for more general cost functions. The second derivation is due to Holevo [14], whereas the previous one is an extended version of the Helstrom's proof [1].

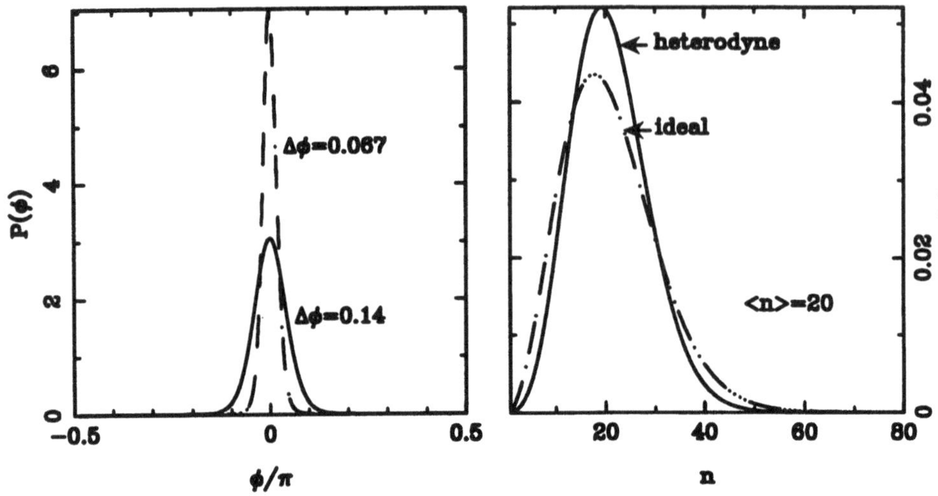

Figure 6. Number and phase probability distribution of optimal phase states with $\langle\hat{n}\rangle = 20$ for ideal and heterodyne phase detection. Optimal states minimize the r.m.s. phase fluctuations (the procedure for deriving such states is explained in Ref. [15]).

Minimizing the cost is equivalent to minimize the corresponding periodicized fluctuations (r.m.s fluctuations in Eq. (149)), or to maximize the likelihood in Eq. (152). Notice that the function $\min\{\phi^2, (\phi-2\pi)^2\}$ does not belong to category (142). As the optimal POM is rather insensitive to the choice of the cost function, it can deserve the name "ideal" POM for the phase.

In Fig. 6 I report a numerical comparison between the ideal and the marginal heterodyne phase detection. For both cases a state with 20 photons is considered that minimizes the r.m.s. phase deviation $\Delta\phi = \sqrt{\langle\Delta\phi^2\rangle}$ of the corresponding detection probability. It is evident that the ideal POM leads to probability distribution sharper than the heterodyne POM. Correspondingly, the number probability of optimal states for ideal detection are slightly broader than the number probability of optimal states for heterodyne detection. A more detailed analysis on marginal phase detection can be found in Ref. [15].

6. Beyond the POM: state reduction and "instrument"

Insofar we have considered only measurements that completely destroy the quantum mechanical description of the system after the interaction with a macroscopic detector: this is the case, for example, of photodetection, where radiation is completely absorbed. We are now interested in a different kind of measurements, which do not destroy the quantum mechanical description of the system, so that in principle a second measurement on the system can be performed after the first one. We call this type of measurements "measurements of the first kind", generalizing a term introduced by W. Pauli [16]. Henceforth, the customary measurement—i.e. those that destroy the state of the system—will be referred to as "measurements of the second kind".[3] More precisely, the definition of measurements

[3]This nomenclature has been used by M. Ozawa in Ref. [17]

of the first and second kind are as follows. For the second kind measurements the quantum mechanical description is provided just by the Born's rule: hence, these measurements are in one-to-one correspondence with POM's. For the first kind measurements, on the other hand, the description provides also the "state reduction" $\hat{\rho} \to \hat{\rho}_\Delta$, which gives the state $\hat{\rho}_\Delta$ immediately after the measurement, for a given experimental event Δ and for state $\hat{\rho}$ immediately before the measurement.[4] The "state reduction" is needed in order to evaluate the statistics of repeated measurements. In the following, we will refer to as "statistics of the measurements" including both the Born's rule and the state reduction.[5] The physical design and the preparation of the measuring apparatus determines the whole statistics of the measurement. In the following we will analyze the mathematical notion of "instrument", which synthetically describes the statistics of a measurement of the first kind.

6.1. INDIRECT MEASUREMENTS

The first kind measurement can be defined as a special type of "indirect" measurement of the second kind. An indirect measurement is a measurement that, instead of being performed directly on the system of interest, is carried out on a different system, which may include also the original system itself. Observables are measured that support informations on the desired quantities (for example, they have the same expectation values), but are different from them, and hence have different statistical distributions. We have already considered this kind of measurement in these lectures, when we analyzed joint measurements of conjugated quadratures of the field by heterodyne detection. The general scheme of this kind of measurements is sketched in Fig. 7. There is a system S and a probe P that interact (but not necessarily) each other; a measurement of the second kind is performed on compatible observables corresponding to operators acting on the whole Hilbert space $\mathcal{H}_S \otimes \mathcal{H}_P$. In this case the state of the system itself is destroyed, and overall the measurement is of the second kind: the probe P is needed only in order to make observables compatible. For example, if $S + P$ have orthogonal projection-valued observables $\mathrm{d}\hat{E}_{SP}(\mathbf{x})$ and preparation $\hat{\rho}\otimes\hat{\rho}_P$ (I drop the subindex S from the system density matrix) and the measurement is performed after the interaction $\hat{U}$, the Born's rule is given by

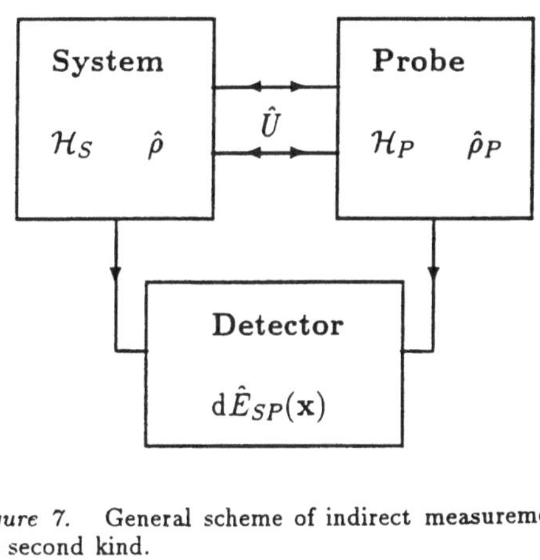

Figure 7. General scheme of indirect measurement of the second kind.

$$\mathrm{d}P(\mathbf{x}) = \mathrm{Tr}_{S+P}[\hat{U}\hat{\rho}\otimes\hat{\rho}_P\hat{U}^\dagger \mathrm{d}\hat{E}_{SP}(\mathbf{x})] = \mathrm{Tr}_S[\hat{\rho}\mathrm{d}\hat{\Pi}(\mathbf{x})] , \tag{153}$$

[4]In this way the nomenclature "first" and "second kind" can be put into correspondence to the "levels of description" of Holevo [18]. Here level I of description is the pure probabilistic one provided by the POM's, level II is the description of state reduction; finally, level III is the complete unitary description of the microscopic apparatus.

[5]Again this nomenclature is due to Ozawa [17].

corresponding to the POM

$$\mathrm{d}\hat{\Pi}(\mathbf{x}) = \mathrm{Tr}_P[\hat{\rho}_P \hat{U}^\dagger \mathrm{d}\hat{E}_{SP}(\mathbf{x})\hat{U}] \,. \tag{154}$$

There is a simple way to change the above scheme in order to make it suited to first kind measurements: just make a second kind measurement only on a probe observable $\mathrm{d}\hat{E}_P(\mathbf{x})$ (it is clear that now the interaction $\hat{U}$ is strictly needed). The resulting measurement scheme is sketched in Fig. 8. The Born's rule is

$$\mathrm{d}P(\mathbf{x}) = \mathrm{Tr}_{SP}[\hat{U}\hat{\rho}\otimes\hat{\rho}_P\hat{U}^\dagger \hat{1}_S \otimes \mathrm{d}\hat{E}_P(\mathbf{x})] = \mathrm{Tr}_S\{\hat{\rho}\mathrm{Tr}_P[\hat{\rho}_P\hat{U}^\dagger \hat{1}_S\otimes \mathrm{d}\hat{E}_P(\mathbf{x})\hat{U}]\} \,, \tag{155}$$

corresponding to the POM

$$\mathrm{d}\hat{\Pi}(\mathbf{x}) = \mathrm{Tr}_P[\hat{\rho}_P \hat{U}^\dagger \hat{1}_S \otimes \mathrm{d}\hat{E}_P(\mathbf{x})\hat{U}] \,. \tag{156}$$

Now, in order to determine the state reduction, one assumes that immediately after this measurement the system S is subjected to another measurement of an arbitrary observable of S, say with spectral resolution $\mathrm{d}\hat{E}_S(\mathbf{y})$.[6] The joint probability for the two combined measurements is

$$\mathrm{d}P(\mathbf{x},\mathbf{y}) = \mathrm{Tr}_{SP}[\hat{U}\hat{\rho}\otimes\hat{\rho}_P\hat{U}^\dagger \mathrm{d}\hat{E}_S(\mathbf{y})\otimes \mathrm{d}\hat{E}_P(\mathbf{x})] \,. \tag{157}$$

It is clear that the result would be exactly the same if the measurement described by $\mathrm{d}\hat{E}_S(\mathbf{y})$ is performed immediately before—instead of immediately after—the first measurement (however after the interaction $\hat{U}$). Hence, there is no causal relation between the "reading of the result $\mathbf{x}$" and the result $\mathbf{y}$ of the second measurement. Now, let us consider the experiment from the point of view of an observer who ignores the apparatus. He asserts that the first measurement has produced result $\mathbf{x}$ according to the Born's rule

$$\mathrm{d}P(\mathbf{x}) = \mathrm{Tr}[\hat{\rho}\,\mathrm{d}\hat{\Pi}(\mathbf{x})] \,. \tag{158}$$

Then, he considers the second measurement as performed "immediately after" the first

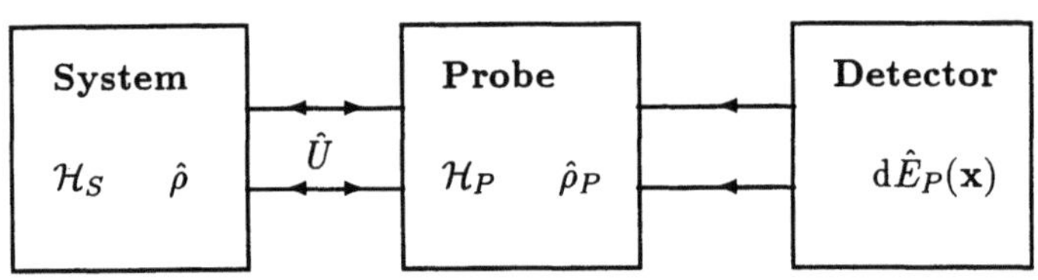

Figure 8. Scheme of an indirect measurement of the first kind.

one, described by the Born's rule with POM $\mathrm{d}\hat{E}_S(\mathbf{y})$, but now the measurement is performed on a different state, say $\hat{\rho}_\mathbf{x}$, that depends on the result $\mathbf{x}$ of the first measurement.

[6] Here I consider that the second measurement is of the second kind. The argument can be easily extended to the case that the second measurement is itself of the first kind: however, for the present purpose, this would create a logical loop.

In formulas, the conditional probability $\mathrm{d}P(\mathbf{x}|\mathbf{y})$ of obtaining $\mathbf{y}$ given the result of the first measurement was $\mathbf{x}$, is given by

$$\mathrm{d}P(\mathbf{x}|\mathbf{y}) = \mathrm{Tr}_S[\hat{\rho}_{\mathbf{x}}\mathrm{d}\hat{E}_S(\mathbf{y})] , \tag{159}$$

and hence the joint probability of obtaining $\mathbf{x}$ and $\mathbf{y}$ can be written as follows

$$\mathrm{d}P(\mathbf{x},\mathbf{y}) = \mathrm{d}P(\mathbf{x}|\mathbf{y})\,\mathrm{d}P(\mathbf{x}) = \mathrm{Tr}_S[\hat{\rho}_{\mathbf{x}}\,\mathrm{d}\hat{E}_S(\mathbf{y})]\mathrm{Tr}_S[\hat{\rho}\,\mathrm{d}\hat{\Pi}(\mathbf{x})] . \tag{160}$$

On the other hand, the probability (160) must be equal to the probability (157): in this way the following identity is obtained

$$\mathrm{Tr}_S[\hat{\rho}_{\mathbf{x}}\mathrm{d}\hat{E}_S(\mathbf{y})] = \frac{\mathrm{Tr}_{SP}[\hat{U}\hat{\rho}\otimes\hat{\rho}_P\hat{U}^\dagger\mathrm{d}\hat{E}_S(\mathbf{y})\otimes\mathrm{d}\hat{E}_P(\mathbf{x})]}{\mathrm{Tr}_{SP}[\hat{U}\hat{\rho}\otimes\hat{\rho}_P\hat{U}^\dagger\hat{1}_S\otimes\mathrm{d}\hat{E}_P(\mathbf{x})]} . \tag{161}$$

The arbitrariness of the choice of the second measurement yields the following relation for any basis $\{|n\rangle\}$

$$\langle n|\hat{\rho}_{\mathbf{x}}|m\rangle = \frac{\mathrm{Tr}_{SP}[\hat{U}\hat{\rho}\otimes\hat{\rho}_P\hat{U}^\dagger|m\rangle\langle n|\otimes\mathrm{d}\hat{E}_P(\mathbf{x})]}{\mathrm{Tr}_{SP}[\hat{U}\hat{\rho}\otimes\hat{\rho}_P\hat{U}^\dagger\hat{1}_S\otimes\mathrm{d}\hat{E}_P(\mathbf{x})]} , \tag{162}$$

namely

$$\begin{aligned}\hat{\rho}_{\mathbf{x}} &= \sum_{nm}|n\rangle\langle m|\frac{\mathrm{Tr}_{SP}[\hat{U}\hat{\rho}\otimes\hat{\rho}_P\hat{U}^\dagger|m\rangle\langle n|\otimes\mathrm{d}\hat{E}_P(\mathbf{x})]}{[\hat{U}\hat{\rho}\otimes\hat{\rho}_P\hat{U}^\dagger\hat{1}_S\otimes\mathrm{d}\hat{E}_P(\mathbf{x})]}\\ &= \sum_{nm}|n\rangle\langle m|\frac{\mathrm{Tr}_S\{\mathrm{Tr}_P[\hat{U}\hat{\rho}\otimes\hat{\rho}_P\hat{U}^\dagger\hat{1}_s\otimes\mathrm{d}\hat{E}_P(\mathbf{x})]|m\rangle\langle n|\}}{\mathrm{Tr}_{SP}[\hat{U}\hat{\rho}\otimes\hat{\rho}_P\hat{U}^\dagger\hat{1}_S\otimes\mathrm{d}\hat{E}_P(\mathbf{x})]}\\ &= \frac{\mathrm{Tr}_P[\hat{U}\hat{\rho}\otimes\hat{\rho}_P\hat{U}^\dagger\hat{1}_s\otimes\mathrm{d}\hat{E}_P(\mathbf{x})]}{\mathrm{Tr}_{SP}[\hat{U}\hat{\rho}\otimes\hat{\rho}_P\hat{U}^\dagger\hat{1}_S\otimes\mathrm{d}\hat{E}_P(\mathbf{x})]} .\end{aligned} \tag{163}$$

Notice that the denominator in Eq. (163) is just the trace of the numerator over $\mathcal{H}_S$. Hence, we can write Eq. (163) as follows

$$\hat{\rho}_{\mathbf{x}} = \frac{\mathrm{d}I(\mathbf{x})\hat{\rho}}{\mathrm{Tr}_S[\mathrm{d}I(\mathbf{x})\hat{\rho}]} , \tag{164}$$

where the mapping $\mathrm{d}I(\mathbf{x})$ is defined as

$$\mathrm{d}I(\mathbf{x})\hat{\rho} = \mathrm{Tr}_P[\hat{U}\hat{\rho}\otimes\hat{\rho}_P\hat{U}^\dagger\hat{1}_S\otimes\mathrm{d}\hat{E}_P(\mathbf{x})] , \tag{165}$$

and is called "instrument".[7]

Thus we have seen that the description of a first kind measurement in terms of Born's rule and state reduction $\hat{\rho}\to\hat{\rho}_{\mathbf{x}}$ pertains an observer who ignores the microscopic description of the apparatus and focuses attention on the system S only. This also makes clear that the state reduction is not a causal evolution: it is not the consequence of the first

[7]For many authors [19] the names "instrument" and "POM" are substituted by "operations" and "effects", respectively.

observation and of "knowing the result", but just the statistical correlation between the results of the two measurements due to the interaction of the system with the probe.[8]

Now, let us consider the properties of the map $\mathrm{d}I(\mathbf{x})$ more abstractly. The result of the measurement is not just a point $\mathbf{x}$: more generally, it is a Borel set, practically an interval $\Delta \in \mathbf{R}$ which the readout of the measurement is known to belong to. The above derivation of the state reduction can be generalized to the following rule (hereafter we drop the subindex S everywhere)

$$\hat{\rho} \to \hat{\rho}_\Delta = \frac{I(\Delta)\hat{\rho}}{\mathrm{Tr}[I(\Delta)\hat{\rho}]}\,, \qquad I(\Delta) = \int_\Delta \mathrm{d}I(\mathbf{x})\,. \tag{166}$$

6.2. REALIZABLE INSTRUMENTS

Mathematically, the map $I(\Delta)$ is a linear transformation of trace class operators with the following rules

$$\begin{aligned} &0 \le \mathrm{Tr}[I(\Delta)\hat{\rho}] \le 1\,, \quad \mathrm{Tr}[I(\mathbf{R})\hat{\rho}] = 1 \\ &\mathrm{Tr}[I(\cup_n \Delta_n)\hat{\rho}] = \sum_n \mathrm{Tr}[I(\Delta)\hat{\rho}]\,, \qquad \{\Delta_n\} \text{ countable disjoint}\,. \end{aligned} \tag{167}$$

Notice that $I(\mathbf{R})\hat{\rho} \neq \hat{\rho}$, in general.[9] It is easy to check that the map defined in Eq. (165) along with Eq. (166) satisfies the above axioms.[10] On the other hand, an abstract map I satisfying Eqs. (167) fully describes the statistics of a measurement of the first kind. It gives both the state reduction and the Born's rule as follows

$$\hat{\rho} \to \hat{\rho}_\Delta = \frac{I(\Delta)\hat{\rho}}{\mathrm{Tr}[I(\Delta)\hat{\rho}]}\,, \qquad P(\Delta) = \int_\Delta \mathrm{d}P(\mathbf{x}) = \mathrm{Tr}[I(\Delta)\hat{\rho}]\,. \tag{168}$$

Now we address the problem if an instrument that satisfies axioms (167) can be physically realized in terms of an indirect measurement involving an interaction with some probe P, and for a suitable preparation of the probe. In other words, we want to know if any "mathematically given" instrument $\mathrm{d}I$ is "physically realizable" with a unitary interaction between S and P, as in Eq. (165). To this purpose, first notice that Eq. (165) leads to

[8]In my knowledge, this point was first clarified by Ozawa [20].

[9]The map $I(\mathbf{R})$ describes a "measurement without reading": this is the evolution of an "open system" S in interaction with an "environment" P.

[10]In particular, let us check positivity of the map, namely

$$\langle v|\mathrm{d}I(\mathbf{x})\hat{\rho}|v\rangle \ge 0 \quad \forall v \in \mathcal{H}\,, \forall \hat{\rho} \text{ traceclass}\,.$$

One has

$$\langle v|\mathrm{d}I(\mathbf{x})\hat{\rho}|v\rangle = \mathrm{Tr}[\hat{U}\hat{\rho}\otimes\hat{\rho}_P\hat{U}^\dagger|v\rangle\langle v|\otimes \mathrm{d}\hat{E}_P(\mathbf{x})]\,.$$

Both density matrices can be written as convex linear combination of pure states. Hence, it is sufficient to prove positivity for pure states only, say $\hat{\rho} \equiv |\psi\rangle\langle\psi|$ and $\hat{\rho}_P \equiv |\varphi\rangle\langle\varphi|$. One has

$$\mathrm{Tr}[\hat{U}|\psi\rangle\langle\psi|\otimes|\varphi\rangle\langle\varphi|\hat{U}^\dagger|v\rangle\langle v|\otimes \mathrm{d}\hat{E}(\mathbf{x})] = \mathrm{d}\mathbf{x}\left\|(|\psi\rangle\langle\psi|\otimes|\varphi\rangle\langle\varphi|)\hat{U}^\dagger|v\rangle|\mathbf{x}\rangle\right\|^2 \ge 0\,.$$

a property for $\mathrm{d}I$ which is stronger than positivity: this is "complete positivity". We say that an instrument is completely positive if it satisfies the following requirement: for any finite sequence of vectors $|v_k\rangle$ and $|w_k\rangle$, $k = 1, \ldots, n$ one has

$$\sum_{k,l=1}^{n} \langle v_k|[\mathrm{d}I(\mathbf{x})|w_k\rangle\langle w_l|]|v_l\rangle \geq 0 \,. \tag{169}$$

From Eq. (165) with $\hat{\rho}_P = |\varphi\rangle\langle\varphi|$ pure state we have

$$\sum_{k,l=1}^{n} \langle v_k|[\mathrm{d}I(\mathbf{x})|w_k\rangle\langle w_l|]|v_l\rangle = \sum_{k,l=1}^{n} \langle w_l|\langle\varphi|\hat{U}^\dagger[|v_l\rangle\langle v_k| \otimes \mathrm{d}\hat{E}_P(\mathbf{x})]\hat{U}|\varphi\rangle|w_k\rangle$$

$$= \sum_{k,l=1}^{n} \langle w_l|\langle\varphi|\hat{U}^\dagger|\mathbf{x}\rangle|v_l\rangle\langle v_k|\langle\mathbf{x}|\hat{U}|\varphi\rangle|w_k\rangle \mathrm{d}\mathbf{x} = \left|\sum_{k=1}^{n}\langle w_l|\langle\varphi|\hat{U}^\dagger|\mathbf{x}\rangle|v_l\rangle\right|^2 \mathrm{d}\mathbf{x} \geq 0 \,. \tag{170}$$

Hence, a realizable instrument is completely positive. Remarkably, Ozawa [20] has proven also the converse assertion, more precisely: Every completely positive instrument $\mathrm{d}I(\mathbf{x})$ is realizable, i.e. there is an extension $\mathcal{H} \otimes \mathcal{H}_P$ of the Hilbert space, a pure state preparation $|\varphi\rangle \in \mathcal{H}_P$ of the probe, a unitary operator $\hat{U}$ acting on $\mathcal{H} \otimes \mathcal{H}_P$, and a selfadjoint operator on $\mathcal{H}_P$ with spectral resolution $\mathrm{d}\hat{E}(\mathbf{x})$, such that

$$\mathrm{d}I(\mathbf{x})\hat{\rho} = \mathrm{Tr}_P[\hat{U}\hat{\rho} \otimes |\varphi\rangle\langle\varphi|\hat{U}^\dagger\hat{1} \otimes \mathrm{d}\hat{E}(\mathbf{x})] \,. \tag{171}$$

For the proof of the theorem the reader is referred to the original work of Ozawa [20].

We have seen that every instrument $\mathrm{d}I(\mathbf{x})$ is associated to a POM $\mathrm{d}\hat{\Pi}(\mathbf{x})$ by trace-duality as follows

$$\mathrm{Tr}[\mathrm{d}\hat{\Pi}(\mathbf{x})\hat{\rho}] = \mathrm{Tr}[\mathrm{d}I(\mathbf{x})\hat{\rho}] \,. \tag{172}$$

It is also true that for every POM $\mathrm{d}\hat{\Pi}(\mathbf{x})$ there is always at least an instrument $\mathrm{d}I(\mathbf{x})$ that satisfies Eq. (172). As a consequence, every POM can be achieved by a measurement of the first kind, and thus the Ozawa's theorem generalizes the Naimark's theorem.

I emphasize that an instrument $\mathrm{d}I(\mathbf{x})$ unambiguously determines a POM $\mathrm{d}\hat{\Pi}(\mathbf{x})$, whereas a POM $\mathrm{d}\hat{\Pi}(\mathbf{x})$ can be generally obtained from many different instruments: in other words, one has the same Born's rule with different state reductions. The POM does not contain sufficient details on the apparatus to describe the back action on the system, whereas the instrument provides a complete description of the statistics of the measurement. On the other hand, there can be still many different apparatus—i.e. different probes, probe preparations, and system-probe interactions—that are described by the same instrument $\mathrm{d}I(\mathbf{x})$.[11]

Now I want to make the relation between instrument and POM more explicit. Let us consider, for simplicity, the case of continuous spectrum and pure state preparation $|\varphi\rangle$ for the probe. One has

$$\begin{aligned}\int_\Delta \mathrm{d}I(\mathbf{x})\hat{\rho} &= \mathrm{Tr}_P\left[\hat{U}(\hat{\rho} \otimes |\varphi\rangle\langle\varphi|)\hat{U}^\dagger\left(\hat{1} \otimes \int_\Delta \mathrm{d}\mathbf{x}|\mathbf{x}\rangle\langle\mathbf{x}|\right)\right] \\ &= \int_\Delta \mathrm{d}\mathbf{x} \int \mathrm{d}\mathbf{x}' \,\langle\mathbf{x}'|\hat{U}|\varphi\rangle\, \hat{\rho}\, \langle\varphi|\hat{U}^\dagger|\mathbf{x}\rangle\delta(\mathbf{x}-\mathbf{x}') \\ &= \int_\Delta \mathrm{d}\mathbf{x}\,\langle\mathbf{x}|\hat{U}|\varphi\rangle\, \hat{\rho}\, \langle\varphi|\hat{U}^\dagger|\mathbf{x}\rangle = \int_\Delta \mathrm{d}\mathbf{x}\,\hat{\Omega}(\mathbf{x})\,\hat{\rho}\,\hat{\Omega}^\dagger(\mathbf{x}) \,, \end{aligned} \tag{173}$$

[11] This is the level III of description of Holevo [18].

where $\hat{\Omega}(\mathbf{x})$ is the (non-unitary) operator acting on the Hilbert space $\mathcal{H}$ of the system only

$$\hat{\Omega}(\mathbf{x}) = \langle \mathbf{x}|\hat{U}|\varphi\rangle \tag{174}$$

which satisfies the completeness relation[12]

$$\int \mathrm{d}\mathbf{x}\, \hat{\Omega}^\dagger(\mathbf{x})\, \hat{\Omega}(\mathbf{x}) = \hat{1}\,. \tag{175}$$

The POM associated with the instrument can be obtained upon substituting Eq. (174) into (172), and using invariance of trace under cyclic permutations. One has

$$\mathrm{d}\hat{\Pi}(\mathbf{x}) = \mathrm{d}\mathbf{x}\, \hat{\Omega}^\dagger(\mathbf{x})\, \hat{\Omega}(\mathbf{x})\,, \tag{176}$$

namely

$$\hat{\Pi}(\Delta) = \int_\Delta \mathrm{d}\mathbf{x}\, \hat{\Omega}^\dagger(\mathbf{x})\hat{\Omega}(\mathbf{x})\,. \tag{177}$$

The generalization of Eq. (177) to the case of discrete spectrum is straightforward. Now we can immediately see that a way to change the instrument without changing the POM is the following "local" unitary transformation of operators $\hat{\Omega}(\mathbf{x})$

$$\hat{\Omega}(\mathbf{x}) \Longrightarrow \hat{V}(\mathbf{x})\hat{\Omega}(\mathbf{x})\,, \qquad \hat{V}^\dagger(\mathbf{x})\hat{V}(\mathbf{x}) = \hat{1}\,. \tag{178}$$

The transformation (178) does not affect the complete positivity of the instrument $\mathrm{d}I(\mathbf{x})$, however it gives a different state reduction (or "back-action") with the same POM.

6.2.1. *Example 1: the standard von Neumann model*

As a first example, we consider the von Neumann model [21] for a first kind measurement. Originally the model was conceived for the measurement of the position of a particle: here I translate it in the language of quantum optics, and I will describe an indirect unsharp measurement of a field quadrature. The interaction Hamiltonian is given by

$$\hat{H} = 2\hbar\kappa\hat{X}\hat{Y}_P\,, \tag{179}$$

where $\hat{X} = \frac{1}{2}(a^\dagger + a)$ and $\hat{Y}_P = \frac{i}{2}(a_P^\dagger - a_P)$ are quadratures of the field modes pertaining the system and the probe, respectively. The quadratures $\hat{X}$ and $\hat{Y}_P$ are the optical equivalent of position and momentum $\hat{q}_1$ and $\hat{p}_2$ of two different interacting particles, as it was considered in the original von Neumann model. Notice that the present optical model is given only for the sake of exemplification, because is would be difficult to achieve the Hamiltonian (179) optically (but also mechanically!) In the impulsive case (i.e. for strong coupling and short interaction time $\tau = \kappa^{-1} \to 0$) the operator $\hat{\Omega}(\mathbf{x})$ in Eq. (174) is given by

$$\hat{\Omega}(x) = \langle x|\exp(-2i\hat{X}\hat{Y}_P)|\varphi\rangle = \int \mathrm{d}x'\langle x|x'\rangle \exp\left[-\hat{X}\frac{\mathrm{d}}{\mathrm{d}x'}\right]\varphi(x') = \varphi(x - \hat{X})\,, \tag{180}$$

[12] I remind that in our notation the domain of the integral, when not specified, is the spectrum of the considered observable. Also one should keep in mind that vectors $|\mathbf{x}\rangle$ and $|\varphi\rangle$ in Eq. (174) belongs to the Hilbert space $\mathcal{H}_P$, so that the matrix element of $\hat{U}$ between them is an operator acting on $\mathcal{H}$ only ($\hat{U}$ is an operator acting on $\mathcal{H}\otimes\mathcal{H}_P$).

and hence the instrument is

$$\mathrm{d}I(x)|\psi\rangle\langle\psi| = \mathrm{d}x\, \varphi(x-\hat{X})|\psi\rangle\langle\psi|\, \varphi(x-\hat{X})\,. \tag{181}$$

To obtain the von Neumann state reduction—i.e. the projection over eigenvectors $|x\rangle$ of $\hat{X}$—let us consider the squeezed-vacuum preparation for P

$$\varphi(x) = \frac{1}{(2\pi\epsilon^2)^{1/4}} \exp\left(-\frac{1}{4}\frac{x^2}{\epsilon^2}\right)\,. \tag{182}$$

In the limit of vanishingly small $\epsilon \to 0$ (ϵ plays the role of the measurement precision) the reduced state will localize on a narrower and narrower Gaussian centered around the value x. Formally, we write the limit as follows

$$\lim_{\epsilon\to 0} \frac{\mathrm{d}I(x)\hat{\rho}}{\mathrm{Tr}[\mathrm{d}I(x)\hat{\rho}]} = |x\rangle\langle x|\,. \tag{183}$$

6.2.2. *Example 2: the Arthurs-Kelly model for joint measurements*

The previous example can be easily generalized to the case of a joint measurement of $\hat{X}$ and $\hat{Y}$. In this case we need two different probe modes that commute each other. Such a measurement model was considered for the first time by Arthurs and Kelly [22]: here I extend their analysis in order to derive also the state reduction of the model.

The impulsive interaction Hamiltonian can be chosen as follows

$$\hat{H} = \hbar(\kappa_1 \hat{X}\hat{Y}_1 - \kappa_2 \hat{Y}\hat{X}_2)\,, \tag{184}$$

where the choice of signs and constants is for later convenience. For simplicity of notation we set the interaction time $\tau = 1$. Let us analyze this model in the Heisenberg picture. One has

$$\hat{X}'_1 = \hat{U}^\dagger \hat{X}_1 \hat{U} = \hat{X}_1 + \frac{1}{2}\kappa_1\hat{X} - \frac{1}{8}\kappa_1\kappa_2\hat{X}_2\,, \tag{185}$$

$$\hat{Y}'_2 = \hat{U}^\dagger \hat{Y}_2 \hat{U} = \hat{Y}_2 + \frac{1}{2}\kappa_2\hat{Y} - \frac{1}{8}\kappa_1\kappa_2\hat{Y}_1\,. \tag{186}$$

It is convenient to require that the indirect measurement of $\hat{X}$ and $\hat{Y}$ be "unbiased", namely that the time evolved expectation values of $\hat{X}_1$ and $\hat{Y}_2$ are equal to those of $\hat{X}$ and $\hat{Y}$ that we want to measure at $t = 0$, i.e. $\langle\hat{X}'_1\rangle = \langle\hat{X}\rangle$ and $\langle\hat{Y}'_2\rangle = \langle\hat{Y}\rangle$. This can be accomplished by choosing $\kappa_1 = \kappa_2 = 2$, and putting the probe modes into the vacuum state. However, considering that the vacuum fluctuations for each quadrature is equal to $\frac{1}{4}$, one has $\langle\Delta\hat{X}'^2_1\rangle = \langle\Delta\hat{Y}'^2_2\rangle = \frac{9}{16}$, which is $\frac{1}{16}$ larger that the minimum noise for a joint measurement. This suggests the further unitary transformation

$$\begin{aligned} &\hat{X}_1 \to \frac{1}{\sqrt{2}}\hat{X}_1\,, \quad \hat{X}_2 \to \sqrt{2}\hat{X}_2\,,\\ &\hat{Y}_1 \to \sqrt{2}\hat{Y}_1\,, \quad \hat{Y}_2 \to \frac{1}{\sqrt{2}}\hat{Y}_2\,, \end{aligned} \tag{187}$$

which minimizes the noise, and can be achieved by the unitary operator

$$\hat{U}_{sq} = e^{i\log 2(\hat{X}_1\hat{Y}_1 - \hat{X}_2\hat{Y}_2)}\,. \tag{188}$$

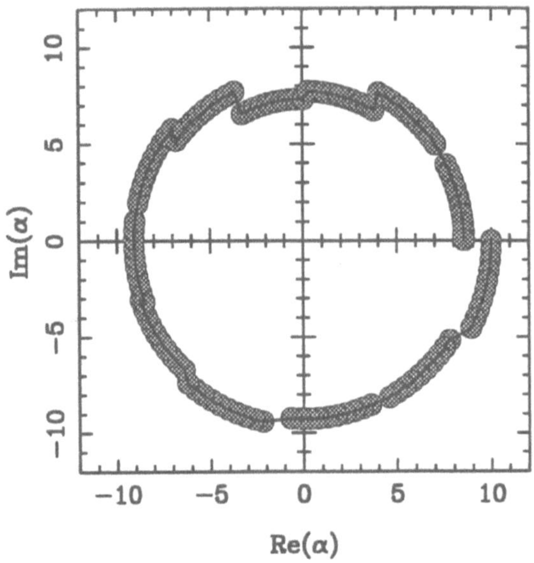

Figure 9. Brownian-motion effect on the free evolution due to the state reduction of the joint first-kind measurement in Eq. (193). This kind of measurement could account for the trajectory description of measurements from cloud or bubble chamber tracks.

The operator (188) squeezes the vacuum state of the probe '1' by a factor two in the $\hat{X}$ direction, and correspondingly unsqueezes the state of the probe '2' by the same factor. The squeezing operator $\hat{U}_{sq}$ acts before the interaction Hamiltonian $\hat{H}$. Hence, in summary, the model corresponds to the Hamiltonian (184) with probe preparation given by

$$\hat{\rho}_P = \hat{U}_{sq}|0,0\rangle\langle 0,0|\hat{U}_{sq}^\dagger \,, \tag{189}$$

and with $\hat{U}_{sq}$ defined in Eq. (188). In order to obtain the instrument corresponding to the present measurement scheme, we evaluate the operators $\hat{\Omega}(\mathbf{x})$ as follows

$$\begin{aligned} \hat{\Omega}(x,y) &= {}_1\langle x|\,{}_2\langle y|\exp[-2i(\hat{X}\hat{Y}_1 - \hat{Y}\hat{X}_2)]\hat{U}_{sq}|0,0\rangle \\ &= {}_1\langle x|\,{}_2\langle y|\hat{D}^\dagger(\hat{X}_2 + i\hat{Y}_1)\hat{U}_{sq}|0,0\rangle \,, \end{aligned} \tag{190}$$

where $\hat{D}(x+iy) \equiv \exp[-2i(x\hat{Y} - y\hat{X})]$ denotes the displacement operator acting on $\mathcal{H}$ [$\hat{Y}_1$ and $\hat{X}_2$ can be treated as c-numbers, because they commute each other and with any system operator]. Using the resolutions of the identity in terms of eigenstates of $\hat{X}_1$ and $\hat{Y}_2$ we obtain

$$\begin{aligned} \hat{\Omega}(x,y) &= \int dx'dy'\,{}_1\langle x|y'\rangle_1\,{}_2\langle y|x'\rangle_2\,{}_1\langle y'|\,{}_2\langle x'|\hat{D}^\dagger(\hat{X}_2 + i\hat{Y}_1)\hat{U}_{sq}|0,0\rangle \\ &= \int dx'dy'\frac{1}{\pi}e^{2i(xy'-yx')}\hat{D}^\dagger(x'+iy')\,{}_1\langle y'|\,{}_2\langle x'|\hat{U}_{sq}|0,0\rangle \\ &= \int \frac{dx'dy'}{\pi}e^{2i(xy'-yx')}\hat{D}^\dagger(x'+iy')\frac{1}{\sqrt{\pi}}e^{-\frac{1}{2}(x'^2+y'^2)} \,, \end{aligned} \tag{191}$$

and introducing the complex variables $\alpha = x + iy$ and $\lambda = x' + iy'$ we recognize in Eq. (191) the coherent state projector

$$\hat{\Omega}(x,y) \equiv \hat{\Omega}(\alpha,\overline{\alpha}) = \frac{1}{\sqrt{\pi}} \int \frac{d^2\lambda}{\pi} e^{\lambda\overline{\alpha}-\overline{\lambda}\alpha} : \hat{D}^\dagger(\lambda) :_A \equiv \frac{1}{\sqrt{\pi}} |\alpha\rangle\langle\alpha| \,. \tag{192}$$

Hence, the state reduction is given by

$$\hat{\rho}_{(\alpha,\overline{\alpha})} = \frac{|\alpha\rangle\langle\alpha|\hat{\rho}|\alpha\rangle\langle\alpha|}{\mathrm{Tr}[|\alpha\rangle\langle\alpha|\hat{\rho}]} \equiv |\alpha\rangle\langle\alpha| \,. \tag{193}$$

In Fig. 9 the effect of the state reduction (193) is illustrated for a freely evolving field mode. The instrument in Eq. (193) reduces the state to a coherent state $|\alpha\rangle\langle\alpha|$ that depends only on the outcome α of the measurement, whatever the starting state $\hat{\rho}$ is. Such kind of measurement—where the reduced state is independent on the input state—is referred to as *Gordon-Louisell*[13] measurement [23]. In general, a Gordon-Louisell measurement has an $\hat{\Omega}$ operator of the form $\hat{\Omega}(\mathbf{x}) = |\psi_{\mathbf{x}}\rangle\langle\varphi_{\mathbf{x}}|$, where $|\psi_{\mathbf{x}}\rangle$ denotes a normalized state vector that depends on the reading $\mathbf{x}$, and $\{|\varphi_{\mathbf{x}}\rangle\}$ is a complete (generally not normalizable) set of vectors in the Hilbert space.

References

1. C. W. Helstrom, *Quantum Detection and Estimation Theory*, Academic Press, (New York, 1976). The reader is also addressed to the original papers: C. W. Helstrom, Found. Phys. **4**, 453 (1974); Int. J. Theor. Phys. **11**, 357 (1974).
2. M. A. Naimark, Iza. Akad. Nauk USSR, Ser. Mat. **4** 277 (1940).
3. N. I. Akhierzer and I. M. Glezman, *Theory of Linear Operators in the Hilbert Space*, (two volumes bound as one), Dover Publ. Inc. N.Y 1993, pag. 121 of Vol. 2.
4. A. Peres, *Quantum Theory: Concepts and Methods*, vol. 57 of *Fundamental Theories of Physics*, Kluwer Academic Publisers (Dordrecht, Boston, London 1993)
5. G. M. D'Ariano, *Measuring Quantum States*, in this volume.
6. L. Mandel, Proc. Phys. Soc. **72** 1037 (1958); *ibid.* **74** 233 (1959)
7. P. L. Kelley, and W. H. Kleiner, Phys. Rev. A **30** 844 (1964)
8. H.Carmichael, *An Open System Approach to Quantum Optics*, Cap. 1 e 2, Springer-Verlag, Heidelberg (1993).
9. G. M. D'Ariano, Int. J. Mod. Phys. B **6** 1291 (1992)
10. G. M. D'Ariano, Nuovo Cimento **107B** 643 (1992).
11. H. P. Yuen and J. H. Shapiro, IEEE Trans. Inf. Theory, **24** 657 (1978); **25** 179 (1979); **26** 78 (1980).
12. H. P. Yuen, Phys. Lett. **91A**, 101 (1982).
13. E. Arthurs and M. S. Goodman, Phys. Rev. Lett. **60** 2447 (1988)
14. A. S. Holevo. *Probabilistic and statistical aspects of quantum theory*, North-Holland, (Amsterdam, 1982).
15. G. M. D'Ariano and M. G. A. Paris, Phys. Rev. A **49** 3022 (1994).
16. See R. Omnès, *The interpretation of Quantum Mechanics*, Princeton Series in Physics, Princeton University Press, (Princeton, 1994), pag. 84. The original reference is: W. Pauli, Handbuch der Physik, **24** 1 (1933)
17. M. Ozawa, in *Squeezed States and Nonclassical Light*, ed. by P. Tombesi and E. R. Pike, pag. 263, Plenum, (New York, 1989).
18. A. S. Holevo, in *Quantum Aspects of Optical Communications*, (Paris 1990), Lecture Notes in Physics **378**, eds. C. Bendjaballah, O. Hirota and S. Reynaud, Springer, Berlin-New York, (1991) pag. 127

[13]This nomenclature was introduced by Yuen

19. G. Ludvig, *Foundations of Quantum Mechanics I and II*, Springer, (Berlin, 1983). K. Kraus, *States, Effects, and Operations: fundamental notions of Quantum theory*, Springer, (Berlin, 1983). E. B. Davies, *Quantum Theory of Open Systems*, Academic, (London, 1976).
20. M. Ozawa, J. Math. Phys. **25** (1984).
21. J. von Neumann. *Mathematical Foundations of Quantum Mechanics*, Princeton UP, (Princeton, NJ, 1955).
22. E. Arthurs and J. L. Kelly, Bell. Syst. Tech. J., **44** 725-729 (1965).
23. J. P. Gordon and W. H. Louisell, in *Physics of Quantum Electronics*, pp. 833-840, McGraw-Hill, (New York, 1966).

MEASURING QUANTUM STATES

G. M. D'ARIANO
Dipartimento di Fisica "A. Volta",
via Bassi 6, I-27100 Pavia, Italy

1. Introduction

The possibility of "measuring" quantum states has been already considered by W. Pauli [1], who wrote in a footnote: "The mathematical problem, as to whether for given functions $W(\mathbf{x})$ and $W(\mathbf{p})$, the wave function ψ, if such function exists, is always uniquely determined, has still not been investigated in all its generality" [by $W(\mathbf{x})$ and $W(\mathbf{p})$ Pauli denoted the probability distributions $W(\mathbf{x}) = |\langle \mathbf{x}|\psi\rangle|^2$ and $W(\mathbf{p}) = |\langle \mathbf{x}|\psi\rangle|^2$, whereas, for "if such function exists" he meant that $W(\mathbf{x})$ and $W(\mathbf{p})$ are compatible]. The answer to Pauli's question is clearly negative [2]. In fact, it is quite obvious that the probabilities $W(\mathbf{x})$ and $W(\mathbf{p})$ at least cannot determine the quantum correlation $\langle\psi|\mathbf{xp}|\psi\rangle$, which can be obtained, for example, from a *joint* measurement of $\mathbf{x}$ and $\mathbf{p}$. On the other hand, as shown in my lectures on "Quantum estimation theory and optical detection" [3], a joint measurement of two conjugated observables would exhibit an additional noise equivalent to an effective quantum efficiency $\eta = 1/2$: as we will see in Sect. 7.5, this is exactly the threshold below which the density matrix cannot be measured.

That more than two observables—actually a complete set of them—would be needed in order to determine the density matrix was clear from the earlier theoretical studies of Fano [4] and remarked in the book of d'Espagnat [5]. However, it is difficult to devise concretely measurable translational observables—other than position, momentum and energy: for this reason a Schrödinger-picture point of view was adopted, in which, instead of measuring varying operators, the state itself is changed in a controlled way, and eventually the energy is measured [6] [for "spin" degrees of freedom a first study has been published only very recently [7]]. Still, such a fundamental problem—measuring the quantum state—has remained at the level of mere speculation for many years [8], and entered the realm of experiments only less than three years ago, after the experiments by Raymer's group [9], in the domain of quantum optics.

What is so special with quantum optics? In quantum optics, differently from quantum mechanics of particles, there is the unique opportunity of measuring all possible linear combinations of position $\hat{q}$ and momentum $\hat{p}$ of a harmonic oscillator, here a mode of the electromagnetic field. As explained in my previous set of lectures [3], such measurement can be achieved by means of a balanced homodyne detector, which measures the quadrature $\hat{x}_\phi = \frac{1}{2}\left(a^\dagger e^{i\phi} + ae^{-i\phi}\right)$ of the field at any desired phase ϕ with respect to the local oscillator (LO) [as usual a denotes the annihilator of the field mode]. The first technique

T. Hakioğlu and A. S. Shumovsky (eds.), Quantum Optics and the Spectroscopy of Solids, 175–202.

to reconstruct the density matrix from homodyne measurements—so called *homodyne-tomography*—originated from the idea of Vogel and Risken [10] that the collection of probability distributions $\{p(x,\phi)\}_{\phi\in[0,\pi)}$ is just the Radon transform (or "tomography") of the Wigner function $W(\alpha,\overline{\alpha})$. Then, by inverting the Radon transform, one obtains the Wigner function, and from the knowledge of $W(\alpha,\overline{\alpha})$ one can recover the matrix elements of the density operator $\hat{\varrho}$. In short, this was the basis of the pioneer method used by Raymer's group.

The first experimental work [9]—which unequivocally demonstrated the feasibility of the method—was, however, affected by some uncontrollable approximations that produce serious systematic errors when the detected state $\hat{\varrho}$ is truly "nonclassical". In fact, as we will see in Sect. 5, strictly speaking the Wigner function cannot be measured, because the inversion of the Radon transform needs an *analytic knowledge* of the probability distributions $p(x,\phi)$. In practice, the inverse Radon transform is achieved by "filtering" data—as in the usual X-ray medical tomography—upon introducing a suitable cut-off parameter that sets a coarse resolution on $W(\alpha,\overline{\alpha})$. Such a cut-off does not correspond to any controllable approximation in terms of density matrix. Setting such resolution in advance makes the measured state "more classical", thus loosing the most interesting quantum features that the experimentalist is looking for! Hence, the density matrix cannot be measured by measuring $W(\alpha,\overline{\alpha})$ as an intermediate step. However, the main idea—namely, homodyning the density matrix—still remains valid: the problem is only related to the method of data-processing, i. e. the use of the inverse Radon transform. With Macchiavello and Paris [11] I derived an exact technique that produces the number-state matrix elements only by averaging functions of data. The rather involved original formulas were simplified later by myself, Leonhardt, and Paul [12]. In the new form, the algorithm was so simple that it was possible also to recognize the feasibility of homodyning the density matrix even for nonideal detector quantum efficiency $\eta < 1$, giving mathematically proved lower bounds for η that depend on the chosen representation for the matrix. After these first results, further theoretical progress has been made, understanding the mechanisms that underly the generation of statistical errors [13] and limit the sensitivity of the method when used for indirectly measuring a single generic observable [14]. The problem of determining the Wigner function was revisited [15], and it was lately recognized [16] that $W(\alpha,\overline{\alpha})$ can be measured at some fixed resolution in polar coordinates, and that fixing such resolution corresponds to truncating the dimension of the density matrix in the Hilbert space. It was also realized that the new method can be profitably used as a novel imaging algorithm—what I call *the quantum Radon transform*—for conventional tomography [16], particularly useful in the extreme situation of low signals and high experimental resolution. Recently, new progress by Richter [17] and Leonhardt *et al.* [18] have made possible to factorize the averaged functions (socalled "pattern functions" [19, 20]) speeding up the algorithm greatly for $\eta = 1$ [an analogous factorization for $\eta < 1$ is still lacking: very recently the factorization has been derived by algebraic methods [21], and we hope that this will eventually provide the ultimate technique]. Currently, the algorithm is so simple, fast, and low-memory, that I believe that after reading these notes the reader will certainly try to implement it on his PC! In the meanwhile there has been an explosion of interest on the subject of *measuring quantum states* from both theoreticians and experimentalists (I apologize for not quoting all the literature): the new exact method has been eventually adopted for real experiments [22, 23], providing a very efficient way for measuring the photon number distribution.

I will review the exact method for homodyning the state of radiation in Sections 6 up to 9, after a premise in Sect. 2 on the central limit theorem, which is at the basis of the method and is needed in order to understand what can be measured and what cannot (for example, the Wigner function cannot be measured: see Sect. 5). A short review on conventional tomographic imaging is given in Sect. 3, and the old homodyne-tomographic method for measuring the state is revisited in Sect. 4. The possibility of using the new method for conventional imaging and the coding of images into density matrices is discussed in Sect. 11. In Sect. 10 a recent debate on the possibility of measuring the density matrix of a *single* quantum system is reviewed. In fact, the notion of "measuring a state" is meaningful only if it refers to many repeated measurements performed on an *ensemble of equally prepared* systems: otherwise, the state of an individual system is unobservable, whatever sequence of measurements is performed on the system.

Appendices contain some relevant proofs and derivations, along with the recently derived factorization formulas. Notice: some parts of the present notes include material that entered my knowledge (or was discovered) after the end of the set of lectures given at Bilkent University. As the field is so rapidly growing, I decided to include this material here in order to make these notes more complete and up to date. In particular the new matter regards the factorization formulas (Refs. [17], [18], and [21]) , the new experiments in Refs. [22] and [23], and finally my recent work with Yuen on the impossibility of measuring the density matrix of a single quantum system [24].

2. Premise on the central limit theorem

The exact method for homodyning the density matrix of the radiation field is based on the fact that the matrix element $\langle\psi|\hat{\varrho}|\varphi\rangle$ can be written as the expectation of a function $f_{\psi\varphi}(I,\phi)$ of the random homodyne photocurrent I at random phase ϕ relative to the LO. The analytic form of the function $f_{\psi\varphi}$ is known in advance, and depends parametrically on both vectors ψ and φ of the matrix element: it will be explicitly given in Sect. 7 for some relevant representations. Hence, the problem of measuring a matrix element resorts to the estimation of a *generalized moment* of the homodyne probability $p(I,\phi)$, or, in other words, to the estimation of what I call the "mathematical expectation" of a given function of I and ϕ. To make things easier, let us consider the case that the random variable is scalar real-valued, denoted by x. Then, the *generalized f-moment* or *mathematical expectation* $\mathbf{E}f(x)$ of the function $f(x)$ is defined as follows

$$F = \mathbf{E}f(x) \equiv \int_{-\infty}^{+\infty} \mathrm{d}x\, p(x) f(x)\,. \tag{1}$$

What we can say in advance on the possibility of "measuring" $\mathbf{E}f(x)$ experimentally? We would be tempted of measuring F by means of the "experimental average"

$$F_N = \frac{1}{N}\sum_{n=1}^{N} f(x_n)\,, \tag{2}$$

where $\{x_n\}_{n=1}^{N}$ denote the outcomes of x in an experiment that consists of N repeated measurements. Is this procedure correct? How reliable is such estimate F_N of F? The answer to these questions are provided by the *central limit theorem* (the proof is reported in Appendix 13.1. According to this theorem, for functions $f(x)$ having moments bounded up to the third order, the random variable F_N will exhibit Gaussian distribution asymptotically for large N, with average F and variance $\varepsilon_N^2 = \mathbf{E}[\Delta f^2]/N$. In such case, we can

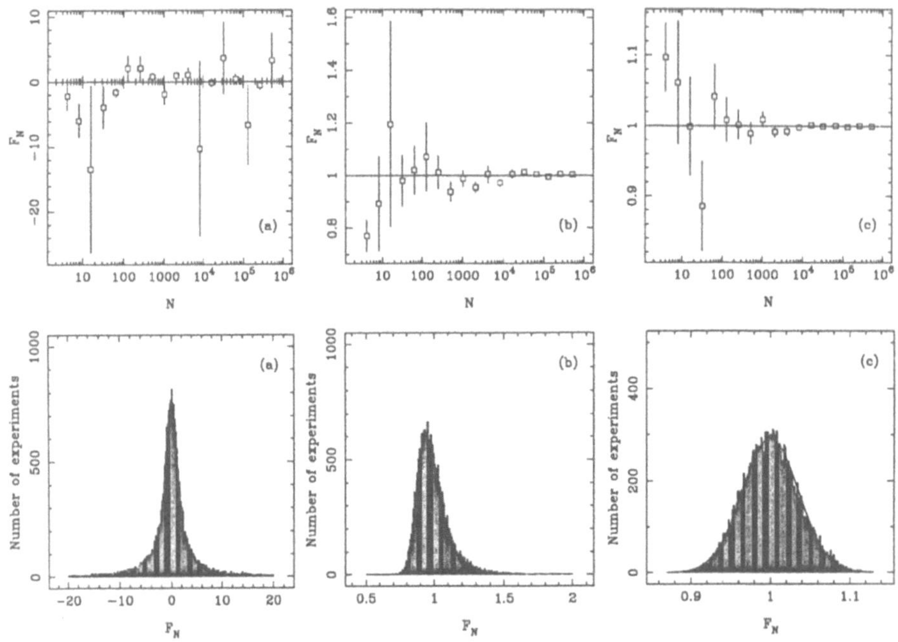

Figure 1. Numerical simulation of experiments for estimating $F = \mathbf{E}f(x)$ in Eq. (1) by means of F_N in Eq. (2). Here, we have uniform probability $p(x) = 1/2$ for $x \in [-1, 1]$, $p(x) = 0$ otherwise. Top figures give F_N within the confidence interval δ_N of Eq. (3) versus N. (a) $f(x) = \frac{1}{x}$: the function is not integrable; (b) $f(x) = \frac{1}{2|x|^{1/2}}$: the function is not square-integrable; (c) $f(x) = \frac{3}{2}|x|^{1/2}$: the function is square-integrable. The three histograms on the bottom give the distribution of F_N for different experiments; every histogram pertains the corresponding case on the top. There are N=20000 measurement for each experiment. For the case (c) the histogram is compared with a Gaussian with the same average and variance.

safely rely on the estimate F_N of F, knowing *a priori* that another similar experiment will produce a result within the *confidence interval* $[F_N - \varepsilon_N, F_N + \varepsilon_N]$ with probability 68%. An estimate of the confidence interval ε_N itself is provided by the well known root mean square deviation

$$\delta_N = \sqrt{\frac{\sum_{n=1}^{N}[f(x_n) - F_N]^2}{N(N-1)}} \,. \tag{3}$$

What happens when the conditions of the central limit theorem are not fulfilled? For example, if the function f is not integrable with respect to the probability measure $p(x)$, the estimate F_N will not approach any definite value for large N. On the other hand, if f is integrable but not square-integrable (i.e. it has unbounded second moment), then δ_N will not rescale correctly as $N^{-1/2}$, and the confidence interval will be utterly unreliable. We can gain some feeling on such pathological situations with the aid of a computer experiment. In the standard Fortran language the random number generator "ran" produces uniform deviates between 0 and 1: by means of translation and rescaling one can generate uniform deviates in any prechosen interval. Let us consider, for example, a uniform probability density: $p(x) = 1/2$ for $x \in [-1, 1]$, $p(x) = 0$ otherwise. In Fig. 1 the result from numerical simulations of the estimate F_N in Eq. (2) with error-bar δ_N in Eq. (3) are given for three kinds of functions: (a) non integrable $f(x) = \frac{1}{x}$; (b) non square-integrable $f(x) = \frac{1}{2|x|^{1/2}}$; (c) square-integrable $f(x) = \frac{3}{2}|x|^{1/2}$. The behavior of F_N versus N is erratic in the non integrable case, whereas it converges nicely to the expected value F in the other two cases.

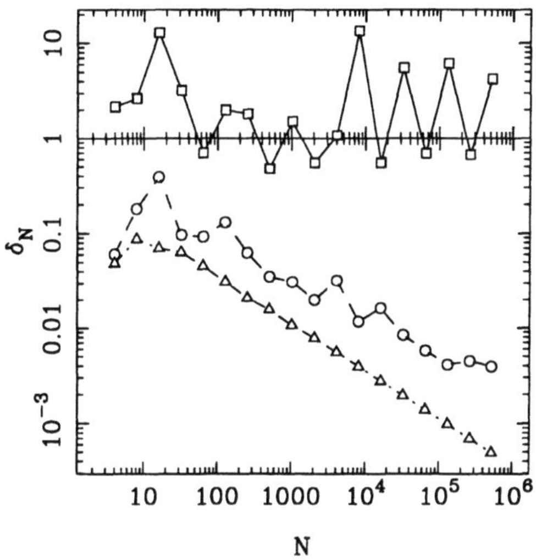

Figure 2. Behavior of the confidence interval δ_N in Eq. (3) versus the number of measurements N per experiment. The squares refer to the case (a) in Fig. 1 [non integrable function], the circles to case (b) [non square-integrable function] and the triangles to case (c) [square-integrable function].

On the other hand, from the same Fig. 1 it is evident that the histogram of the values of F_N from different experiments is Gaussian only in the square-integrable case, and from Fig. 2 we can see that the this is the only case leading to a $\delta_N \sim N^{-1/2}$.

What we can say if the probability $p(x)$ is completely unknown *a priori*—as it is the case in practice? Then, in order to assure in advance that $f(x)$ satisfies the conditions imposed by the central limit theorem, the function $f(x)$ must be square-integrable for any possible probability measure $p(x)$. It is easy to see that this requirement is fulfilled only if $f(x)$ is bounded by some constant $|f(x)| < C < \infty$. In fact, if $f(x)$ is unbounded, then it is always possible to find a probability measure $p(x)$ for which $f(x)$ is not even integrable. On the other hand, if $f(x)$ is bounded, then every moment of f is bounded—in particular the first three ones—and the central limit theorem holds. Thus we conclude that for all possible (a priori unknown) probabilities $p(x)$, the generalized f-moment in Eq. (1) can be estimated by means of the experimental average (2) if and only if the function $f(x)$ is bounded. In such case we will also say that F is *statistically sampled* by F_N, and Eqs. (1) and (2) will be referred to as *sampling formulas* for F.

3. What is a tomography?

The essential problem of tomographic imaging is to recover a distribution of mass $m(x, y)$ in a 2-d slab from a finite collection of one dimensional projections $r(s, \phi)$ at different angles ϕ. The situation is schematically sketched in Fig. 3, where $m(x, y)$ describes two

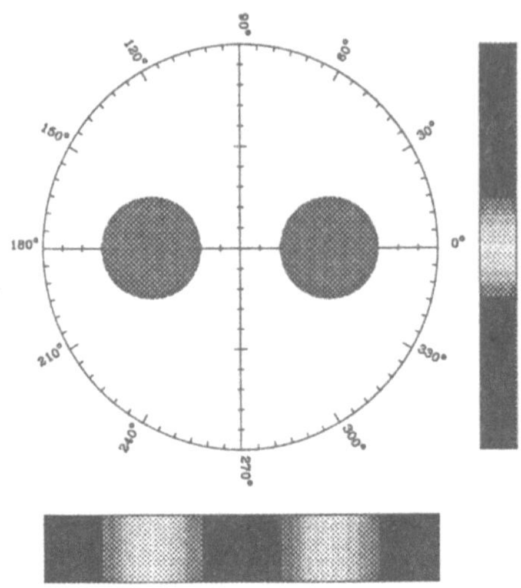

Figure 3. Illustration of the tomographic reconstruction of a 2-d image (here two holes in a uniform background) from its 1-d transmission profiles at different angles ϕ.

circular holes in a uniform background. The tomographic machine—-for example, an X-ray equipment—collects many stripe photos of the sample from various directions ϕ, and then numerically performs a mathematical transform in order to reconstruct $m(x, y)$ from its radial profiles $r(s, \phi)$ at different ϕ's. The word "tomography" is customary to denote such imaging procedure starting from radial projections. Mathematically, the radial profiles $r(s, \phi)$ are just the integrals of the image $m(x, y)$ along different directions ϕ, namely

$$r(s,\phi) = \int_{-\infty}^{+\infty} \frac{\mathrm{d}t}{\pi}\, m\left(s\cos\phi - t\sin\phi, s\sin\phi + t\cos\phi\right) , \tag{4}$$

with s denoting the current coordinate along the direction orthogonal to the projection, and t the coordinate along the projection direction. The choice of constants in Eq. (4) gives $r(s, \phi)$ normalized to unit versus s for $m(x, y)$ also normalized to unit. In the following I will adopt the complex notation $m(x, y) \to m(\alpha, \bar{\alpha})$, with $\alpha = x + iy$. Then Eq. (4) is rewritten as follows

$$r(x,\phi) = \int_{-\infty}^{+\infty} \frac{\mathrm{d}y}{\pi}\, m\left((x+iy)e^{i\phi}, (x-iy)e^{-i\phi}\right) , \tag{5}$$

The collection of all projections $r(x, \phi)$ at different ϕ's is called *Radon transform.*

The reconstruction of the "image" $m(\alpha, \bar{\alpha})$ from its "projections" $r(x, \phi)$—this reconstruction is also called "back-projection"—is given by the inverse Radon transform [25]

$$m(\alpha,\bar{\alpha}) = \int_{-\infty}^{+\infty} \frac{\mathrm{d}k\,|k|}{4} \int_0^{\pi} \frac{\mathrm{d}\phi}{\pi} \int_{-\infty}^{+\infty} \mathrm{d}x\;\, r(x,\phi)\exp\left[ik\left(x-\alpha_\phi\right)\right] , \tag{6}$$

where $\alpha_\phi = \mathrm{Re}(\alpha e^{-i\phi})$. The derivation of (6) from (5) is given in Appendix 13.2. Upon exchanging integrals over k and ϕ, the image $m(\alpha, \bar{\alpha})$ is written in double integral form

$$m(\alpha,\bar{\alpha}) = \int_0^{\pi} \frac{d\phi}{\pi} \int_{-\infty}^{+\infty} \mathrm{d}x\, r(x,\phi)\, K(x-\alpha_\phi) , \tag{7}$$

with the kernel $K(z)$ given by

$$K(z) = -\frac{1}{2}\,\mathrm{P}\frac{1}{z^2} \equiv -\lim_{\varepsilon\to 0^+}\frac{1}{2}\mathrm{Re}\frac{1}{(z+i\varepsilon)^2}\,, \tag{8}$$

P denoting the Cauchy principal value. Integrating Eq. (7) by parts, one obtains the familiar "filtering" procedure

$$m(\alpha,\bar{\alpha}) = \int_0^{\pi}\frac{\mathrm{d}\phi}{2\pi}\,\mathrm{P}\int_{-\infty}^{+\infty}\mathrm{d}x\,\frac{\partial r(x,\phi)/\partial x}{x-\alpha_\phi}\,, \tag{9}$$

which is commonly used in conventional tomographic imaging [26]. Notice that because of the principal value and of the x-derivative in Eq. (9) one can recover $m(\alpha,\bar{\alpha})$ only when $r(x,\phi)$ is given analytically. An analytical form for projections $r(x,\phi)$ can be achieved from raw data either by "splining" the experimental profiles, or by filtering them with a high-k cutoff in the Fourier transform (6), or finally by setting a nonzero ε in the kernel (8).

4. Why quantum homodyne tomography?

The homodyne-tomography technique originated from the simple idea [10] that the probability distributions $p(x,\phi)$ of the outcomes x of the quadrature $\hat{x}_\phi$ at phase ϕ with respect to the LO is just the marginal probability of the Wigner function $W(\alpha,\overline{\alpha})$ along the line $\alpha = re^{i\phi}$, $r\in(-\infty,+\infty)$ in the complex plane: in other words, the collection of probability densities $p(x,\phi)$ for $0\le\phi\le\pi$ is the Radon transform of $W(\alpha,\overline{\alpha})$. Hence, $W(\alpha,\overline{\alpha})$ can be evaluated by tomographic imaging (inverse Radon transform) from $p(x,\phi)$, namely

$$W(\alpha,\overline{\alpha}) = \int_{-\infty}^{+\infty}\frac{\mathrm{d}k\,|k|}{4}\int_0^{\pi}\frac{\mathrm{d}\phi}{\pi}\int_{-\infty}^{+\infty}\mathrm{d}x\,p(x,\phi)\exp\left[ik(x-\alpha_\phi)\right]\,, \tag{10}$$

where $\alpha_\phi = \mathrm{Re}(\alpha e^{-i\phi})$. Eq. (10) was the basis of the first tomographic method for detecting the density matrix of the radiation field. By means of the filtering or splining procedures discussed in Section 3, an analytic form for $p(x,\phi)$ is recovered from histograms of many collected data, and the Wigner function is obtained. Then, the density matrix in the number representation is evaluated using the following integral transforms [9]

$$\langle x+x'|\hat{\varrho}|x-x'\rangle = \int_{-\infty}^{\infty}\frac{\mathrm{d}y}{\pi}\,e^{2ix'y}W(x+iy,x-iy)\,, \tag{11}$$

$$\langle n|\hat{\varrho}|m\rangle = \frac{1}{\sqrt{\pi 2^{n+m-1}n!m!}}\int_{-\infty}^{\infty}\mathrm{d}x\int_{-\infty}^{\infty}\mathrm{d}x'\,e^{-(x^2+x'^2)}H_n(\sqrt{2}x)H_m(\sqrt{2}x')\,\langle x|\hat{\varrho}|x'\rangle\,, \tag{12}$$

where $H_n(x)$ denote Hermite polynomials. The choice of the k-cutoff or of the maximum degree of splining should be done carefully as a function of the number of data. However, both filtering parameters do not correspond to any meaningful kind of approximation in terms of the matrix elements $\langle n|\hat{\varrho}|m\rangle$: thus, this method is quantitatively unreliable, and can only give qualitative informations on the Wigner function for large numbers of data and nearly unit quantum efficiency. As noticed in Ref.[11], *over-smoothing* produces systematic errors equivalent to make the state "more classical", thus preventing from detecting the most interesting quantum features, such as, for example, the oscillations of the photon number probability of a squeezed state [27]. On the other hand, *under-smoothing* would lead to systematic negative probabilities (i. e. diagonal matrix elements). In Section 6 I will derive the exact sampling formulas of Refs. [11] and [12] for measuring the density matrix without filtering.

5. On the impossibility of measuring the Wigner function

Upon exchanging the "averaging" integrals over ϕ and x with respect to the outer integral over k, Eq. (10) can be written in a form similar to Eq. (7) as expectation over the random couple (x, ϕ). One has

$$W(\alpha,\overline{\alpha}) = \int_0^\pi \frac{\mathrm{d}\phi}{\pi} \int_{-\infty}^{+\infty} \mathrm{d}x\, p(x,\phi)\, K(x-\alpha_\phi)\,, \tag{13}$$

where, as in Eq. (8), the kernel $K(x)$ is given by

$$K(x) = -\frac{1}{2}\mathrm{P}\frac{1}{x^2} \equiv -\lim_{\varepsilon\to 0^+} \frac{1}{2}\mathrm{Re}\frac{1}{(x+i\varepsilon)^2}\,. \tag{14}$$

From Eq. (14) it is apparent that the kernel $K(x)$ is unbounded, and thus, as discussed in Sect. 2, the Wigner function cannot be "measured" (i.e. statistically sampled), unless boundness is artificially introduced by fixing a nonzero value of ε.

6. Exact method for measuring the density matrix

We want to obtain a general matrix element $\langle\psi|\hat{\varrho}|\varphi\rangle$ in form of expectation, namely as in Eq. (13), but with $\langle\psi|\hat{\varrho}|\varphi\rangle$ in place of $W(\alpha,\overline{\alpha})$ and with a function $f_{\psi\varphi}(x,\phi)$ in place of the kernel $K(x-\alpha_\phi)$. Then, if and only if the function $f_{\psi\varphi}$ is bounded, every moment of the function will be bounded for all possible probability distributions $p(x,\phi)$ and, according to the central-limit theorem discussed in Sect. 2, the matrix element can be sampled over a sufficiently large set of data, and the average values for different experiments will be Gaussian distributed, allowing estimation of the confidence intervals (3).

In order to obtain the integral kernel $f_{\psi\varphi}$ for the matrix element $\langle\psi|\hat{\varrho}|\varphi\rangle$, one can start from the operator identity

$$\hat{\varrho} = \int \frac{\mathrm{d}^2\alpha}{\pi}\, \mathrm{Tr}(\hat{\varrho} e^{-\overline{\alpha}a+\alpha a^\dagger})\, e^{-\alpha a^\dagger+\overline{\alpha}a} \tag{15}$$

which, by changing to polar variables $\alpha = (i/2)ke^{i\phi}$, becomes

$$\hat{\varrho} = \int_{-\infty}^{+\infty} \frac{\mathrm{d}k\,|k|}{4} \int_0^\pi \frac{\mathrm{d}\phi}{\pi}\, \mathrm{Tr}(\hat{\varrho} e^{ik\hat{x}_\phi})\, e^{-ik\hat{x}_\phi}\,. \tag{16}$$

Equation (15) is nothing but the operator form of the Fourier-transform relation between Wigner function and characteristic function: the reader can easily check it by himself. The trace-average in Eq. (16) can be evaluated in terms of $p(x,\phi)$, using the complete set $\{|x\rangle_\phi\}$ of eigenvectors of $\hat{x}_\phi$. After exchanging the integrals over x and ϕ with respect to the outer integral over k, one obtains the quantum equivalent of Eq. (13), namely

$$\hat{\varrho} = \int_0^\pi \frac{\mathrm{d}\phi}{\pi} \int_{-\infty}^{+\infty} \mathrm{d}x\, p(x,\phi) K(x-\hat{x}_\phi)\,, \tag{17}$$

where K is still given by Eq. (14), but now is a function of the quadrature operator $\hat{x}_\phi$. Taking matrix elements of both sides of Eq. (17) between vectors ψ and φ, we obtain the desired sampling formula

$$\langle\psi|\hat{\varrho}|\varphi\rangle = \int_0^\pi \frac{\mathrm{d}\phi}{\pi}\int_{-\infty}^{+\infty} \mathrm{d}x\, p(x,\phi)\langle\psi|K(x-\hat{x}_\phi)|\varphi\rangle\,. \tag{18}$$

Hence the averaged function is $f_{\psi\varphi}(x,\phi) \equiv \langle\psi|K(x-\hat{x}_\phi)|\varphi\rangle$. Now the reader could object that Eq. (18) is useless, because $K(x)$ is unbounded. But, he must carefully consider that for some vectors ψ and φ the matrix element $\langle\psi|K(x-\hat{x}_\phi)|\varphi\rangle$ can be bounded, although the function $K(x)$ is not! As an example, one can consider the expectation value $\langle\alpha|\delta(x-\hat{x}_\phi)|\alpha\rangle$ of $\delta(x-\hat{x}_\phi)$ between coherent states $|\alpha\rangle$, which is just the Gaussian $|\langle x|\alpha\rangle|^2 = \sqrt{2/\pi}\exp[-2(x-\alpha_\phi)^2]$, and hence is clearly bounded.

Before analyzing specific matrix representations, I generalize the sampling formula (18) to the case of nonunit quantum efficiency. As shown in my lectures on "Quantum estimation theory and optical detection" [3], low efficiency detectors in a homodyne scheme simply produce a probability $p_\eta(x,\phi)$ which is a Gaussian convolution of the ideal probability $p(x,\phi)$ for $\eta = 1$. In terms of the generating functions of the $\hat{x}_\phi$-moments one has

$$\int_{-\infty}^{+\infty} \mathrm{d}x\, p_\eta(x,\phi)e^{ikx} = \exp\left(-\frac{1-\eta}{8\eta}k^2\right)\int_{-\infty}^{+\infty} \mathrm{d}x\, p(x,\phi)e^{ikx}\,. \tag{19}$$

Upon substituting Eq. (19) into Eq. (16) and following the same lines that lead us to Eq. (17), one obtains the operator identity

$$\hat{\varrho} = \int_0^\pi \frac{\mathrm{d}\phi}{\pi}\int_{-\infty}^{+\infty} \mathrm{d}x\, p_\eta(x,\phi)K_\eta(x-\hat{x}_\phi)\,, \tag{20}$$

where now the kernel is given by

$$K_\eta(x) = \frac{1}{2}\mathrm{Re}\int_0^{+\infty} \mathrm{d}k\, k\,\exp\left(\frac{1-\eta}{8\eta}k^2 + ikx\right)\,. \tag{21}$$

Again the desired sampling formula for $\langle\psi|\hat{\varrho}|\varphi\rangle$ is obtained by taking matrix elements of both sides of Eq. (21). Notice that now the kernel $K_\eta(x)$ is generally not even a tempered distribution: however, as we will see in the Sect. 7, the matrix elements of $K_\eta(x-\hat{x}_\phi)$ are bounded for some representations, depending on the value of η.

7. Measurable representations and bounds for quantum efficiency

The matrix element $\langle\psi|K_\eta(x-\hat{x}_\phi)|\varphi\rangle$ is bounded if $\langle\psi|e^{-ir\hat{x}_\phi}|\varphi\rangle$ decays faster than $\exp\left[-(1-\eta)k^2/8\eta\right]$ in Eqs. (20) and (21). We can determine the balancing of Gaussian functions in Eqs. (20) and (21) upon introducing the concept of "resolution" $\varepsilon_\psi^2(\phi)$ of the vector ψ in the $\hat{x}_\phi$-representation. This is given by

$$|_\phi\langle x|\psi\rangle|^2 \simeq \exp\left[-\frac{x^2}{2\varepsilon_\psi^2(\phi)}\right] \qquad \text{(leading term)}\,. \tag{22}$$

The resolution $\varepsilon_\psi^2(\phi)$ represents the Gaussian decay length of the probability $|_\phi\langle x|\psi\rangle|^2$ versus x, with the rule that $\varepsilon_\psi^2(\phi) = 0$ and $\varepsilon_\psi^2(\phi) = \infty$ for functions decaying faster and slower than the Gaussian one, respectively. Now, let us define the "reduced" resolution averaged on the two vectors ψ and φ

$$\frac{2}{\varepsilon^2(\phi)} \doteq \frac{1}{\varepsilon^2_\psi(\phi)} + \frac{1}{\varepsilon^2_\varphi(\phi)}\,. \tag{23}$$

In terms of $\varepsilon^2(\phi)$ it is easy to state the balancing between the decay of $\langle\psi|e^{-ir\hat{x}_\phi}|\varphi\rangle$ and the growing of the anti-Gaussian in Eq. (21). Using the resolution of the identity in terms of the eigenvectors $\{|x\rangle_\phi\}$ of $\hat{x}_\phi$ we obtain

$$\begin{aligned}\langle\psi|K_\eta(x-\hat{x}_\phi)|\varphi\rangle &= \frac{1}{2}\mathrm{Re}\int_0^{+\infty} dk\, k\, e^{ikx}\int_{-\infty}^{+\infty} dx'\, e^{-ikx'}\exp\left(\frac{1-\eta}{8\eta}k^2\right)\langle\psi|x'\rangle_{\phi\phi}\langle x'|\varphi\rangle \\ &= \frac{1}{2}\mathrm{Re}\int_0^{+\infty} dk\, k\, e^{ikx}\exp\left[-\frac{1}{2}k^2\left(\varepsilon^2(\phi)-\frac{1-\eta}{4\eta}\right)\right] \qquad \text{(leading term)}\,,\end{aligned} \tag{24}$$

namely, we can readily assert that the matrix element $\langle\psi|K(x-\hat{x}_\phi)|\varphi\rangle$ is bounded if the following inequality is satisfied for $\phi\in[0,\pi)$

$$\eta > \frac{1}{1+4\varepsilon^2(\phi)}\,, \tag{25}$$

or, in terms of $\varepsilon(\phi)$,

$$\varepsilon(\phi) > \frac{1}{2}\sqrt{\eta^{-1}-1}\,. \tag{26}$$

Hence, we conclude that:

The matrix element $\langle\psi|\hat{\varrho}|\varphi\rangle$ can be measured if the minimum reduced resolution of vectors ψ and φ in the quadrature representations satisfies the bound

$$\varepsilon = \min_{0\le\phi\le\pi}\{\varepsilon(\phi)\} > \frac{1}{2}\sqrt{\eta^{-1}-1}\,. \tag{27}$$

We now consider some particular representations of interest for applications.

7.1. QUADRATURE REPRESENTATION

The scalar product of two quadrature eigenvectors at the same phase ϕ is a delta-function, and the square of the delta-function in Eq. (22) is not well defined. We can easily check that the kernel is unbounded, even for $\eta = 1$, where it is given by

$${}_\phi\langle x_1|K(x-\hat{x}_\phi)|x_2\rangle = \frac{|x_1-x_2|}{\sin^2\phi}\exp\left\{-2i\frac{x_1-x_2}{\sin\phi}\left[x-\frac{1}{2}(x_1+x_2)\cos\phi\right]\right\}\,. \tag{28}$$

Hence, one concludes that it is not possible to measure the density matrix in the quadrature representation, even for $\eta = 1$.

7.2. COHERENT-STATE REPRESENTATION

The resolution is $\varepsilon = \frac{1}{2}$: according to Eq. (27) the density matrix can be measured for $\eta > \frac{1}{2}$. The integral kernel is given by

$$\langle\alpha|K_\eta(x-\hat{x}_\phi)|\beta\rangle = 2\kappa^2\langle\alpha e^{i\phi}|\beta e^{i\phi}\rangle e^{-2(\kappa x - w_\phi)^2}\,\Phi\left(-\frac{1}{2},\frac{1}{2};2(\kappa x-w_\phi)^2\right)\,, \tag{29}$$

with $\kappa = \sqrt{\eta/(2\eta-1)}$, $w_\phi = \frac{1}{2}(\beta e^{i\phi} + \overline{\alpha} e^{-i\phi})$, and $\Phi(\alpha,\beta;z)$ denoting the degenerate hypergeometric function.

7.3. NUMBER–STATE REPRESENTATION

The resolution is $\varepsilon = \frac{1}{2}$: the matrix element can be measured for $\eta > \frac{1}{2}$. The integral kernel is given by

$$\langle n|K_\eta(x-\hat{x}_\phi)|n+d\rangle = e^{-id\phi} 2\kappa^{d+2} \sqrt{\frac{n!}{(n+d)!}} e^{-\kappa^2 x^2} \tag{30}$$
$$\times \sum_{\nu=0}^{n} \frac{(-)^\nu}{\nu!} \binom{n+d}{n-\nu} (2\nu+d+1)!\kappa^{2\nu} \mathrm{Re}\left\{(-i)^d D_{-(2\nu+d+2)}(-2i\kappa x)\right\} .$$

In Eq. (30) $D_\sigma(z)$ denotes the parabolic cylinder function. Notice that the analytic expression of the kernel does not contain any dimensional truncation of the Hilbert space.

7.4. SQUEEZED–STATE REPRESENTATION

The leading term in Eq. (22) comes only from the squeezed vacuum (the coherent part just shifts the quadrature). For a squeezed vacuum with squeezing parameter s, the probability of the quadrature $\hat{x}_\phi$ is the Gaussian $|_\phi\langle x|\psi\rangle|^2 = \sqrt{\frac{2s_\phi^2}{\pi}} \exp\left(-2s_\phi^2 x^2\right)$, where $s_\phi = |s^{\frac{1}{2}} \sin\phi - is^{-\frac{1}{2}} \cos\phi|^{-2}$. Hence, the smallest resolution is $\varepsilon = \frac{1}{2}\min(s, s^{-1})$ and the matrix element can be measured for $\eta > [1 + \min(s, s^{-1})]^{-1}$.

7.5. ABSOLUTE BOUND FOR QUANTUM EFFICIENCY

We have seen that for the coherent and number-state representations the density matrix can be measured for quantum efficiency $\eta > 1/2$. On the other hand, for the squeezed state representation one has $\eta > [1 + \min(s, s^{-1})]^{-1}$, which is even greater than 1/2. Therefore, one could wonder if 1/2 is an absolute bound for η, as conjectured in Ref. [12]. This is clearly true for all representations with minimum uncertainty product

$$\varepsilon(\phi)\varepsilon(\phi + \pi/2) \geq \frac{1}{4} , \tag{31}$$

and either the number (complete) or the coherent/squeezed (over-complete) representations satisfy the Heisenberg inequality (31) with the equal sign. Now the question is if it is possible to find (over)-complete basis for the Fock space having uncertainty product larger than $\frac{1}{4}$: in my knowledge, there are no such "spread" basis. Notice that it is difficult to obtain "spread" basis as unitary transformations of either the number or the coherent basis. In fact, any unitary operator that is the exponential of a Hermitian bilinear polynomial in a and $a^\dagger$ preserves the uncertainty product: on the other hand, the exponential of a Hermitian polynomial of degree higher than 2 is generally not analytical on the Fock basis (apart from trivial functions of the number operator) [28].

8. Statistical errors

In this section we study in more detail the analytic form of the integral kernel, in order to understand the mechanisms producing statistical errors in the measured matrix elements.

We only analyze the case of the number representation, with the kernel given by Eq. (30): the same arguments can be easily extended to other representations.

In Fig. 4 the kernel $\langle n|K_\eta(x-\hat{x}_\phi)|n+d\rangle$ for $\eta=1$ is plotted versus x for $\phi=0$ at different values of n, and d. One can see that for $d=0$—along the diagonal of the

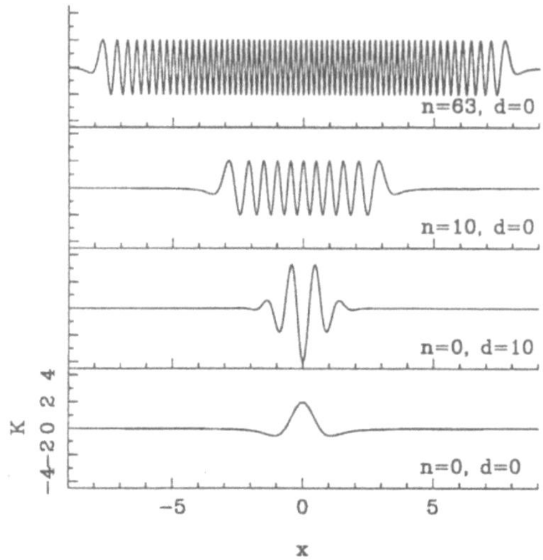

Figure 4. The kernel $\langle n|K_\eta(x-\hat{x}_\phi)|n+d\rangle$ of Eq. (30) for $\eta=1$ (the y-scale is fixed).

matrix—the range of the kernel is bounded between -2 and 2, and increases slowly versus distance d from the diagonal. For increasing n and d the kernel oscillates fast, with an increasing number of nodes. Fast oscillations make the average of the kernel—hence the measured value $\langle n|\hat{\varrho}|m\rangle$—more sensitive to fluctuations of the random outcomes x of the quadrature, producing confidence intervals that increase versus n and d. On the other hand, the bounded range makes errors themselves bounded, so they saturate at large n's.

For $\eta<1$ the behavior of the kernel changes dramatically, with the range increasing versus n more and more rapidly as η approaches the lower bound $\eta=0.5$ (see Fig.5). In this case the resulting errors increase rapidly versus n, and more data will be needed to "clean out" the additional noise due to nonunit quantum efficiency.

The above mechanisms for errors are well illustrated in Figs. 6 and 7, where the photon number probability distribution is plotted from computer-simulated experiments. The behavior of the confidence intervals (represented by the gray-shaded thickness of horizontal lines) is qualitatively very different for $\eta=1$ (Fig. 6) and for $\eta=.8$ (Figs. 7). Fig. 6 shows the Wheeler-Schleich oscillations [27] of a squeezed state. Fig. 7a gives the probability for a squeezed vacuum measured for $\eta=0.8$, whereas Fig. 7b gives the probability obtained from the same data at $\eta=.8$, but using the kernel with $\eta=1$: this shows the "smearing" effect of improper accounting for quantum efficiency. From both Figs. 6 and 7 it is evident that homodyne detection of the density matrix allows to recover delicate interference oscillations of nonclassical states.

The statistical reliability of both measured values and confidence intervals can be checked by means of Monte Carlo simulations. A typical check histogram for the devia-

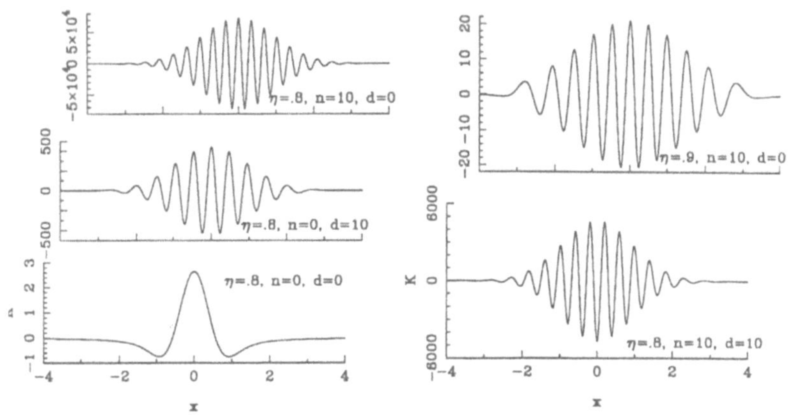

Figure 5. The kernel $\langle n|K_\eta(x-\hat{x}_\phi)|n+d\rangle$ in Eq. (30) for $\eta<1$. Notice thevery different ranges.

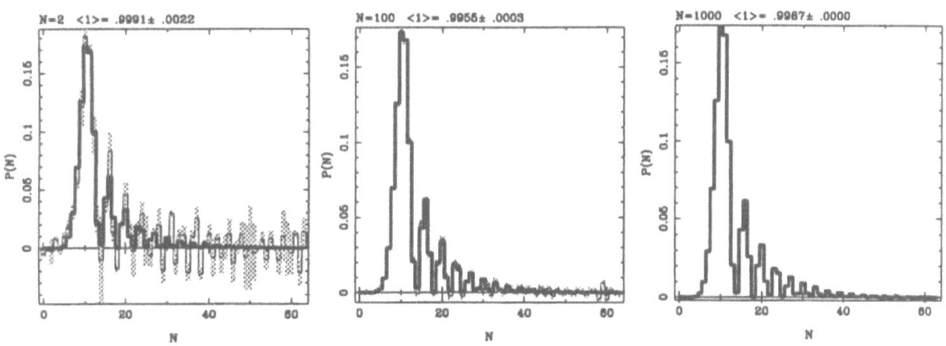

Figure 6. Tomographic reconstruction of the Wheeler-Schleich oscillations of a squeezed state with $\langle\hat{n}\rangle = 13.5125$ and 5.51 squeezing photons. Confidence intervals are represented by the gray-shaded thickness of horizontal lines. Detection efficiency is $\eta = 1$. Homodyne data are computer-simulated. From the left to the right we have N=2, 100 and 1000 blocks of data, respectively, with 5000 data each at totally random phases. On the top of each figure a normalization check is reported.

tions from theoretical values $\Delta\varrho_{n,m} \equiv \left(\varrho_{n,m}^{(exper)} - \varrho_{n,m}^{(theor)}\right)/\delta_{n,m}$ normalized by confidence intervals $\delta_{n,m}$ is given in Fig. 8 for all matrix elements of a 65×65 density matrix: both histograms for real and imaginary parts of $\varrho_{n,m}$ compare very well with a standard Gaussian.

Finally, in Fig. 9 a typical result for convergence of normalization and mixing is given for a coherent state.

In conclusion of this section a remark is in order on homodyning the density matrix as a tool for measuring any desired observable. In fact, from the measured density matrix one can evaluate the expected value of any operator (which is sufficiently well behaved as a function of a and $a^\dagger$). However, nonzero confidence intervals for the matrix elements correspond to additional noise for the indirectly detected single observable. In Ref. [14] it has been estimated that homodyning the density matrix adds several dB's of noise

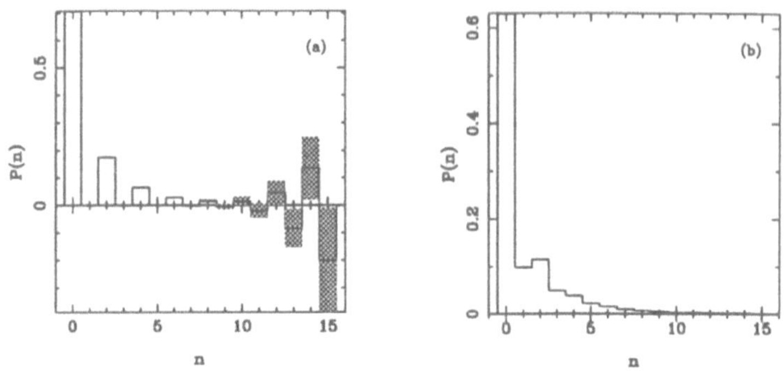

Figure 7. Homodyne detection with quantum efficiency $\eta = .8$ of the photon-number probability of a squeezed vacuum with $\langle \hat{n} \rangle = 1$. Homodyne data are computer-simulated. Here result are obtained from 200 blocks of 5×10^5 data with 27 phases each. Confidence intervals are represented by the gray-shaded thickness of horizontal lines. Fig. 7a: reconstruction using the kernel (30) with $\eta = .8$. Fig. 7b: reconstruction from the same data without taking into account quantum efficiency, namely using the kernel (30) with $\eta = 1$.

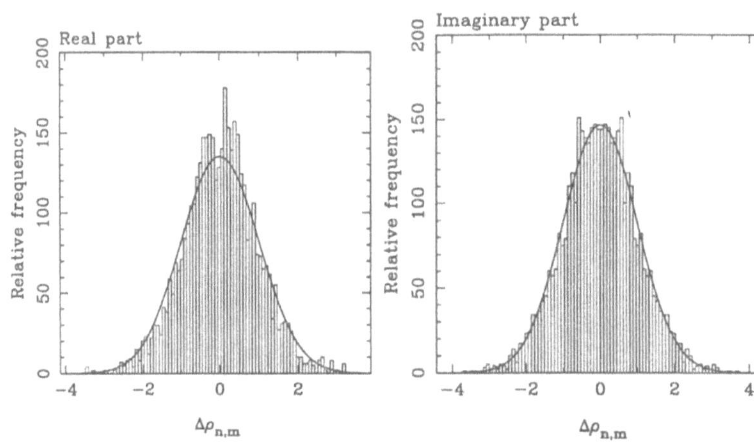

Figure 8. Distribution of the normalized deviations from the theoretical values $\Delta\varrho_{n,m} \equiv \left(\varrho_{n,m}^{(exper)} - \varrho_{n,m}^{(theor)}\right)/\delta_{n,m}$ for the first 65×65 matrix elements. Detection efficiency is $\eta = 1$. The quantum state is coherent with $\langle n \rangle = 4$. The histograms pertain the deviations of all matrix elements (real and imaginary parts separately) detected during 32 experiments with 26 scanning phases each, and 40 measurements for each phase. A standardized Gaussian curve is superimposed to each histogram.

with respect to the ideal measurement of the single observable (it is particularly noisy for detection of the phase of the field). Therefore, ideal detection of a single observable—which provides only a partial information on the quantum state—is always less noisy than homodyning the density matrix—which, however, gives the most complete knowable information.

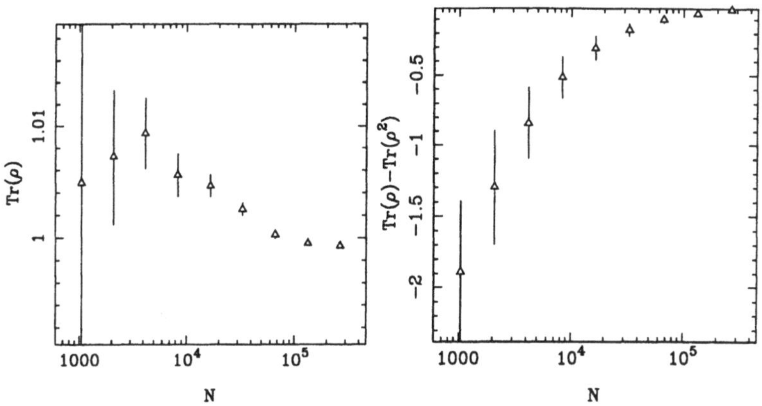

Figure 9. Normalization and mixing of the homodyne-detected density matrix versus the total number of data of the experiment (i.e. number of phases × number of data per phase) for a coherent state with $\langle n \rangle = 4$. Detection efficiency is $\eta = 1$. The matrix is 22×22. Error bars are estimated by grouping data into blocks of 20 data of 26 phases each. The experiments are computer-simulated.

9. Reconstruction of the Wigner function at finite resolution

As explained in Sect. 5, the Wigner function cannot be measured directly from homodyne data: however, it can be reconstructed from the measured density matrix in a dimensionally-truncated Hilbert space. This is achieved using the formula

$$W(\alpha, \bar{\alpha}) = \frac{2}{\pi} \mathrm{Tr} \left[\hat{\varrho} e^{2\left(\alpha a^{\dagger} - \bar{\alpha} a\right)} e^{i\pi a^{\dagger} a} \right] , \tag{32}$$

which in the number representation is rewritten as the Fourier transform

$$W(\alpha, \bar{\alpha}) = \mathrm{Re} \sum_{d=0}^{\infty} e^{i d \arg(\alpha)} \sum_{n=0}^{\infty} \Lambda(n, d; |\alpha|^2) \varrho_{n,n+d} , \tag{33}$$

with $\varrho_{n,m} \doteq \langle n | \hat{\varrho} | m \rangle$, and

$$\Lambda(n, d; |\alpha|^2) = (-)^n 2(2 - \delta_{d0}) |2\alpha|^d \sqrt{\frac{n!}{(n+d)!}} e^{-2|\alpha|^2} L_n^d(|2\alpha|^2) , \tag{34}$$

$L_n^d(x)$ denoting Laguerre polynomials. The Hilbert space truncation of the sums in (33) sets the *resolution in polar coordinates* for the reconstructed Wigner function: this will be illustrated in more detail in Section 11. Laguerre polynomials in Eq. (34) produce radial oscillations, with the result that despite the truncated dimension of the matrix is relatively small, the natural quantum oscillations in the Wigner functions are easily recovered. In Fig. 10 the Wigner function of Schrödinger cat states are given as reconstructed from homodyne computer-simulated data. As compared to the exact method, the old filtered technique (revisited in Sect. 4) would have required very small cutoffs to recover such Wigner oscillations, hence needing much larger sets of data.

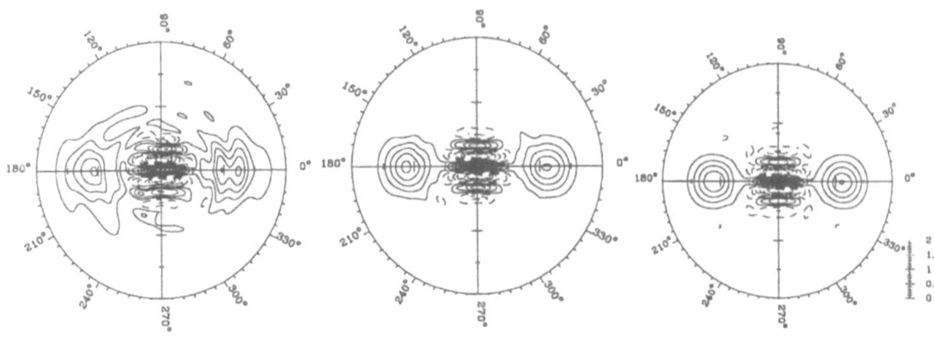

Figure 10. Reconstructed Wigner function of a Schrödinger-cat superposition of two symmetrical coherent states with $\langle n \rangle = 5$ photons each. From the left to the right: 1, 10, 100 blocks of data each with 52 phases and 1000 data for each phase. The dimension of the density matrix is truncated at 16 photons max. Detection efficiency is $\eta = 1$. Notice the oscillations near the origin (dashed lines represent negative values).

10. On the impossibility of measuring the density matrix of a single system

The notion of "measuring a state" is meaningful only if it refers to many repeated measurements performed on an *ensemble of equally prepared* systems: otherwise, the state of an individual system is unobservable, whatever sequence of measurements is performed on the system. In fact, a single measurement cannot yield sufficient information to reconstruct the state; on the other hand, it would uncontrollably disturb the state, thereby ruling out the possibility of extracting the remaining information from successive measurements.

Recently, the possibility of determining the wave function of a single quantum system has been reconsidered by several authors [29, 30, 31, 32, 33], exploring special measurement schemes, based essentially on vanishingly weak quantum nondemolition measurements [29, 30], or "reversible" measurements [31, 32, 33]. In each of these schemes the conclusion is that it is practically impossible to measure the wave function of a single system, either because the weakness of the measuring interaction prevents one from gaining information on the wave function [29], or because the scheme essentially measures only orthogonal states [30], or because quantum measurements can be physically reverted only with probability of success equal to 1/2 [33]. Only very recently Yuen and I[24] proved the general impossibility of determining the state of a single quantum system for arbitrary measuring schemes, including any kind of succession of measurements, which we proved to have the same probability distribution of a single joint measurement with output state independent on the input. We have derived this proof because it seems that, despite its fundamental relevance in the logical framework of quantum mechanics, it has been never proved in general. Here, from Ref. [24] I simply report a scheme for making repeated homodyne measurements on a single copy of the radiation field, in order to show in which way the system state is perturbed however weak the system-apparatus interaction is. This is due to the fact that in order to overcome the low effective quantum efficiency corresponding to the weak interaction, the apparatus must be prepared in a highly squeezed state, as also shown by Tombesi in his set of lectures [34] and in Ref. [35].

Insofar the tomographic scheme that we have considered is based on "second kind" measurements, which completely destroy the quantum state of the system. Many measurements are performed, but the system is prepared in the same state $\hat{\varrho}$ before each measurement. We want now to consider a "first kind" version of the above scheme that in principle allows to measure the density matrix without destroying it. Similarly to any first kind measurement, this goal can be achieved using an indirect-measurement, namely by performing second-kind measurements on another "probe" mode b that interacts with a via a unitary operator $\hat{U}$: this was already explained in the previous set of lectures on Quantum estimation theory and optical detection [3]. An easy way to achieve a first-kind measurement of the quadrature $\hat{a}_\phi \equiv \hat{x}_\phi$ is by measuring the quadrature $\hat{b}_\phi = \frac{1}{2}(b^\dagger e^{i\phi} + be^{-i\phi})$ of another mode b. Without loss of generality, we consider that before every single measurement the probe is prepared in a pure state $|v(\phi)\rangle$, which is generally optimized as a function of the observable $\hat{b}_\phi$ that is measured. The generating function of the moments of $\hat{b}_\phi$ after the interaction with a is given by

$$X(\lambda,\phi) = \mathrm{Tr}\left(e^{i\lambda\hat{b}_\phi}\hat{U}\hat{\varrho}\otimes|v(\phi)\rangle\langle v(\phi)|\hat{U}^\dagger\right)=\mathrm{Tr}\left[\exp\left(i\lambda\hat{U}^\dagger\hat{b}_\phi\hat{U}\right)\hat{\varrho}\otimes|v(\phi)\rangle\langle v(\phi)|\right], \quad (35)$$

and is just the Fourier transform of the probability distribution of the experimental outcomes. We consider the following interaction

$$\hat{U} = \exp\left[\kappa\left(ab^\dagger - a^\dagger b\right)\right] , \quad (36)$$

which describes the linear field transformation at a beam splitter [3]

$$\hat{U}^\dagger b\hat{U} = \sin\kappa\, a + \cos\kappa\, b \equiv \vartheta^{1/2} a + (1-\vartheta)^{1/2} b , \quad (37)$$

where ϑ is the mirror transmissivity, and b is the field mode of the unused port at the same frequency of a. Thus, the present scheme physically corresponds to let the field mode a shine over a long chain of low-transmissivity mirrors, detecting the quadrature of the weak transmitted field at each mirror. Due to the linearity of Eq. (37), the moment generating function factorizes in the following way

$$X(\lambda,\phi) = \chi_a(\vartheta^{1/2}\lambda,\phi)\chi_b((1-\vartheta)^{1/2}\lambda,\phi) , \quad (38)$$

where $\chi_a(\lambda,\phi) = \mathrm{Tr}[\exp(i\lambda\hat{a}_\phi)\hat{\varrho}]$ is the generating function for the noninteracting mode a only, and, analogously, $\chi_b(\lambda,\phi)$ for the mode b. Using the operator identity (15) the density operator $\hat{\varrho}$ is written in terms of the generating function $\chi_a(\lambda,\phi)$

$$\hat{\varrho} = \int_0^\pi \frac{d\phi}{\pi}\int_{-\infty}^{+\infty}\frac{d\lambda\,|\lambda|}{4}e^{-i\lambda\hat{a}_\phi}\chi_a(\lambda,\phi) , \quad (39)$$

and using Eq. (38) one obtains

$$\hat{\varrho} = \int_0^\pi \frac{d\phi}{\pi}\int_{-\infty}^{+\infty} dx\, p_\vartheta(x,\phi)\Xi_\vartheta(x-\hat{a}_\phi) , \quad (40)$$

where $p_\vartheta(x,\phi)$ is the probability of the measured quadrature $\hat{b}_\phi$ rescaled by $\vartheta^{1/2}$ (as for customary homodyning, where the output photocurrent is rescaled by $\eta^{1/2}$[3])

$$p_\vartheta(x,\phi) = \vartheta^{1/2}\int_{-\infty}^{+\infty}\frac{d\lambda}{2\pi}e^{-i\lambda\vartheta^{1/2}x}X(\lambda,\phi) . \quad (41)$$

The kernel $\Xi_\vartheta(x)$ in Eq. (40) is given by

$$\Xi_\vartheta(x) = \frac{1}{2}\mathrm{Re}\int_0^{+\infty} d\lambda\, \lambda e^{i\lambda x}[\chi_b(\lambda\sqrt{(1-\vartheta)/\vartheta}, \phi)]^{-1}, \tag{42}$$

and generally depends on the coupling parameter ϑ ($\equiv \sin\kappa$) and on the probe state $|v(\phi)\rangle$. One can easily see that when the probe mode b is in the vacuum state the kernel (42) is identical to $K_\eta(x)$ in Eq. (21) with $\vartheta \equiv \eta$, namely, the transmissivity ϑ plays the role of the overall quantum efficiency of the indirect measurement. However, as noticed in Ref. [35], the effective quantum efficiency can be decreased at will by squeezing the probe mode b in the direction of the quadrature $\hat{b}_\phi$. More precisely, one prepares the probe in the squeezed vacuum

$$|v(\phi)\rangle \equiv \hat{S}_\phi|0\rangle\,, \tag{43}$$

where

$$\hat{S}_\phi = e^{ib^\dagger b\phi}\hat{S}_0 e^{-ib^\dagger b\phi}\,, \quad \hat{S}_0 = \exp\left[-\frac{r}{2}\left(b^{\dagger 2} - b^2\right)\right], \tag{44}$$

$r > 0$ denoting the squeezing parameter. One has

$$\hat{S}_0\hat{b}_0\hat{S}_0^\dagger = e^r\hat{b}_0\,, \qquad \hat{S}_0^\dagger|x\rangle = e^{r/2}|e^r x\rangle\,, \tag{45}$$

with $|x\rangle$ denoting the eigenvector of $\hat{b}_0$ for eigenvalue x: the rescaling (45) more generally holds for the quadrature $\hat{b}_\phi$ and its eigenvectors $e^{ib^\dagger b\phi}|x\rangle$ when using the rotated squeezing operator $\hat{S}_\phi$ in place of $\hat{S}_0$. With the help of transformations (44) and (45) it is easy to check that the kernel $\Xi_\vartheta(x)$ in Eq. (42) coincides with $K_\eta(x)$ in Eq. (21) with effective efficiency

$$\eta \equiv \frac{e^{2r}\vartheta}{e^{2r}\vartheta + 1 - \vartheta}\,, \tag{46}$$

as also shown in Refs. [34] and [35]. Therefore, by increasing the squeezing parameter r it is possible to enhance the effective quantum efficiency η beyond the allowed bound for measuring the density matrix ($\eta > 1/2$ for number and coherent states). At this point one may think that squeezing the vacuum of b allows one to consider weaker and weaker interactions with $\vartheta \to 0$, with the possibility of performing repeated measurements on the same system with vanishing perturbation at each measurement. However, as we will show immediately, this cannot be attained, because the squeezing needed to keep η as constant also amplifies the perturbation back to a finite extent. This can be seen upon analyzing the limiting behavior of the transition operator $\hat{\Omega}(x,\phi)$, that gives the state reduction after each measurement [3]

$$\hat{\varrho}' = \frac{\hat{\Omega}(x,\phi)\hat{\varrho}\hat{\Omega}^\dagger(x,\phi)}{\mathrm{Tr}[\hat{\varrho}\hat{\Omega}^\dagger(x,\phi)\hat{\Omega}(x,\phi)]}\,. \tag{47}$$

Here, the transition operator $\hat{\Omega}(x,\phi)$ is given by

$$\hat{\Omega}(x,\phi) = \vartheta^{1/4}\langle\vartheta^{1/2}x|e^{-ib^\dagger b\phi}e^{\kappa(ab^\dagger - a^\dagger b)}\hat{S}_\phi|0\rangle\,, \tag{48}$$

where the powers of ϑ account for quadrature rescaling, and one should keep in mind that the matrix element is evaluated between vectors $|\vartheta^{1/2}x\rangle$ and $|0\rangle$ in the Hilbert space of mode b only, so that $\hat{\Omega}(x,\phi)$ is an operator on the Hilbert space of mode a. One has

$$\hat{\Omega}(x,\phi) = \vartheta^{1/4}\langle\vartheta^{1/2}x|\hat{S}_0 e^{-ib^\dagger b\phi} e^{\kappa(a_r b^\dagger - a_r^\dagger b)}|0\rangle \ , \tag{49}$$

where

$$a_r = \cosh r\, a + e^{2i\phi}\sinh r\, a^\dagger \ . \tag{50}$$

Using Eq. (45) and normal ordering the interaction operator with respect to b (by the BCH formula for the $SU(2)$ group [36]), one obtains

$$\hat{\Omega}(x,\phi) = (e^{2r}\vartheta)^{1/4}\langle(e^{2r}\vartheta)^{1/2}x|e^{-ib^\dagger b\phi}e^{\tan\kappa a_r b^\dagger}|0\rangle|\cos\kappa|^{a_r^\dagger a_r} \ . \tag{51}$$

Eq. (51) can be arranged in the following way

$$\hat{\Omega}(x,\phi) = \exp\left[\tan\kappa\, a_r \frac{d}{d\lambda}\bigg|_{\lambda=0}\right] e^{\frac{1}{2}\lambda^2}(e^{2r}\vartheta)^{1/4}\langle(e^{2r}\vartheta)^{1/2}x|\lambda e^{-i\phi}\rangle|\cos\kappa|^{a_r^\dagger a_r} \ , \tag{52}$$

where λ is a real running parameter, and $|\lambda e^{-i\phi}\rangle$ is a coherent state. Upon evaluating the derivative with respect to λ one obtains the explicit operator form of $\hat{\Omega}(x,\phi)$

$$\begin{aligned}\hat{\Omega}(x,\phi) &= \left(\frac{2e^{2r}\vartheta}{\pi}\right)^{1/4} \exp\left[-\left(e^{2r}\vartheta x - \tan\kappa e^{-i\phi}a_r\right)^2\right] \\ &\times \exp\left[\frac{1}{2}\tan^2\kappa e^{-2i\phi}a_r^2\right]|\cos\kappa|^{a_r^\dagger a_r} \ .\end{aligned} \tag{53}$$

We can now finally evaluate the asymptotic form of the transition operator in Eq. (53) in the simultaneous limits of vanishing transmission coefficient $\vartheta \to 0$ and infinite squeezing parameter $r \to \infty$ keeping the effective quantum efficiency η as constant according to Eq. (46). In Eq. (53), for $\vartheta \to 0$ using Eq. (37) one obtains $e^{2r}\vartheta \to \Delta_\eta^{-2}$, where

$$\Delta_\eta^2 \equiv \frac{1-\eta}{\eta} \ . \tag{54}$$

Moreover, one has $e^{-r}e^{-i\phi}a_r \to \hat{a}_\phi$, and $|\cos\kappa|^{a_r^\dagger a_r} \to \exp[-\hat{a}_\phi^2/(2\Delta_\eta^2)]$. Therefore, the asymptotic form of the transition operator is simply given by

$$\hat{\Omega}(x,\phi) = \left(\frac{2}{\pi\Delta_\eta^2}\right)^{1/4} \exp\left[-\frac{(x-\hat{a}_\phi)^2}{\Delta_\eta^2}\right] \ . \tag{55}$$

Equation (55) is the typical form of a von Neumann "reduction" of the state (see my other set of lectures in this volume [3]). Therefore, one concludes that despite the interaction has been tuned as vanishingly small, the state is in fact "reduced" at each measuring step.

In conclusion of this section I want to emphasize that the need of a lower bound for quantum efficiency η is of fundamental relevance. In fact, if the density matrix could be homodyne detected for vanishingly small quantum efficiency, then there would be no need for squeezing the vacuum of the measuring apparatus in order to enhance the efficiency beyond the bound, and the density matrix of a single system could be detected by vanishingly weak measurements.

11. Quantum and classical Radon transform

In this section I show how the new exact algorithm for homodyning the density matrix can be used also for ordinary imaging. The situation of interest is when the radial profiles are not well resolved digitalized functions, but actually represent the density distribution of random points—as if in our X-ray machine the beam is so weak that radial photos are just the collection of many small spots, each from a single X-ray photon (this situation is sketched in Fig. 11. It is obvious that this case can be reduced to the one of conventional imaging by counting all points falling in a predetermined 1-d mesh, and giving radial profiles in form of histograms (this is what actually happens in a real machine that uses arrays of photodetectors). However, we want to utilize the whole available information from each "event"—i.e. the exact 1-d location of each spot—in a way that is independent on any predetermined mesh. In practice, this extreme situation happens when the signal is so weak and the machine resolution is so high (i.e. the mesh-step is so tiny) that only zero or one photon at most can be collected in each channel.

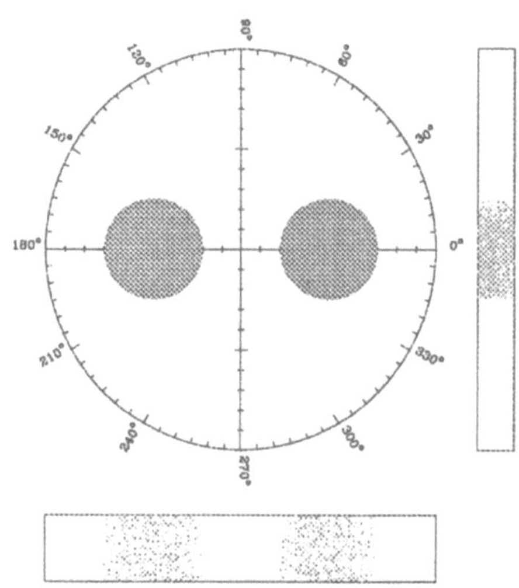

Figure 11. Illustration of imaging in presence of very weak signals. The image is the same as in Fig. 3, but here the transmission profiles are given in terms of random points on a photographic plate.

The above low-signal/high-resolution case naturally brings the imaging problem into the quantum domain. Here, we have at disposal our technique for measuring the density matrix of the field in terms of homodyne outcomes (the equivalent of our radial spots) for different phases of the local oscillator (the equivalent of our ϕ angle). In general an image $W(\alpha, \bar{\alpha})$ does not correspond to a Wigner function of a physical state, due to the fact that the Heisenberg relations unavoidably produce only smooth Wigner functions, whereas a conventional image can have very sharp edges. However, if one allows the density matrix to be no longer positive definite (but still trace class), a correspondence between density matrices and images is obtained that holds in general, and a convenient truncation of the matrix dimension $d_{\mathcal{H}}$ can be chosen to set the imaging resolution. In summary, the connection between images and density matrix $\varrho_{n,m}$—what I call either the *quantum Radon transform* or the *quantum imaging*—is given by Eqs. (18), (30), (33) and (34) with the sums truncated at maximum photon number $d_{\mathcal{H}} - 1$. I emphasize again that the density matrix does not correspond to any physical state of radiation, and this is why the method is named *Fictitious Photons Tomography* [16].

For suitably chosen dimension $d_{\mathcal{H}}$ the quantum Radon transform also provides a procedure for image compression, with $d_{\mathcal{H}}$ setting the resolution of imaging. The kind of resolution can be understood by studying the behavior of the kernels $\langle n|K(x - \hat{x}_\phi)|n + d\rangle$ in Eq. (30) which are averaged over the experimental data in order to obtain the matrix elements $\varrho_{n,n+d}$. As shown in Fig. 4, outside a region that is increasing versus the indices n and $n + d$, the functions $\langle n|K(x - \hat{x}_\phi)|n + d\rangle$ decrease as x^{-d-2}, whereas inside they oscillate with a number of nodes that increases with $2n + d$. This behavior produces the effects illustrated in Fig. 12 where for increasing dimension $d_{\mathcal{H}}$ I report the tomographic reconstruction of an image that describes a chess-board shaped step-function $W(\alpha, \bar{\alpha})$ in the plane. The plot is obtained by numerically integrating Eq. (18) with the kernel (30) from analytically given transmission profiles $p(x, \phi)$ and then using Eqs. (33) and (34). From Fig. 12 one can see that both radial and angular resolutions improve versus $d_{\mathcal{H}}$, making the details of the image sharper and sharper.

In Fig. 13 a Monte Carlo simulation is reported of the quantum imaging of the font "ψ" for weak signals. The reconstruction is quite accurate already for $5 * 10^6$ data with resolution $d_{\mathcal{H}} = 32$. The situation occurring for small numbers of data is given in the first plot, where the highly resolved image exhibits the natural statistical fluctuations due to the limited number of data. For larger samples the image starts appearing sharper from the random background. For comparison the analytical imaging (as in Fig. 12) is also given for resolution $d_{\mathcal{H}} = 128$.

12. Concluding remarks on experimental applications

I conclude these notes with some remarks on experimental applications of the method of homodyning the density matrix. It is important to realize that although photodetectors themselves in principle seem to provide the easiest way for measuring the photon number distribution, it is very difficult to achieve a photon counting that is ideal for a large range of photon numbers starting from the vacuum. In fact, it is well known that if the detector ideally has no threshold (as for an avalanche photo-diode) then it has a low quantum efficiency, or vice versa, if it has a good quantum efficiency, then it has a high sensitivity threshold (as for a linear photo-diode). Homodyning has the advantage of amplification from the LO, which is highly excited and coherent, such that the difference photocurrent at the output is very intense and can be detected by high-efficiency linear photo-diodes (the overall quantum efficiency of the homodyne is originated by incomplete mode-matching between the signal and the LO). In this way, homodyning the density matrix becomes a perfect method for measuring the photon statistics. In addition, as for the diagonal matrix elements the phase of the LO is averaged out [the kernel in Eq. (30) does not depend on ϕ for $d = 0$], it is sufficient to collect data without worrying for the phase, with the only caution of having negligible phase-correlation between signal and LO: this simplifies the

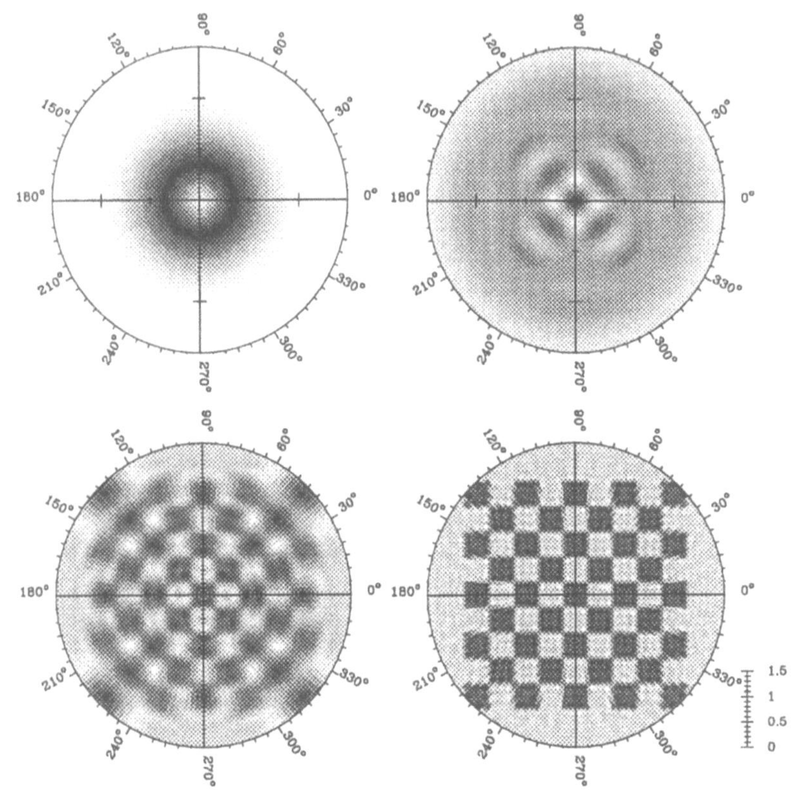

Figure 12. Tomographic reconstruction of a chess-board shaped step function $W(\alpha,\overline{\alpha})$ for increasing dimension of the truncated matrix, $d_{\mathcal{H}} = 2, 8, 32, 128$ from the top-left to the bottom-right. The plots are obtained by integrating numerically Eq. (18) with kernel (30), from analytically assigned transmission profiles $p(x,\phi)$, and then using Eqs. (33) and (34).

experimental method greatly. However, by no means such "homodyne photodetector" can be used to detect single events, but only statistics. That homodyning is a good tool to monitor delicate oscillations in photon statistics was already suggested in Refs. [12, 19, 20]. More recently, the new exact method has been adopted in real experiments: in Refs. [22, 23] it has been applied to measure ultrafast (subpicosecond) time-resolved photon statistics of weak fields.

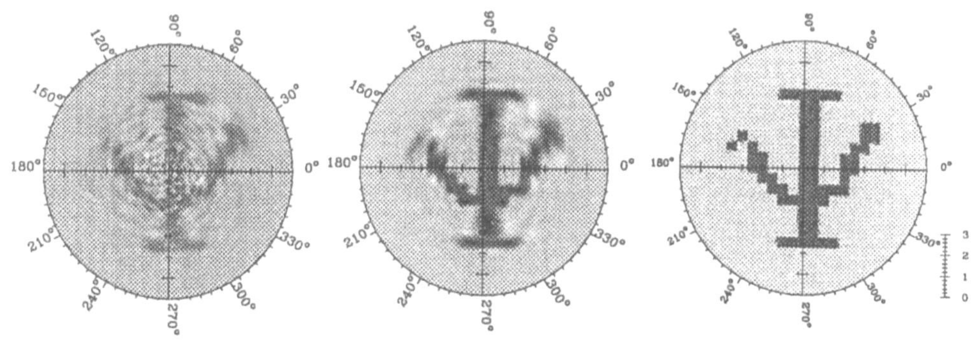

Figure 13. Monte Carlo simulation of a quantum imaging for weak signals of the font "ψ". The truncation dimension is fixed at $d_{\mathcal{H}} = 32$. The first plot on the left is the reconstruction from only 50000 data (50 phases and 1000 data each); the second plot is for $5 * 10^6$ data (100 blocks as before). The last plot is a reconstruction obtained by integrating numerically Eq. (18) with kernel (30) from analytically assigned transmission profiles $p(x, \phi)$, as in Fig. 12 (here $d_{\mathcal{H}} = 128$).

13. Appendices

13.1. PROOF OF THE CENTRAL LIMIT THEOREM

Generally speaking, the central limit theorem says that the probability distribution of the sum of a large number N of random variables $\{x_n\}_{n=1}^N$ approaches a Gaussian function. For application to the scheme of repeated measurements, we only need a simplified version of the theorem, with all variables x_n having the same probability distribution. In this case the proof of the theorem is relatively straightforward and is next given in detail.

Theorem: *Let $\{x_n\}_{n=1}^N$ be statistically independent random variables having the same probability distribution, with average $\overline{x}$, variance σ^2 and bounded third moment*

$$\mathbf{E}|x_n - \overline{x}|^3 < B < +\infty \; . \tag{56}$$

Then the random variable $X_N = \frac{1}{N}\sum_{n=1}^N x_n$ has average $\overline{x}$ and variance $\varepsilon_N^2 = \frac{\sigma^2}{N}$, and has a probability distribution approaching a Gaussian function asymptotically for $N \to \infty$.

Proof: The first part of the theorem is straightforward. One has that

$$\mathbf{E}X_N = \frac{1}{N}\sum_{n=1}^N \mathbf{E}x_n = \frac{1}{N}N\overline{x} \equiv \overline{x} \; . \tag{57}$$

Moreover, as a consequence of statistical independence of x_n one has

$$\begin{aligned}\mathbf{E}X_N^2 &= \frac{1}{N^2}\sum_{n=1}^N\sum_{m=1}^N \mathbf{E}[x_n x_m] = \frac{1}{N^2}\sum_{n\neq m=1}^N \mathbf{E}x_n\,\mathbf{E}x_m + \frac{1}{N^2}\sum_{n=1}^N \mathbf{E}x_n^2 \\ &= \frac{N-1}{N}\overline{x}^2 + \frac{1}{N}(\sigma^2 + \overline{x}^2) = \frac{1}{N}\sigma^2 + \overline{x}^2 \; ,\end{aligned} \tag{58}$$

namely $\mathbf{E}[\Delta X_N^2] = \frac{\sigma^2}{N}$. The second part of the proof consists of showing that for $N \to \infty$ the characteristic function of the standardized variable $\chi_N = \frac{X_N - \overline{x}}{\sigma/\sqrt{N}}$ is that of a zero-mean/unit-variance Gaussian distribution, namely

$$\lim_{N\to\infty} \mathbf{E}\exp(i\kappa\chi_N) = \exp\left(-\frac{1}{2}\kappa^2\right) . \tag{59}$$

In fact, the standardized Gaussian probability distribution is the Fourier transform of the characteristic function (59), and the Fourier transform is unique, and this will prove the last part of the theorem. From the statistical independence and equivalence of the random variables $\{x_n\}_{n=1}^N$ we have

$$\mathbf{E}\exp(i\kappa\chi_N) = \mathbf{E}\exp\left[\frac{i\kappa}{\sqrt{N}\sigma}\sum_{n=1}^N (x_n - \overline{x})\right] = \left\{\mathbf{E}\exp\left[i\frac{\kappa}{\sqrt{N}\sigma}(x_1 - \overline{x})\right]\right\}^N$$
$$= \left\{\mathbf{E}\left[1 + i\frac{\kappa}{\sqrt{N}\sigma}(x_1 - \overline{x}) + \frac{1}{2!}\left(i\frac{\kappa}{\sqrt{N}\sigma}\right)^2 (x_1 - \overline{x})^2 + \frac{R_N}{N^{3/2}}\right]\right\}^N$$
$$= \left\{1 - \frac{\kappa^2}{2N} + \frac{\mathbf{E}R_N}{N^{3/2}}\right\}^N , \tag{60}$$

where by Taylor's formula $|\mathbf{E}R_N| \le C\frac{|\kappa|^3}{\sigma^3}\mathbf{E}|x_1 - \overline{x}|^3$ for some constant C. Forming the natural logarithm we have

$$\log \mathbf{E}\exp(i\kappa\chi_N) = N\log\left(1 - \frac{\kappa^2}{2N} + \frac{1}{N^{3/2}}\mathbf{E}R_N\right) = -\frac{\kappa^2}{2} + \mathcal{O}(N^{-1/2}) , \tag{61}$$

and hence Eq. (59). The theorem can be easily extended to more general situations, for example when the random variables $\{x_n\}_{n=1}^N$ are not statistically equivalent, but all of them have finite average and variance and satisfy the bound (56) for some common constant B. In this case for large N X_N has still Gaussian distribution, with mean value $\mathbf{E}X_N = \frac{1}{N}\sum_{n=1}^N \mathbf{E}x_n$ and variance $\mathbf{E}\Delta X_N^2 = \frac{1}{N}\sum_{n=1}^N \mathbf{E}[\Delta x_n^2]$. It is also easy to see that the theorem can be proven using, in place of the bound (56), the more general Lyapunov condition

$$\lim_{N\to\infty} \frac{\sum_{n=1}^N \mathbf{E}|x_n - \mathbf{E}x_n|^3}{\left\{\sum_{n=1}^N \mathbf{E}[\Delta x_n^2]\right\}^{3/2}} = 0 . \tag{62}$$

In fact, in this case the same asymptotic expansion (61) holds (see Ref. [37], pag. 78). Finally, the central limit theorem is known to apply to some cases of statistically dependent variables [38].

Necessary conditions for the theorem's validity are difficult to state: however, it is obvious that the result of the theorem does not hold true at least when the first two moments of the random variables $\{x_n\}_{n=1}^N$ are not bounded.

13.2. DERIVATION OF THE INVERSE RADON TRANSFORM FORMULA

The Radon transform of a function $m(x,y)$ in the complex plane is given by the collection of functions $r(x,\phi)$ $\phi \in [0,\pi]$, which are the marginal integrals of $m(x,y)$ along the direction with polar angle ϕ, namely

$$r(x,\phi) = \int_{-\infty}^{+\infty} \frac{dy}{\pi}\, m\left((x+iy)e^{i\phi}, (x-iy)e^{-i\phi}\right) . \tag{63}$$

In order to derive the inverse formula of (63), we start from the definition of the Dirac's delta in the complex plane

$$\delta^{(2)}(\alpha) = \int \frac{\mathrm{d}^2\gamma}{\pi^2} e^{\gamma\bar{\alpha}-\bar{\gamma}\alpha} , \tag{64}$$

and change to polar coordinates $\gamma = -\frac{i}{2}e^{-i\phi}k$, obtaining

$$\delta^{(2)}(\alpha) = \int_0^{\infty} \frac{\mathrm{d}k\, k}{4} \int_0^{2\pi} \frac{\mathrm{d}\phi}{\pi^2} e^{-ik\alpha_\phi} , \tag{65}$$

where $\alpha_\phi = \mathrm{Re}(\alpha e^{-i\phi})$. Using the symmetry $\alpha_{\phi+\pi} = -\alpha_\phi$ we can rewrite (65) as follows

$$\delta^{(2)}(\alpha) = \int_{-\infty}^{+\infty} \frac{\mathrm{d}k\,|k|}{4} \int_0^{\pi} \frac{\mathrm{d}\phi}{\pi^2}\, e^{-ik\alpha_\phi} . \tag{66}$$

The last formula is *in nuce* the inverse Radon transform. In fact, it is just sufficient to write $m(\alpha,\bar{\alpha})$ identically as follows

$$m(\alpha,\bar{\alpha}) = \int \mathrm{d}^2\beta\, \delta^{(2)}(\alpha-\beta)\, m(\beta,\overline{\beta}) . \tag{67}$$

Then, using the polar representation (66) of the Dirac delta and exchanging the integral over β with those defining $\delta^{(2)}(\alpha)$ we obtain

$$m(\alpha,\bar{\alpha}) = \int_{-\infty}^{+\infty} \frac{\mathrm{d}k\,|k|}{4} \int_0^{\pi} \frac{\mathrm{d}\phi}{\pi^2} \int \mathrm{d}^2\beta\, e^{-ik(\alpha_\phi-\beta_\phi)}\, m(\beta,\overline{\beta}) . \tag{68}$$

Finally, upon rewriting $\beta \doteq (x+iy)e^{i\phi}$ and using Eq. (63), we obtain the inverse Radon transform

$$m(\alpha,\bar{\alpha}) = \int_{-\infty}^{+\infty} \frac{\mathrm{d}k\,|k|}{4} \int_0^{\pi} \frac{\mathrm{d}\phi}{\pi} \int_{-\infty}^{+\infty} \mathrm{d}x\;\; r(x,\phi) \exp\left[ik\left(x-\alpha_\phi\right)\right] . \tag{69}$$

13.3. FACTORIZATION FORMULA FOR THE INTEGRAL KERNEL

The numerical evaluation of the integral kernel as is written in Eq. (30) involves the calculation of parabolic cylinder functions, which can be connected to each other iteratively. However, for sufficiently large n and d (around 10) the iteration becomes numerically unstable, and the cylinder functions must be evaluated by series summation, slowing down the numerical calculation considerably. On the other hand, the series over ν in Eq. (30) has terms with oscillating signs and absolute value that increases dramatically with n and d. Again for sufficiently large n and d (20) the calculation becomes unstable, and in order to reach values as $n = 40$ one is forced to use extended precision. Recently, elegant factorization formulas have been found by Richter [17], Leonhardt *et al.* [18], and more recently derived by D'Ariano *et al.* in a completely algebraic way [21]. Such formulas provide an algorithm that is fast, low-memory, and stable for large n and d. Here I only report the recipe for numerical calculation: for the derivation of formulas the reader is addressed to the above quoted references.

The kernel functions—also called "pattern functions" in Refs. [17, 19, 20] or "sampling functions" in Ref. [18]—can be factorized as follows

$$\begin{aligned}&\langle n|K(x-\hat{x})|m\rangle = \langle m|K(x-\hat{x})|n\rangle \\ &= 2xu_n(x)v_m(x) - \sqrt{n+1}u_{n+1}(x)v_m(x) - \sqrt{m+1}u_n(x)v_{m+1}(x)\,, \end{aligned} \tag{70}$$

for $m \geq n$. In Eq. (70) only the case of $\phi = 0$ ($\hat{x} \equiv \hat{x}_\phi$) has been considered, as for $\phi \neq 0$ the trivial rotation factor $\exp(-id\phi)$ is just needed [see Eq. (30)]. The functions $u_n(x)$ (energy eigenfunctions of the harmonic oscillator) are connected each other by the (stable) iteration

$$u_n(x) = \frac{1}{\sqrt{n}}[2xu_{n-1}(x) - \sqrt{n-1}u_{n-2}(x)]\,, \tag{71}$$

with starting values

$$u_{-1}(x) = 0\,, \qquad u_0(x) = \exp\left(-x^2\right)\,. \tag{72}$$

The functions $v_m(x)$ ("irregular" harmonic oscillator eigenfunctions [18]) also satisfy recursion (71), but now this would be unstable, because it depends too critically on accuracy of $v_0(x)$, which would involve an imaginary error function. Instead, Ref. [18] suggests using the backward recursion

$$v_{m-2}(x) = [2xv_{m-1}(x) - \sqrt{m}v_m(x)]/\sqrt{m-1}\,, \tag{73}$$

starting from the semiclassical approximation for large numbers

$$v_m(x) = \left(\frac{8\pi}{\alpha_m^2 \sin^2\kappa_m}\right)^{1/4} \sin\left\{\frac{1}{2}\alpha_m^2\left[\sin(2\kappa_m) - 2\kappa_m\right] + \frac{\pi}{4}\right\}\,, \tag{74}$$

where

$$\alpha_m = \sqrt{m+1/2}\,, \qquad \kappa_m = \arccos\frac{x}{\alpha_m}\,. \tag{75}$$

For a density matrix with maximal number of photons N it is suggested to start the iteration (73) from $m = 4N$ and $m = 4N-1$. The classically allowed region for such quantum number is $|x| < \alpha_{4N} - \frac{1}{2}\alpha_{4N}^{-1/3}$, otherwise one can use the asymptotic expression for $v_m(x)$ which are provided by the forward recursion

$$v_m(x) = \frac{\sqrt{m}}{2x}v_{m-1}(x)\,, \tag{76}$$

with the starting condition

$$v_0(x) = \exp\left(x^2\right)/x\,. \tag{77}$$

All tomographic reconstructions in the present paper for $\eta = 1$ have been obtained using the above numerical recipe. For $\eta < 1$ a factorization formula is not yet known, and the results here presented have been obtained by summing the series (30). In Ref. [39] the following Bernoulli transformation has been suggested

$$\langle n|\hat{\varrho}|m\rangle = \eta^{-\frac{1}{2}(n+m)} \sum_{k=0}^{\infty} \langle n+k|\hat{\varrho}|m+k\rangle^{(\eta)} \left[\binom{n+k}{n} \binom{m+k}{m} \right]^{1/2} , \qquad (78)$$

to connect the desired density matrix elements $\langle n|\hat{\varrho}|m\rangle$ to the "smeared" ones $\langle n|\hat{\varrho}|m\rangle^{(\eta)}$ measured by inefficient homodyne detection with the kernel for unit efficiency (such "smeared" matrix elements are Bernoulli convolutions of the true matrix elements: an example has been given in Fig. 7). However, in practice the deconvolution (78) can be used only for efficiency η very near to one, and for small indices n and m, because the convergence of the series in Eq. (78) is very slow (geometric convergence with finite convergence radius). Moreover, this method is not numerically efficient, because one needs to sample many smeared matrix elements $\langle n|\hat{\varrho}|m\rangle^{(\eta)}$ in order to obtain only few true elements $\langle n|\hat{\varrho}|m\rangle$.

References

1. W. Pauli, in *Encyclopedia of Physics* (Springer, Berlin, 1958), Vol. V, p. 17.
2. For example, the spherical-harmonic eigenstates $f(r)P_l^m(\theta)e^{im\phi}$ of the angular momentum (whatever is the radial function) have probabilities $W(\mathbf{x})$ and $W(\mathbf{p})$ that are independent on the sign of m.
3. G. M. D'Ariano, *Quantum estimation theory and optical detection*, in this volume.
4. U. Fano, Rev. Mod. Phys. **29**, 74 (1957)
5. B. d'Espagnat, *Conceptual Foundations of Quantum Mechanics*, W. A. Benjamin, Massachusetts (1976)
6. A. Royer, Found. Phys. **19**, 3 (1989)
7. U. Leonhardt, Phys. Rev. Lett. **74**, 4101 (1995)
8. For a historical review on theoretical work and proposals for measuring quantum states see Ref. [6]. For more recent references see also Refs. [7] and [20].
9. D. T. Smithey, M. Beck, M. G. Raymer, and A. Faridani, Phys. Rev. Lett. **70**, 1244 (1993). See also D. T. Smithey, M. Beck, J. Cooper, and M. G. Raymer, Phys. Rev. A **48**, 3159 (1993) and references therein.
10. K. Vogel and H. Risken, Phys. Rev. A **40**, 2847 (1989). This Radon-transform connection between homodyne probability distributions and Wigner function was lately generalized to Fourier-transform relations between different representations of the density matrix in: H. Kühn, D.-G. Welsch, W. Vogel, J. Mod. Optics **41**, 1607 (1994).
11. G. M. D'Ariano, C. Macchiavello and M. G. A. Paris, Phys. Rev. A**50**, 4298 (1994).
12. G. M. D'Ariano, U. Leonhardt and H. Paul, Phys. Rev. A **52**, R1801 (1995)
13. G. M. D'Ariano, Quantum Semiclass. Opt. **7**, 693 (1995)
14. G. M. D'Ariano, C. Macchiavello and M. G. A. Paris, Phys. Lett. A **195**, 31 (1994)
15. G. M. D'Ariano, C. Macchiavello, and M. G. A. Paris, Nuovo Cimento **110** B, 237 (1995)
16. G. M. D'Ariano, C. Macchiavello and M. G. A. Paris, Opt. Commun. (1996 in press)
17. Th. Richter, Phys. Lett. A (in press)
18. U. Leonhardt, M. Munroe, T. Kiss, M. G. Raymer and Th. Richter, Opt. Commun. (1996 in press)
19. H. Paul, U. Leonhardt, and G. M. D'Ariano, Acta Phys. Slov. **45**, 261 (1995)
20. U. Leonhardt, H. Paul and G. M. D'Ariano, Phys. Rev. A **52** 4899 (1995)
21. G. M. D'Ariano, C. Macchiavello and N. Sterpi, unpublished.
22. M. Munroe, D. Boggavarapu, M. E. Anderson, and M. G. Raymer, Phys. Rev. A **52**, R924 (1995)
23. M. Munroe, D. Boggavarapu, M. E. Anderson, U. Leonhardt, and M. G. Raymer, in *Coherence and Quantum Optics VII*, ed. by J. Eberly, L. Mandel, and E. Wolf (in press)
24. G. M. D'Ariano and H. P. Yuen, Phys. Rev. Lett. **76** 2832 (1996)
25. F. Natterer, *The mathematics of computerized tomography*, (Wiley, 1986).
26. P. Mansfield and P. G. Morris, *NMR Imaging in Biomedicine*, (Academic Press, 1982).
27. W. Schleich and J. A. Wheeler, Nature **326**, 574 (1987).
28. R. A. Fisher, M. M. Nieto, and V. D. Sandberg, Phys. Rev. D**29**, 110 (1984)
29. O. Alter, and Y. Yamamoto, Phys. Rev. Lett. **74**, 4106 (1995).
30. Y. Aharonov, J. Anandan, L. Vaidman, Phys. Rev. A **47**, 4616 (1993); Y. Aharonov and L. Vaidman, Phys. Lett. A **178**, 38 (1993).

31. M. Ueda and M. Kitagawa, Phys. Rev. Lett. **68**, 3424 (1992)
32. A. Imamoglu, Phys. Rev. A **47**, R4577 (1993)
33. A. Royer, Phys. Rev. Lett. **73**, 913 (1994).
34. P. Tombesi, in this volume.
35. G. M. D'Ariano, S. Mancini, P. Tombesi, unpublished
36. G. M. D'Ariano, Int. J. Mod. Phys. B **6**, 1291 (1992).
37. Y. A. Rozanov, *Probability Theory*, reprinted by Dover Publ. Inc., N. Y. 1977.
38. H. Cramér, *Mathematical Methods of Statistics*, Princeton Univ. Press, Princeton, NJ, 1951.
39. T. Kiss, U. Herzog, and U. Leonhardt, Phys. Rev. A **52**, 2433 (1995)

OPTICAL TOMOGRAPHY AND MACROSCOPIC COHERENCE

P. TOMBESI
Dipartimento di Matematica e Fisica, Università di Camerino, I-62032
and
Istitito Nazionale di Fisica della Materia, Italy

1. Wigner Function and Marginal Probabilities

It is well known that the complete information about all the quantum statistical properties of a quantum mechanical system are contained in the density operator $\hat{\rho}$. Various representations of the density matrix operator were introduced in quantum optics as valuable mathematical tools. These are the Glauber-Sudarshan P-representation [1], [2], the Husimi Q-function [3] and the Wigner function [4]. All these representations are called quasi-probability distributions. The Wigner function is the first quasi-probability density introduced in Quantum Mechanics. It may be defined as the Fourier transform of the symmetrically ordered characteristic function $\chi(\eta)$ of the density operator $\hat{\rho}$. The density operator is indeed uniquely determined by its characteristic function

$$\chi(\eta) = Tr(\hat{\rho}e^{\eta a^+ - \eta^* a}) \tag{1}$$

where a,a^+ are boson operators of the quantum system with the canonical commutation $[a, a^+] = 1$. The Wigner function is its complex Fourier transform

$$W(\alpha, \alpha^*) = \int \frac{d^2\eta}{\pi^2} e^{\eta^*\alpha - \eta\alpha^*} \chi(\eta). \tag{2}$$

that is why it is called a quasi-probability. The W-function gives the symmetrically-ordered moments of boson operators a and a^+. Cahill and Glauber [5] introduced the s-parametrized quasi-probability distribution. This distribution may be defined as the Fourier transform of the s-parametrized characteristic function

$$\chi(\eta; s) = Tr(\hat{\rho}e^{\eta a^+ - \eta^* a + s\frac{|\eta|^2}{2}}) \tag{3}$$

i.e.

$$W(\alpha, \alpha^*; s) = \int \frac{d^2\eta}{\pi^2} e^{\eta^*\alpha - \eta\alpha^*} \chi(\eta; s). \tag{4}$$

Hence, $W(\alpha, \alpha^*; 1)$ is the Glauber-Sudarshan P-function, $W(\alpha, \alpha^*; 0)$ is the Wigner W-function and $W(\alpha, \alpha^*; -1)$ is the Husimi Q-function. It turns out that the quantum character of a physical system is better identified by considering its Wigner function instead of other representations, because the Q-function is always too smooth and the P-function

T. Hakioğlu and A. S. Shumovsky (eds.), Quantum Optics and the Spectroscopy of Solids, 203–224.

becomes highly singular for particular quantum states. One may notice that the characteristic function $\chi(\eta)$ in Eq.(1) is the mean value of the displacement operator

$$\hat{D}(\eta) = e^{\eta a^+ - \eta^* a} \tag{5}$$

which creates the coherent state $|\eta\rangle$ from the vacuum [1]; i.e.

$$|\eta\rangle = \hat{D}(\eta)|0\rangle. \tag{6}$$

If we now choose

$$\eta = i\frac{r}{2}e^{i\phi} \tag{7}$$

then

$$\chi(\eta) = \chi(r,\phi) = Tr(\hat{\rho}e^{ir\frac{ae^{-i\phi}+a^+e^{i\phi}}{2}}) \tag{8}$$

holds. Defining the Hermitian rotated quadrature

$$\hat{X}_\phi = \frac{ae^{-i\phi} + a^+e^{i\phi}}{2} \tag{9}$$

we finally obtain that the characteristic function of the density matrix $\hat{\rho}$ is by definition the mean value of the exponential of the rotated quadrature operator $\hat{X}_\phi$

$$\chi(r,\phi) = Tr(\hat{\rho}e^{ir\hat{X}_\phi}) = \langle e^{ir\hat{X}_\phi}\rangle. \tag{10}$$

This is nothing else that the characteristic function of the probability distribution $p(x,\phi)$ of the rotated quadrature operator at fixed phase angle ϕ. This distribution may be defined as the Fourier transform of its characteristic function

$$p(x,\phi) = \int \frac{dr}{2\pi}\chi(r,\phi)e^{-irx}. \tag{11}$$

Because $\hat{X}_\phi$ in Eq.(9) is a Hermitian operator this probability is always well behaved and positive everywhere. The distribution $p(x,\phi)$ is the so-called marginal distribution of the Wigner function and represents the projection of the W-function onto a plane pependicular to the complex plane (α,α^*) at a given phase angle. The distribution $p(x,\phi)$ represents thus the probability distribution of the outcomes x, for a fixed phase angle ϕ, of the quadrature observable $\hat{X}_\phi$ which is the rotated quadrature operator.

Vogel and Risken [6] showed that the s-parametrized Wigner function in Eq.(4) may be obtained from an ensemble of repeated measurements of the rotated quadrature operator $\hat{X}_\phi$ at various phases ϕ relative to the local oscillator.

2. Quadrature Phase Measurements

Let us first of all consider how the measurement process in a homodyne detection gives the value of the quadrature operator. Various schemes to make quadrature phase measurements were discussed by Shapiro and Yuen [7]. These involve homodyning the signal field with a local oscillator as a reference field before photodetection. Consider the two impinging fields in a beam splitter with transmissivity coefficient ξ. The two field $E_S(\mathbf{r},t)$ and $E_{LO}(\mathbf{r},t)$ have the same frequencies and the same polarizations; thus

$$E_S(\mathbf{r},t) = i\left(\frac{\hbar\omega}{2V\epsilon_0}\right)^{1/2}\left(ae^{i(\mathbf{k}\cdot\mathbf{r}-\omega t)} - h.c.\right) \tag{12}$$

and

$$E_{LO}(\mathbf{r},t) = i\left(\frac{\hbar\omega}{2V\epsilon_0}\right)^{1/2}\left(be^{i(\mathbf{k}\cdot\mathbf{r}-\omega t)} - h.c.\right) \tag{13}$$

where a and b are boson operators and we only consider one mode for each field. The total field, after recombination at the beamsplitter, is

$$E_T(\mathbf{r},t) = i\left(\frac{\hbar\omega}{2V\epsilon_0}\right)^{1/2}\left(ce^{i(\mathbf{k}\cdot\mathbf{r}-\omega t)} - h.c.\right). \tag{14}$$

For a phase shift of $\pi/2$ between the reflected and transmitted beam we have

$$c = a\sqrt{\xi} + ib\sqrt{1-\xi}. \tag{15}$$

The photons are detected when, impinging on the surface of a photodetector, are converted into electric pulses. Not all photons, however, are converted in electric pulses and hence detected. This effect introduces the quantum efficiency η of detection. For $\eta = 1$, the ideal case, all photons are detected while, in reality, is always $\eta < 1$. We shall show later on that the quantum efficiency may inhibits the complete reconstruction of any density matrix elements. It is possible to show [8] that the probability of detecting m photons, when the detector has quantum efficiency η and only one mode of the field is considered, is given by

$$P_m^\eta = Tr\left[\hat{\rho}\ :\ \frac{(\eta c^+ c)^m}{m!}e^{-\eta c^+ c}\ :\right], \tag{16}$$

where : ... : denotes normal ordering. When $\eta = 1$ one gets

$$P_m^1 = \langle m|\hat{\rho}|m\rangle, \tag{17}$$

while for $\eta < 1$

$$P_m^\eta = \sum_{k=m}^{\infty} \rho_{kk}\begin{pmatrix} k \\ m \end{pmatrix}\eta^m(1-\eta)^{k-m} \tag{18}$$

which represents the convolution of number distribution with the binomial distribution. The non ideal photodetector with quantum efficiency $\eta < 1$ can be represented as the action of a beam splitter with transmission coefficient η which is placed before an ideal photodetector with unit quantum efficiency. In the following we will show the effect of quantum efficiency. For the moment we will choose ideal photodetectors. The photodetector gives a photocurrent which is integrated in a counter that counts the number $\hat{N} = c^+c$ of photons impinging the photodetector unit surface in a given time interval Δt. It represents a point process [9], thus

$$m = \langle c^+c\rangle = \xi\langle a^+a\rangle + (1-\xi)\langle b^+b\rangle + i\sqrt{\xi(1-\xi)}\langle(a^+b - ab^+)\rangle. \tag{19}$$

If one chooses the local oscillator field in a coherent state of large amplitude β we may neglect the first term in the right hand side of Eq.(19) and write

$$m = \langle c^+c\rangle \simeq (1-\xi)|\beta|^2 + 2\sqrt{\xi(1-\xi)}|\beta|\langle\frac{(ae^{-i\phi} - a^+e^{i\phi})}{2i}\rangle \tag{20}$$

or

$$m \simeq (1-\xi)|\beta|^2 + 2|\beta|\sqrt{\xi(1-\xi)}\langle\hat{X}_{\phi+\pi/2}\rangle, \tag{21}$$

where the phase

$$\phi = \arg\beta \tag{22}$$

is controlled by the experimenter through the local oscillator. In the limits $|\beta| \to \infty$ (classical field), $\Delta t \to 0$ and $\xi \to 1$, so that $|\beta|\sqrt{(1-\xi)}$ remains finite, the underlaying point process becomes a Gaussian stochastic process [10]. Thus, when the classical intensity from the local oscillator is dropped off, the random outcomes of the photodetector directly give the quadrature phase values.

3. Wigner Function from Repeated Measurements

In a series of repeated measurements the marginal probability density $p(x,\phi)$ may be constructed at each phase ϕ as an istogram of the random outcomes of quadrature operator. Vogel and Risken [6] obtained the following expression connecting the measured marginal probability to the Wigner function

$$W(\alpha,\alpha^*) = \frac{1}{4}\int_{-\infty}^{+\infty}|r|\,dr\int_0^\pi\frac{d\phi}{\pi}\int_{-\infty}^{+\infty}dx p(x,\phi)e^{ir[x-Re(\alpha e^{i\phi})]}. \tag{23}$$

Now we will derive a more generalized expression of Vogel and Risken result [6] by using not only the rotated quadrature phase $\hat{X}_\phi$ but the shifted and rotated quadrature operator $\hat{X}$ [11] defined as

$$\hat{X} = \mu\hat{q} + \nu\hat{p} + \delta. \tag{24}$$

For

$$\hat{q} = \frac{a\,+\,a^+}{2} \tag{25}$$

$$\hat{p} = i\frac{(a^+\,-\,a)}{2} \tag{26}$$

$$\mu = \cos\phi, \quad \nu = -\sin\phi, \quad \delta = 0 \tag{27}$$

we recover the usual rotated quadrature operator in Eq.(9). We will then extend the results of Ref. [6] to the case of a generic linear transform of both quadratures belonging to the Lie group $ISp(2,R)$. We are then aloud to introduce extra parameters which include squeezing and shifting of quadratures.

As we have already seen, the characteristic function of a given density matrix of a quantum system is, by definition, the mean value of the exponential of the linear form operator $\hat{X}$ in position $\hat{q}$ and momentum $\hat{p}$ (we use now different notations to evidentiate the generalization)

$$F(k) = \langle e^{ik\hat{X}}\rangle = Tr(\hat{\rho}e^{ik\hat{X}}). \tag{28}$$

The observable $\hat{X}$ is considered to be measured by some experimental device. The Fourier transform of the characteristic function is the marginal distribution for the variable x (eigenvalue of the observable $\hat{X}$). Such a marginal distribution will depend on the parameters of the linear transform (24)

$$p(x,\mu,\nu,\delta) = \frac{1}{2\pi}\int_{-\infty}^{+\infty} dkF(k)\, e^{-ikx} \tag{29}$$

The direct calculation for $F(k)$ gives

$$\begin{aligned} F(k) &= \int_{-\infty}^{+\infty} dy\langle y|\hat{\rho}e^{ik\hat{X}}|y\rangle \\ &= e^{ik\delta}\int_{-\infty}^{+\infty} dy\langle y|\hat{\rho}e^{ik(\mu\hat{q}+\nu\hat{p})}|y\rangle \\ &= e^{\frac{ik^2}{2}\mu\nu+ik\delta}\int_{-\infty}^{+\infty} dy\langle y|\hat{\rho}e^{ik\nu\hat{p}}e^{ik\mu\hat{q}}|y\rangle \end{aligned} \tag{30}$$

finally

$$F(k) = e^{\frac{ik^2}{2}\mu\nu+ik\delta}\int_{-\infty}^{+\infty} dy\rho(y+\nu k,y)\, e^{iky\mu}, \tag{31}$$

in obtaining Eq.(31) the Baker-Hausdorff formula [12] has been used to disentangle the exponential operator and we defined $\rho(y+\nu k,y) = \langle y+k\nu|\hat{\rho}|y\rangle$ being $\hat{p}$ the displacement operator in coordinate basis $|y\rangle$. Thus, the marginal distribution $p(x,\mu,\nu,\delta)$ is related to the density matrix in coordinale representation by the expression

$$p(x,\mu,\nu,\delta) = \frac{1}{2\pi}\int_{-\infty}^{+\infty} dk\int_{-\infty}^{+\infty} dy\; e^{-ikx+\frac{i}{2}\mu\nu k^2+ik\mu y+ik\delta}\rho(y+\nu k,y). \tag{32}$$

Now, the Wigner function is related to the density matrix in coordinate representation by the expression [13]

$$W(q,p) = \int_{-\infty}^{+\infty} du\; e^{-ipu}\rho(q+u/2,q-u/2). \tag{33}$$

This is a Fourier transform and its inverse is

$$\rho(q,q') = \frac{1}{2\pi}\int_{-\infty}^{+\infty} dp\; e^{ipu}W(\frac{q+q'}{2},p) \tag{34}$$

being $u = q-q'$. In Ref.[5] it was shown that the characteristic function is expressed in terms of the Wigner function (which is the s-parametrized quasi-probability of Eq.(4) with $s=0$) by the expression

$$\langle e^{ik\hat{X}}\rangle = \frac{1}{2\pi}\int dq\int dp\, W(q,p)\, e^{ik(\mu q+\nu p+\delta)}, \tag{35}$$

then the marginal distribution, which is its Fourier transform, is connected to the Wigner function by the relation

$$p(x,\mu,\nu,\delta) = \int\int\int e^{-ik(x-\mu q-\nu p-\delta)}W(q,p)\frac{dk\,dq\,dp}{(2\pi)^2}. \tag{36}$$

Our aim is to represent the Wigner function $W(q,p)$ if we know, from given measurements, the marginal distribution $p(x,\mu,\nu,\delta)$. The case in Eq.(27) was first considered by Smithey *et al.* [14] who showed the possibility of reconstructing from experimental data the Wigner function of the radiation field. They reconstructed the Wigner function by using the Vogel and Risken result in Eq.(23), which allows the evaluation of the Wigner function only when the probability density $p(x,\phi)$ is given in analytical form. However, $p(x,\phi)$ is obtained from experimental data, thus in order to obtain an analytical expression, a smoothing procedure is required. This was accomplished by means of backprojection techniques which are standard in classical tomographic imaging [15]. Recently D' Ariano *et al.* [16] devised a method which does not need any backprojection. An analogous result was also reported by Kühn *et al.* [17].

4. Marginal Distribution for Shifted and Squeezed Quadratures

Our generalization will permit the tomographic reconstruction of the Wigner function from an ensemble of measurements taken for all real values of parameters μ, ν and δ which produce the marginal distribution $p(x,\mu,\nu,\delta)$. Our parameters are general and we could choose for example

$$\mu = e^{-\zeta}\cos\phi \quad \nu = -e^{\zeta}\sin\phi \quad \delta \neq 0 \tag{37}$$

that might be considered as the influence of the squeezing parameter ζ and shift δ on the rotated quadrature. Our aim is then to express the Wigner function in terms of the marginal distribution $p(x,\mu,\nu,\delta)$ in the domain of linear transformations. We first calculate the Fourier transform of $p(x,\mu,\nu,\delta)$ with respect to the shift parameter δ as

$$\tilde{p}(x,\mu,\nu,d) = \frac{1}{2\pi}\int d\delta\, e^{-i\delta d} p(x,\mu,\nu,\delta) \tag{38}$$

and by using Eq.(36) we obtain

$$\begin{aligned}\tilde{p}(x,\mu,\nu,d) &= \frac{1}{(2\pi)^3}\int\int\int\int d\delta\, dk\, dq\, dp W(q,p) e^{-ik(x-\mu q-\nu p-\delta)-i\delta d} \\ &= \frac{1}{(2\pi)^3}\int\int\int dk\, dq\, dp W(q,p)(2\pi)\delta(d-k) e^{-ik(x-\mu q-\nu p)} \\ &= \frac{1}{(2\pi)^3}\int\int dq\, dp W(q,p) e^{-id(x-\mu q-\nu p)}.\end{aligned} \tag{39}$$

We now introduce the Fourier transform in the (μ,ν) plane as

$$\tilde{\tilde{p}}(x,m,n,d) = \frac{1}{(2\pi)^2}\int\int \tilde{p}(x,\mu,\nu,d) e^{-im\mu-in\nu} d\mu\, d\nu \tag{40}$$

thus in terms of the Wigner function we get

$$\begin{aligned}\tilde{\tilde{p}}(x,m,n,d) &= \frac{1}{(2\pi)^4}\int\int\int\int dq\, dp W(q,p) e^{-ixd-i\mu(m-qd)-i\nu(n-pd)} d\mu\, d\nu \\ &= \frac{1}{(2\pi)^4}\int\int dq\, dp W(q,p) e^{-ixd}(2\pi)^2\delta(m-qd)\delta(n-pd)\end{aligned} \tag{41}$$

and using

$$\delta(m-qd)\delta(n-pd)=\frac{1}{d^2}\delta(q-\frac{m}{d})\delta(p-\frac{n}{d}) \tag{42}$$

we obtain

$$\tilde{\tilde{p}}(x,m,n,d)=\frac{1}{(2\pi)^2}e^{-ixd}\frac{1}{d^2}W(\frac{m}{d},\frac{n}{d}) \tag{43}$$

or better

$$W(\frac{m}{d},\frac{n}{d})=(2\pi)^2d^2e^{ixd}\,\tilde{\tilde{p}}(x,m,n,d). \tag{44}$$

By setting

$$q=\frac{m}{d}\quad,\quad p=\frac{n}{d} \tag{45}$$

we get

$$W(q,p)=(2\pi)^2d^2e^{ixd}\,\tilde{\tilde{p}}(x,qd,pd,d), \tag{46}$$

and the Wigner function is reconstructed from the knowledge of

$$(2\pi)^2\tilde{\tilde{p}}(0,q,p,1)=W(q,p). \tag{47}$$

Because this one is expressed as a particular Fourier component of the marginal distribution, it means that the marginal distribution contains overcomplete information about the system under study.

5. The Harmonic Oscillator Ground State

As an useful example let us consider the ground state of the harmonic oscillator

$$\psi_0(x)=\frac{1}{\sqrt[4]{\pi}}e^{-\frac{x^2}{2}}, \tag{48}$$

its density matrix in coordinate representation is

$$\rho_0(x,x')=\frac{1}{\sqrt{\pi}}e^{-\frac{x^2}{2}-\frac{x'^2}{2}}, \tag{49}$$

and the W-function is

$$\begin{aligned} W(q,p) &= \int du\; e^{-ipu}\frac{1}{\sqrt{\pi}}e^{-\frac{1}{2}(q+\frac{u}{2})^2-\frac{1}{2}(q-\frac{u}{2})^2} \\ &= e^{-(q^2+p^2)}\int\frac{du}{\sqrt{\pi}}\;e^{-\frac{1}{2}(ip\sqrt{2}+\frac{u}{\sqrt{2}})^2} \\ &= 2e^{-(q^2+p^2)}. \end{aligned} \tag{50}$$

where the well known Gaussian integral trick was used. We wish now obtain the same result starting from the characteristic function of the density matrix for the given state

$$\begin{aligned} \langle e^{ik\hat{X}}\rangle &= e^{ik\delta}\int dx\,\langle x|\hat{\rho}\,e^{ik(\mu\hat{q}+\nu\hat{p})}|x\rangle \\ &= e^{ik\delta+\frac{i}{2}k^2\mu\nu}\int\frac{dx}{\sqrt{\pi}}e^{ikx\mu-\frac{1}{2}(x+k\nu)^2-\frac{x^2}{2}} \\ &= e^{ik\delta+\frac{i}{2}k^2\mu\nu-\frac{k^2\nu^2}{2}+\frac{k^2}{2}(\nu-i\mu)^2}\int\frac{dx}{\sqrt{\pi}}e^{-(x+\frac{k(\nu-i\mu)}{2})^2} \\ &= e^{ik\delta}e^{-\frac{k^2}{4}(\mu^2+\nu^2)}. \end{aligned} \tag{51}$$

Its Fourier transform gives the marginal distribution

$$p(x,\mu,\nu,\delta) = \int \frac{dk}{\sqrt{2\pi}} e^{-ikx} e^{ik\delta - \frac{k^2}{4}(\mu^2+\nu^2)}$$
$$= [\pi(\mu^2+\nu^2)]^{-1/2} \exp[-\frac{(x-\delta)^2}{\mu^2+\nu^2}]. \quad (52)$$

The Fourier transform in δ, μ and ν finally is

$$\tilde{\tilde{p}}(x,m,n,d) = \frac{1}{2\pi^2} \frac{e^{-ixd}}{d^2} e^{-[(\frac{m}{d})^2+(\frac{n}{d})^2]}, \quad (53)$$

which obviously gives

$$\tilde{\tilde{p}}(0,q,p,1) = \frac{1}{2\pi^2} e^{-(q^2+p^2)} = \frac{1}{(2\pi)^2} W(q,p) \quad (54)$$

or

$$W(q,p) = 2e^{-(q^2+p^2)} \quad (55)$$

as we have seen above.

6. Vogel and Risken Result

We have a direct way of connecting the marginal distribution to the Wigner function, however, we again need an analytical expression for the marginal distribution because we have to perform its Fourier transform in the parameters space. The Vogel and Risken result (23) is obtained by the following steps. Let us start from the result obtained in Eq.(44)

$$W(\frac{m}{d},\frac{n}{d}) = (2\pi)^2 d^2 e^{ixd} \tilde{\tilde{p}}(x,m,n,d). \quad (56)$$

and by using Eqs.(40), (39) we get

$$W(\frac{m}{d},\frac{n}{d}) = \frac{d^2}{2\pi} \int\int\int p(x,\mu,\nu,\delta) e^{id(x-\delta)-im\mu-in\nu} d\mu\, d\nu\, d\delta. \quad (57)$$

We now set the parameters μ,ν as in Eq.(27) with $\delta \neq 0$ thus we have

$$d\mu\, d\nu = -r dr\, d\phi \quad (58)$$

and

$$x = r x_\phi + \delta \quad (59)$$

so, being r, ϕ and δ independent variables one can consider $d\delta = dx$ in Eq.(57) hence, setting

$$\delta = 0 \ , \ \frac{m}{d} = q \ , \ \frac{n}{d} = p \ , \ d = 1, \quad (60)$$

Eq.(57) becomes

$$W(q,p) = -\frac{1}{2}\int_0^{+\infty} r dr \int_0^{\pi} \frac{d\phi}{\pi} \int_{-\infty}^{+\infty} dx p(x,\phi) e^{irx-irq\cos\phi+irp\sin\phi}$$
$$= \frac{1}{4}\int_{-\infty}^{\infty} |r| dr \int_0^{\pi} \frac{d\phi}{\pi} \int_{-\infty}^{\infty} dx p(x,\phi) e^{ir[x-(q\cos\phi-p\sin\phi)]}. \quad (61)$$

Setting now

$$q = \frac{\alpha + \alpha^*}{2} \quad p = \frac{\alpha - \alpha^*}{2i} \tag{62}$$

we finally get

$$W(\alpha, \alpha^*) = \frac{1}{4} \int_{-\infty}^{+\infty} |r| dr \int_0^{\pi} \frac{d\phi}{\pi} \int_{-\infty}^{+\infty} dx p(x, \phi) e^{ir[x - Re(\alpha e^{i\phi})]}, \tag{63}$$

which is the Vogel and Risken result for the Wigner function (i.e. $s = 0$). This expression was used by Smithey *et al.* [14] to experimentally reconstruct the Wigner function. However, as was shown by D'Ariano *et al.* [16] this procedure is not convergent because of the integral in the parameter r, and a cut-off should be used which introduces systematic errors. With a completely different procedure, by using the s-parametrized quasi-probability $W(\alpha, s)$ and considering $s = -1$, which coincides with the Husimi Q-function, they showed that one can directly obtain the density matrix elements in the number representation by only averaging over a sub-ensemble of experimental data for a fixed quadrature phase ϕ. The main point of Ref. [16] is that it is possible to obtain the density matrix elements from the marginal distribution avoiding the evaluation of the Wigner function as an intermediate step.

7. Direct Reconstruction of Density Matrix Elements

We now show how that result can be obtained. To this goal we start from another well known representation of the density operator [5]

$$\hat{\rho} = \int \frac{d^2\alpha}{\pi} W(\alpha, \alpha^*) \hat{T}(\alpha) \tag{64}$$

where the operator $\hat{T}(\alpha)$ is defined as the complex Fourier transform of the displacement operator $\hat{D}(\xi) = \exp(a^+\xi - a\xi^*)$, i.e.

$$\hat{T}(\alpha, \alpha^*) = \int \frac{d^2\xi}{\pi} \hat{D}(\xi) e^{(\alpha\xi^* - a^+\xi)} \tag{65}$$

and $W(\alpha, \alpha^*)$ in Eq.(64) is the usual Wigner function which represents the weight function for the expansion of the density operator in terms of the operator $\hat{T}(\alpha)$. We now use the canonical coordinates (q, p) connected to (α, α^*) by

$$\alpha = \frac{q + ip}{\sqrt{2}} \ , \ \alpha^* = \frac{q - ip}{\sqrt{2}} \tag{66}$$

and by using Eq.(46) we can write the Wigner function in terms of the marginal distribution for the generic shifted and rotated quadrature $\hat{X}$ in Eq.(24) as

$$W(q, p) = (2\pi)^2 d^2 e^{ixd} \, \tilde{\tilde{p}}(x, qd, pd, d), \tag{67}$$

where qd, pd and d are conjugate variables, in the Fourier transform, to μ, ν and δ respectively; thus

$$\hat{\rho} = \int\int \frac{dqdp}{(2\pi)^2} \hat{T}(q, p) \int\int\int d\mu \, d\nu \, d\delta \; d^2 e^{ixd} \, p(x, \mu, \nu, \delta) e^{-i\mu qd - i\nu pd - i\delta d} \tag{68}$$

where we used the inverse Fourier transforms. We may write Eq.(68) in a compact form

$$\hat{\rho} = \int\int\int d\mu\, d\nu\, d\delta\; p(x,\mu,\nu,\delta)\hat{K}_{\mu,\nu,\delta} \tag{69}$$

where the Kernel operator $\hat{K}_{\mu,\nu,\delta}$ is given by

$$\hat{K}_{\mu,\nu,\delta} = \int\int \frac{dqdp}{(2\pi)^2}\, d^2\hat{T}(q,p)e^{-i\mu qd - i\nu pd - id(\delta - x)}. \tag{70}$$

By using the Baker-Hausdorf formula we finally get

$$\hat{K}_{\mu,\nu,\delta} = \frac{1}{2\pi}d^2 e^{-id(\delta-x)} e^{\frac{d}{\sqrt{2}}(\nu - i\mu)a^+} e^{-\frac{d}{\sqrt{2}}(\nu+i\mu)a} e^{-\frac{d^2}{4}(\nu^2+\mu^2)}. \tag{71}$$

This result is obtained by using Eqs.(65) and (66) i.e.

$$\begin{aligned}
\hat{K}_{\mu,\nu,\delta} &= \int\int \frac{dqdp}{(2\pi)^2} d^2 e^{-i\mu qd - i\nu pd - id(\delta-x)} \int \frac{d^2\xi}{\pi}\hat{D}(\xi)e^{(\alpha\xi^* - \alpha^*\xi)} \\
&= \frac{d^2}{\pi}e^{id(\delta-x)} \int\int\int\int \frac{dq\, dp\, d\xi_1\, d\xi_2}{(2\pi)^2} e^{-i\mu qd - i\nu pd - \xi^* a + \xi a^+} e^{\frac{q+ip}{\sqrt{2}}(\xi_1 - i\xi_2) - \frac{q-ip}{\sqrt{2}}(\xi_1 + i\xi_2)} \\
&= \frac{d^2}{\pi}e^{id(\delta-x)} \int\int d\xi_1\, d\xi_2 e^{-(\xi_1 - i\xi_2)a + (\xi_1 + i\xi_2)a^+} \delta(\nu d - \sqrt{2}\xi_1)\delta(\mu d + \sqrt{2}\xi_2) \\
&= \frac{d^2}{2\pi}e^{id(\delta-x)} e^{-\frac{1}{\sqrt{2}}(\nu d + i\mu d)a + \frac{1}{\sqrt{2}}(\nu d - i\mu d)a^+}.
\end{aligned} \tag{72}$$

We will now express the kernel operator $\hat{K}_{\mu,\nu,\delta}$ in various representations: i) *coherent states basis*

$$\begin{aligned}
\langle\alpha|\hat{K}_{\mu,\nu,\delta}|\beta\rangle &= \frac{d^2}{2\pi}e^{id(\delta-x) - \frac{d^2}{4}(\nu^2+\mu^2)} \langle\alpha| e^{\frac{d}{\sqrt{2}}(\nu - i\mu)a^+} e^{-\frac{d}{\sqrt{2}}(\nu + i\mu)a}|\beta\rangle \\
&= \frac{d^2}{2\pi}e^{id(\delta-x) - \frac{d^2}{4}(\nu^2+\mu^2)} e^{\frac{d}{\sqrt{2}}(\nu - i\mu)\alpha^*} e^{-\frac{d}{\sqrt{2}}(\nu + i\mu)\beta}\langle\alpha|\beta\rangle \\
&= \frac{d^2}{2\pi}e^{id(\delta-x) - \frac{d^2}{4}(\nu^2+\mu^2)} e^{\frac{d}{\sqrt{2}}(\nu - i\mu)\alpha^*} e^{\frac{d}{\sqrt{2}}(\nu + i\mu)\beta} e^{-\frac{|\alpha|^2}{2} - \frac{|\beta|^2}{2} + \alpha^*\beta},
\end{aligned} \tag{73}$$

ii) *number states basis*

$$\langle n+s|\hat{K}_{\mu,\nu,\delta}|n\rangle = \frac{d^2}{2\pi}e^{id(\delta-x) - \frac{d^2}{4}(\nu^2+\mu^2)} \langle n+s| e^{\frac{d}{\sqrt{2}}(\nu - i\mu)a^+} e^{-\frac{d}{\sqrt{2}}(\nu + i\mu)a}|n\rangle. \tag{74}$$

Since we have

$$e^{-\frac{d}{\sqrt{2}}(\nu+i\mu)a}|n\rangle = \sum_{k=0}^{n} \frac{\left[-\frac{d}{\sqrt{2}}(\nu + i\mu)\right]^k}{k!} a^k|n\rangle \tag{75}$$

and

$$a^k|n\rangle = \sqrt{(n-k)!}|n-k\rangle \tag{76}$$

we get

$$\langle n+s| e^{\frac{d}{\sqrt{2}}(\nu - i\mu)a^+} e^{-\frac{d}{\sqrt{2}}(\nu + i\mu)a}|n\rangle = \sum_{l=0}^{n} \frac{\sqrt{n!(n+s)!}}{(n-l)!} \frac{[\frac{d}{\sqrt{2}}(\nu - i\mu)]^l}{l!} \frac{[-\frac{d}{\sqrt{2}}(\nu + i\mu)]^{(l+s)}}{(l+s)!} \tag{77}$$

which should be inserted in Eq.(74). iii) *coordinate representation*

$$
\begin{aligned}
\langle x|\hat{K}_{\mu,\nu,\delta}|y\rangle &= \frac{d^2}{2\pi}e^{id(\delta-x)}\langle x|e^{\frac{d}{\sqrt{2}}[\nu(a^+-a)-i\mu(a^++a)]}|y\rangle \\
&= \frac{d^2}{2\pi}e^{id(\delta-x)}\langle x|e^{-id[\nu\hat{p}+i\mu\hat{q}]}|y\rangle \\
&= \frac{d^2}{2\pi}e^{id(\delta-x)}\langle x|e^{-i\mu\hat{q}d}e^{-i\nu\hat{p}d}|y\rangle e^{\frac{i}{2}\mu\nu d^2} \\
&= \frac{d^2}{2\pi}e^{id(\delta-x)}e^{-i\mu x d}\langle x|y+\nu d\rangle e^{\frac{i}{2}\mu\nu d^2} \\
&= \frac{d^2}{2\pi}e^{id(\delta-x)}e^{-i\mu x d}e^{\frac{i}{2}\mu\nu d^2}\delta(y+\nu-x).
\end{aligned} \tag{78}
$$

It is now evident that the kernel operator is only bounded in coherent and number states bases while is unbounded in coordinate representation. As a consequence every moment is bounded, for all possible distributions $p(x,\mu,\nu,\delta)$, for those representations which give bounded values of the kernel only. According to the central limit theorem the density matrix elements can be sampled, on a sufficiently large set of data, in those bases which give finite expectation values. This result was first obtained by D'Ariano *et al.* [18]. To recover their result we will start from another representation of the density matrix [5]

$$\hat{\rho} = \int \frac{d^2\alpha}{\pi} Tr[\hat{\rho}\hat{D}(\alpha)]\hat{D}^{-1}(\alpha) \tag{79}$$

where $\hat{D}(\alpha)$ is the displacement operator in Eq.(5) and $\hat{D}^{-1}(\alpha)$ its inverse. By changing the variables

$$\mu = \sqrt{2}Im(\alpha) \quad , \quad \nu = -\sqrt{2}Re(\alpha) \tag{80}$$

it can be written as

$$\hat{\rho} = \int \frac{d\mu\, d\nu}{2\pi} Tr[\hat{\rho}e^{i\hat{X}}]e^{-i\hat{X}}. \tag{81}$$

To evaluate the trace in Eq.(81) we use the complete set of eigenvectors $\{|X\rangle\}$ of the shifted and rotated quadrature operator $\hat{X}$ whose eigenvalue is X which, of course, is a linear function of μ, ν and δ. Thus

$$Tr[\hat{\rho}e^{i\hat{X}}] = \int dX\, p(X,\mu,\nu,\delta)e^{iX}. \tag{82}$$

We can now replace the integration on the variable X with δ because δ represents the shift of the variable X. Thus, using Eq.(82), we can write Eq.(81) as

$$\hat{\rho} = \int \frac{d\mu\, d\nu\, d\delta}{2\pi} p(X,\mu,\nu,\delta)e^{iX}e^{-i\hat{X}} \tag{83}$$

or better

$$\hat{\rho} = \int d\mu\, d\nu\, d\delta\, p(X,\mu,\nu,\delta)\hat{K}_{\mu\nu\delta} \tag{84}$$

with

$$\hat{K}_{\mu\nu\delta} = \frac{1}{2\pi}e^{-i(\hat{X}-X)}, \tag{85}$$

which is the kernel previously given in Eq.(71) with $d = 1$ and $\hat{X} = \mu\hat{q} + \nu\hat{p} + \delta$ with $\hat{q} = (a+a^+)/\sqrt{2}$, $\hat{p} = (a-a^+)/i\sqrt{2}$. In order to reconstruct the tomographic formula for the homodyne detection [18], one needs to go in the polar variable , i.e. $\mu = r\cos\phi$, $\nu = r\sin\phi$ setting $\delta = 0$, then

$$\hat{X}_{\mu,\nu,\delta} \to \hat{X}_{\mu,\nu} = r\hat{X}_\phi = r[\hat{q}\cos\phi + \hat{p}\sin\phi]. \tag{86}$$

Furthermore

$$\mathrm{Tr}\{\hat{\rho}e^{i\hat{X}_{\mu,\nu}}\} = \mathrm{Tr}\{\hat{\rho}e^{ir\hat{X}_\phi}\} = \int dX\, p(X,\phi)e^{irX} \tag{87}$$

and thus

$$wher\hat{\rho} = \int d\phi dX\, p(X,\phi)\hat{K}_\phi \tag{88}$$

with

$$\hat{K}_\phi = \frac{1}{2\pi}\int dr\, re^{ir(X-\hat{X}_\phi)} \tag{89}$$

which is the same kernel of Ref. [18]. Substantially the kernel of Eq. (89) is given by the radial integral of the kernel of Eq. (85), and this is due to the fact that one goes from a general transformation, with two free parameters, to a particular transformation (homodyne rotation) with only one free parameter and then one needs to integrate over the other one. The extension of the previuos results to the multimode case is straightforward [19]. We only present here the idea. The characteristic function of the multimode case can be written as (we use the convention that boldface letters represent multidimensional vectors)

$$\chi(\boldsymbol{\eta}) = \mathrm{Tr}(\hat{\rho}e^{i\boldsymbol{\eta}\cdot\hat{\mathbf{X}}}) = \langle e^{i\boldsymbol{\eta}\cdot\hat{\mathbf{X}}}\rangle \tag{90}$$

where $\hat{\rho}$ is the multimode density operator and the vectorial operator $\hat{\mathbf{X}}$ is the mult-dimensional extension of Eq.(24); it is defined toghether with its conjugate operator $\hat{\mathbf{Y}}$ by

$$\begin{pmatrix}\hat{\mathbf{X}}\\ \hat{\mathbf{Y}}\end{pmatrix} = \Lambda\begin{pmatrix}\hat{\mathbf{q}}\\ \hat{\mathbf{p}}\end{pmatrix} + \begin{pmatrix}\boldsymbol{\delta}\\ \boldsymbol{\delta}'\end{pmatrix}, \tag{91}$$

with $\hat{\mathbf{q}}$ and $\hat{\mathbf{p}}$ the multimode extension of position and momentum operators and the matrix Λ a symplectic matrix [19]. With a procedure similar to the above it is possible to obtain the multimode version of results previously obtained for the one mode case [19].

8. The Effect of Detector's Quantum Efficiency

Until now we have completely disregarded the efficiency of the detection process. Leonhardt and Paul [20] recently analysed the deteriorating effect of low efficiency detectors. Their conclusion is that the optical homodyne tomography does not allow, in principle, the complete reconstruction of the Wigner function when the quantum efficiency η of the detector is less than unit. To obtain this result we will use the heuristic model for the photodetector efficiency already introduced in Sec.(2). We consider thus a beam splitter with transmission coefficient η, equal to the quantum efficiency of the detector, followed by a photodetector with unit quantum efficiency. We assume that into the two input ports of the beam splitter are impinging the mode a, whose density matrix we want to reconstruct, and the mode b, which represents the vacuum fluctuations always present in any physical process and which mimic the effect of quantum efficiency or the loss of detected photons. The two output modes are the reflected mode c and transmitted mode d. They are obtained by the unitary transformation

$$\begin{pmatrix} c \\ d \end{pmatrix} = \mathbf{M} \begin{pmatrix} a \\ b \end{pmatrix} , \tag{92}$$

with the unitary scattering matrix $\mathbf{M}$ (containing reflection and trasmission coefficients) written as

$$\mathbf{M} = \begin{pmatrix} \cos\kappa & -\sin\kappa \\ \sin\kappa & \cos\kappa \end{pmatrix} . \tag{93}$$

In Eq. (93) essentially $\cos^2\kappa = 1 - \eta$ is the reflection coefficient (and $\sin^2\kappa = \eta$ is the trasmission one). This transformation can be achieved with the following unitary operator

$$\hat{U} = \exp\left[\kappa\left(ab^+ - a^+b\right)\right] , \tag{94}$$

so that $c = \hat{U}^+ a \hat{U}$ and $d = \hat{U}^+ b \hat{U}$. This can now be normally ordered with respect to mode b using the Baker-Campbell-Hausdorff formula (BCH) for angular momenta ($SU(2)$ group) as follows [21]

$$\hat{U} = e^{\tan\kappa ab^+} |\cos\kappa|^{a^+a-b^+b} e^{-\tan\kappa a^+b}. \tag{95}$$

If $\hat{\rho}_a$ is the density matrix of the mode we want to measure and $\hat{\rho}_b = |0\rangle_{bb}\langle 0|$ is the vacuum density matrix, the joint density matrix of the output mode d becomes

$$\hat{R}_{cd} = \hat{U}\hat{\rho}_a \otimes |0\rangle_{bb}\langle 0|\hat{U}^+ , \tag{96}$$

and any desired expectation value of operators corresponding only to the mode a can be obtained with the reduced density matrix

$$\hat{\rho}_d = \mathrm{Tr}_b[\hat{R}_{cd}]. \tag{97}$$

By using the BCH formula (95) we easily obtain

$$\begin{aligned} \hat{\rho}_d &= \mathrm{Tr}_b[e^{ab^+\tan\kappa}|\cos\kappa|^{a^+a}\hat{\rho}_a \otimes |0\rangle_{bb}\langle 0|\,|\cos\kappa|^{a^+a}e^{-a^+b\tan\kappa}] \\ &= \mathrm{Tr}_b[e^{ab^+\tan\kappa}|\cos\kappa|^{a^+a}\hat{\rho}_a|\cos\kappa|^{a^+a}|0\rangle_{bb}\langle 0|e^{-a^+b\tan\kappa}] \\ &= \sum_{k=0}^{\infty}(-)^k\frac{(\tan\kappa)^{2k}}{k!}a^k|\cos\kappa|^{a^+a}\hat{\rho}_a|\cos\kappa|^{a^+a}a^{+k}. \end{aligned} \tag{98}$$

We now use the number representation

$$\langle n|\hat{\rho}_d|m\rangle = (\cos\kappa)^{n+m}\sum_{k=0}^{\infty}(-)^k(\sin\kappa)^{2k}\left[\begin{pmatrix} m+k \\ m \end{pmatrix}\begin{pmatrix} n+k \\ n \end{pmatrix}\right]^{\frac{1}{2}}\langle n+k|\hat{\rho}_a|m+k\rangle \tag{99}$$

and the transformation can be inverted as follows

$$\langle n|\hat{\rho}_a|m\rangle = (\cos\kappa)^{-n-m}\sum_{k=0}^{\infty}(\tan\kappa)^{2k}\left[\begin{pmatrix} m+k \\ m \end{pmatrix}\begin{pmatrix} n+k \\ n \end{pmatrix}\right]^{\frac{1}{2}}\langle n+k|\hat{\rho}_d|m+k\rangle. \tag{100}$$

This is the way to connect the desired density matrix $\langle n|\hat{\rho}_a|m\rangle$ in the number representation to the measured one $\langle n|\hat{\rho}_d|m\rangle$. Equation (100) represents a convolution of the number distribution with the binomial probability, which produces the undesired effect of smearing the number distribution because of vacuum fluctuations (i.e. the detector quantum efficiency). We can do, however, things in a more general way. The generating function $\chi_d(r,\phi)$ of quadrature moments at the output

$$\hat{d}_\phi = \frac{1}{2}(d\, e^{i\phi} + d^+\, e^{-i\phi}) \tag{101}$$

is given in the two equivalent (Schrödinger or Heisenberg picture) ways

$$\chi_d(r,\phi) = \langle e^{ir\hat{d}_\phi}\rangle = \mathrm{Tr}\left(e^{ir\hat{b}_\phi}\hat{R}_{cd}\right) = \mathrm{Tr}\left(e^{ir\hat{d}_\phi}\hat{\rho}_a \otimes |0\rangle_{bb}\langle 0|\right) . \tag{102}$$

The nth derivative of $\chi_d(r,\phi)$ with respect to the parameter r is connected with the nth moment of the output quadrature $\hat{d}_\phi$. We can easily factor out $\chi_d(r,\phi)$ into two components, each pertaining a single input mode. One has

$$\begin{aligned}\chi_d(r,\phi) &= {}_b\langle 0|\mathrm{Tr}_a[e^{ir(\sin\kappa\hat{a}_\phi+\cos\kappa\hat{b}_\phi)}\hat{\rho}_a]|0\rangle_b \\ &= \chi_a(r\sqrt{\eta},\phi)\chi_b(r\sqrt{1-\eta},\phi)\,,\end{aligned} \tag{103}$$

where

$$\chi_a(\lambda,\phi) = \mathrm{Tr}_a[e^{i\lambda\hat{a}_\phi}\hat{\rho}_a], \tag{104}$$

and

$$\chi_b(\lambda,\phi) = \langle 0|e^{i\lambda\hat{b}_\phi}|0\rangle \tag{105}$$

and we set $\eta = \sin^2\kappa$ which represents the mirror transmissivity and in our case the detector quantum efficiency. The Fourier transform of $\chi_d(r,\phi)$ gives the marginal probability distribution $p_d(x,\phi)$ of the quadrature $\hat{d}_\phi$ of the detected output field

$$p_d(x,\phi) = \int_{-\infty}^{+\infty}\frac{dr}{2\pi}e^{-irx}\chi_d(r,\phi)\,, \tag{106}$$

and, similarly, $\chi_a(\lambda,\phi)$ and $\chi_b(\lambda,\phi)$ are related to the quadrature probability distributions of their respective modes. Our aim is now to connect the density matrix elements of $\hat{\rho}_a$ with the detected probability distribution $p_d(x,\phi)$; this will give us the desired tomographic formula. We start from the operator identity [18]

$$\hat{\rho}_a = \int_0^\pi\frac{d\phi}{\pi}\int_{-\infty}^{+\infty}\frac{dr|r|}{4}e^{-ir\hat{a}_\phi}\chi_a(r,\phi)\,, \tag{107}$$

and we substitute Eq. (103) into Eq. (107), thus obtaining

$$\begin{aligned}\hat{\rho}_a &= \int_0^\pi\frac{d\phi}{\pi}\int_{-\infty}^{+\infty}\frac{dr|r|}{4}e^{-ir\hat{a}_\phi} \\ &\times\ \chi_d\left(\frac{r}{\sqrt{\eta}},\phi\right)\chi_b^{-1}\left(r\sqrt{\frac{1-\eta}{\eta}},\phi\right) .\end{aligned} \tag{108}$$

Using Eq. (106) one has

$$\hat{\rho}_a = \int_0^\pi \frac{d\phi}{\pi} \int_{-\infty}^{+\infty} \frac{dr|r|}{4} e^{-ir\hat{a}_\phi} \int_{-\infty}^{+\infty} dx\, e^{irx/\sqrt{\eta}} \times p_d(x,\phi)\chi_b^{-1}\left(r\sqrt{\frac{1-\eta}{\eta}},\phi\right) , \tag{109}$$

which provides the desired relation

$$\hat{\rho}_a = \int_0^\pi \frac{d\phi}{\pi} \int_{-\infty}^{+\infty} dx\, p_d(x,\phi) K\left(\frac{x}{\sqrt{\eta}} - \hat{a}_\phi\right) , \tag{110}$$

where

$$K(x) = \frac{1}{2}\mathrm{Re} \int_0^\infty dr\, r\, e^{irx} \chi_b^{-1}\left(r\sqrt{\frac{1-\eta}{\eta}},\phi\right) . \tag{111}$$

For b in the vacuum state, one has

$$\chi_b\left(r\sqrt{\frac{(1-\eta)}{\eta}},\phi\right) = \exp\left(-\frac{1-\eta}{8\eta}r^2\right) , \tag{112}$$

which is giving the same kernel obtained in Ref. [18] for homodyne tomography with quantum efficiency η, i.e.

$$K(x) = \frac{1}{2}\mathrm{Re} \int_0^\infty dr\, r\, \exp\left(\frac{1-\eta}{8\eta} + irx\right) . \tag{113}$$

Hence, our present scheme is equivalent to a direct measurement in presence of nonunit efficiency. Therefore, we can use the theorem of Ref. [18], namely: for the number representation the matrix elements of $K(x - \hat{a}_\phi)$ are bounded for $\eta > 0.5$, and in this case the density matrix can be statistically sampled from experimental data.

9. Macroscopic Quantum Coherence

Recently there has been a growing interest in the possibility of observing linear superpositions of macroscopically distinguishable states, the so called "Schrödinger cat" states. Good candidates for these macroscopic states are the coherent states of an electromagnetic field mode and a large number of different schemes has been proposed for the generation of linear superpositions of two coherent states with opposite phases [22], [23]

$$|\psi\rangle = c_+|\alpha\rangle + c_-|-\alpha\rangle \tag{114}$$

where $|\alpha\rangle$ is the coherent state and c_s the probability amplitude. Since the seminal work of Yurke and Stoler [22] it becomes clear that a Kerr medium could be used to generate such states at optical level. They showed, indeed, that the unitary evolution of an initial coherent state, interacting with a Kerr medium with a well defined length, will produce a superposition of coherent states. For instance an initial state $|\alpha_0>$ will generate the superposition

$$|\alpha_0>= e^{-i\pi/4}|\alpha_0> +e^{i\pi/4}|-\alpha_0> . \quad (115)$$

after an interaction time $t_0 = \pi/(2\Omega)$ where Ω is the strength of Kerr nonlinearity. At well defined shorter times three or more coherent states could also be generated [22]. This, of course, requires the precise knowledge of the length of the medium (or interaction time). It is also well known, and was shown in great details by Daniel and Milburn [24], that as soon as one takes into account the loss in the Kerr medium the generation of those states is suddenly inhibited. Thus, the best should be to have a Kerr medium with high nonlinearity to loss ratio. Recently [25], quadrature squeezed light was observed in semiconductors at frequencies less tahn half of band gap, where larger ratios of nonlinearity to loss could be obtained [26]. Furthermore, it has been recently shown [27] that a quasi-superposition of macroscopic states, with interference fringes still present, could be generated in a Kerr medium with the above ratio of 10 when one uses a squeezed bath to model the loss, even though at present this ratio is far from being realizable. In this context it was also shown that a squeezed bath could be realized by a suitable feedback [28]. Moreover, it was also shown [29] that by using a time modulation of the Kerr nonlinearity one can obtain the coherent superposition without the precise knowledge of the length of the medium (or interaction time) by only adjusting the phase of the time modulation.

Anyway, the delicate point in the observation of optical Schrödinger cats is not only the generation of these states, but also their experimental detection. In fact the signature of linear superpositions of two states is the existence of interference fringes, which allows one to distinguish them from their statistical mixture. These fringes are extremely sensitive to the coupling with any external environment, such as the macroscopic apparatus employed to detect them, and they disappear exponentially fast with the mean number of photons. Therefore it is crucial to set up a detection scheme which is able to slow down as much as possible the unavoidable destruction of the interference pattern caused by the measurement. Kennedy and Walls [30] presented a general model for dissipation with squeezed quantum fluctuations and showed that, with such a bath as model for the experimental apparatus, one is able to preserve macroscopic quantum coherence, confirming thus previous heuristic results [31]. In [28] we showed that a squeezed bath can be simulated not only by shining the cavity with a squeezed vacuum but, more generally, by applying a suitable feedback of the output signal corresponding to a quantum non-demolition measurement (QND) of an observable. However, it is not clear whether this feedback scheme could be easily experimentally implemented or not and we have proposed [32] a way to obtain the same result without the feedback. To this end we introduce the following Hamiltonian model which was proposed some years ago by Alsing, Milburn and Walls [33] as a model for quantum nondemolition (QND) measurements.

10. The Detection Model

We consider a cavity supporting two different modes, with annihilation operators a and b. The two modes are coupled by a nonlinear crystal so that (in the interaction picture)

$$H_{int} = \hbar\chi X_\xi Y_\varphi , \quad (116)$$

where

$$X_\xi = \frac{\left(ae^{i\xi} + a^{+ger}e^{-i\xi}\right)}{2}, \tag{117}$$

and

$$Y_\varphi = \frac{(be^{i\varphi} + b^{+ger}e^{-i\varphi})}{2}. \tag{118}$$

This interaction could be achieved by, for example, a crystal with a $\chi^{(2)}$ nonlinearity in which two processes driven by classical fields, amplification at the frequency $\omega_s = \omega_a + \omega_b$, and frequency conversion at the frequency $\omega_d = \omega_a - \omega_b$, have equal strengths [33]. Because of the QND coupling, when the "meter" mode b is heavily damped at rate k_b, one can monitor the quadrature X_ξ of the signal mode a just by performing a homodyne measurement of a quadrature Y_δ of the mode b. In fact, when $k_b \gg k_a$ (damping rate of the a mode) the homodyne photocurrent $I(t)$ can be directly expressed in terms of the "instantaneous" mean value $\langle X_\xi(t)\rangle_c$, conditioned on the result of the measurement [34, 35], as

$$I(t) = \eta\chi\left[2\sin(\delta - \varphi)\langle X_\xi(t)\rangle_c + \sqrt{\frac{2k_b}{\eta\chi^2}}\xi(t)\right], \tag{119}$$

where η is the efficiency of the homodyne detection and $\xi(t)$ is a Gaussian white noise with $\langle\xi(t)\xi(t')\rangle = \delta(t-t')$. We detune the two modes in the cavity, so that their uncoupled evolution is no more driven by the standard vacuum bath term alone, but by

$$\mathcal{L}_a\rho = k_a\left(2a\rho a^{+ger} - a^+a\rho - \rho a^+a\right) - i[\delta_a a^{+ger}a, \rho] \tag{120}$$

and an analogous expression holds for the b mode. The effect of the two nonzero detunings δ_a and δ_b can be intuitively described in terms of an "internal feedback" mechanism, because the detunings mix the two quadratures X_ξ and Y_φ with their respective $\pi/2$ out of phase quadratures, so that any variation of X_ξ is "fed back" to the X_ξ dynamics itself by the joint action of the detunings and the nonlinear coupling. Provided that the adiabatic condition $k_b \gg k_a$ is satisfied, the homodyne measurement of the quadrature Y_δ allows monitoring the a mode quadrature X_ξ also in the presence of nonzero detunings. In fact, when $\delta_b \neq 0$, Eq. (119) generalizes to

$$I(t) = \eta\chi\Big\{\Big[\frac{2k_b^2}{k_b^2+\delta_b^2}\sin(\delta-\varphi) - \frac{2k_b\delta_b}{k_b^2+\delta_b^2}\cos(\delta-\varphi)\Big]\langle X_\xi(t)\rangle_c + \sqrt{\frac{2k_b}{\eta\chi^2}}\xi(t)\Big\}, \tag{121}$$

so that from the homodyne photocurrent it is still possible to reconstruct the marginal probability distribution of the quadrature X_ξ, which is the quantity usually considered for revealing the interference fringes associated to an optical Schrödinger cat. We have therefore the model defined by the following master equation for the density matrix D of the two modes

$$\dot{D} = \mathcal{L}_a D + \mathcal{L}_b D - \frac{i}{\hbar}[H_{int}, D], \tag{122}$$

where the superoperator $\mathcal{L}_i$ $(i = a, b)$ is given by Eq. (120). Because the Wigner function of the two modes evolves according to the Fokker-Planck equation for a four-dimensional

Ornstein-Uhlenbeck process [36], Eq. (122) can be exactly solved [34]. Anyway, the analytical expressions in the general case are very cumbersome and therefore we shall explicitely discuss only the adiabatic limit $k_b \gg k_a$, where the meter mode b can be adiabatically eliminated, and which, as we have seen above, is the most interesting case for our purposes. After the adiabatic elimination of the b mode, one gets the following master equation for the a mode reduced density matrix ρ

$$\dot{\rho} = \mathcal{L}_a\rho - \frac{\Gamma}{2}\left[X_\psi,\left[X_\psi,\rho\right]\right] + iF\left[X_\psi,\left\{X_\psi,\rho\right\}\right] \;, \tag{123}$$

where

$$\Gamma = \chi^2 k_b / 2(k_b^2 + \delta_b^2), \tag{124}$$

and

$$F = \chi^2 \delta_b / 4(k_b^2 + \delta_b^2). \tag{125}$$

11. Detection of Optical Schrödinger Cats

We will now focus on the detection of optical Schrödinger cats rather than on their generation, and therefore we shall assume that at $t = 0$ a superposition of coherent states of the a mode has been already prepared, i.e., we consider an initial condition $\rho(0) = \sum_{\alpha,\beta} N_{\alpha,\beta}|\alpha\rangle\langle\beta|$. The exact time evolution from this initial state can be obtained with the same method of [34] and it is better expressed in terms of the normally ordered characteristic function $\chi(\lambda,\lambda^*;t) = \mathrm{Tr}\left\{\rho(t)\exp(\lambda a^{+ger})\exp(-\lambda^* a)\right\}$

$$\begin{aligned} \chi(\lambda,\lambda^*;t) &= \sum_{\alpha,\beta} N_{\alpha,\beta}\langle\beta|\alpha\rangle \exp\Big\{B^*(t)\lambda - A(t)\lambda^* \\ &- \nu(t)|\lambda|^2 + \frac{\mu(t)}{2}\lambda^{*\,2} + \frac{\mu(t)^*}{2}\lambda^2\Big\}, \end{aligned} \tag{126}$$

where

$$\begin{aligned} A(t) &= \left[\frac{\alpha}{2}\frac{2\Delta - iF + 2i\delta_a}{2\Delta} - \beta^*\frac{iFe^{-2i\xi}}{4\Delta}\right]e^{-(k_a+\Delta)t} \\ &+ \left[\frac{\alpha}{2}\frac{2\Delta + iF - 2i\delta_a}{2\Delta} + \beta^*\frac{iFe^{-2i\xi}}{4\Delta}\right]e^{-(k_a-\Delta)t} \end{aligned} \tag{127}$$

$$\begin{aligned} B^*(t) &= \left[\frac{\beta^*}{2}\frac{2\Delta + iF - 2i\delta_a}{2\Delta} + \alpha\frac{iFe^{2i\xi}}{4\Delta}\right]e^{-(k_a+\Delta)t} \\ &+ \left[\frac{\beta^*}{2}\frac{2\Delta - iF + 2i\delta_a}{2\Delta} - \alpha\frac{iFe^{2i\xi}}{4\Delta}\right]e^{-(k_a-\Delta)t} \end{aligned} \tag{128}$$

$$\begin{aligned} \nu(t) &= \frac{F}{16}\left(\frac{\Gamma\delta_a}{\Delta^2} - \frac{2F}{\Delta}\right)\left(\frac{1 - e^{-2(k_a+\Delta)t}}{2\,(k_a+\Delta)}\right) \\ &- \frac{\Gamma\delta_a(2\delta_a - F)}{8\Delta^2}\left(\frac{1 - e^{-2k_a t}}{2k_a}\right) \\ &+ \frac{F}{16}\left(\frac{\Gamma\delta_a}{\Delta^2} + \frac{2F}{\Delta}\right)\left(\frac{1 - e^{-2(k_a-\Delta)t}}{2\,(k_a-\Delta)}\right) \end{aligned} \tag{129}$$

$$\begin{aligned}\mu(t) &= \frac{F^2 e^{-2i\xi}}{16\,(2\delta_a - F + 2i\Delta)}\left(\frac{\Gamma\delta_a}{\Delta^2} - \frac{2F}{\Delta}\right)\frac{1-e^{-2(k_a+\Delta)t}}{2\,(k_a+\Delta)} \\ &- \frac{F\Gamma e^{-2i\xi}\delta_a}{8\Delta^2}\frac{1-e^{-2k_a t}}{2k_a} \\ &+ \frac{F^2 e^{-2i\xi}}{16\,(2\delta_a - F - 2i\Delta)}\left(\frac{\Gamma\delta_a}{\Delta^2} + \frac{2F}{\Delta}\right)\frac{1-e^{-2(k_a-\Delta)t}}{2\,(k_a-\Delta)}\end{aligned} \tag{130}$$

$$\Delta = \sqrt{\delta_a F - \delta_a^2}\,. \tag{131}$$

We see that the system is stable and reaches a steady state if and only if

$$\Delta < k_a \quad \text{i.e.} \quad \chi^2\delta_a\delta_b < 4\left(k_b^2+\delta_b^2\right)\left(k_a^2+\delta_a^2\right)\,. \tag{132}$$

In the stable case, the stationary state is described by a Gaussian density operator of the form

$$\rho_{st} = Z^{-1}\exp\left\{-na^{+ger}a - \frac{m}{2}a^{+2} - \frac{m^*}{2}a^2\right\}\,, \tag{133}$$

where Z is a normalization constant and the equilibrium parameters m and n can be written as

$$\begin{aligned} n &= \frac{\nu_\infty + 1/2}{\sqrt{(\nu_\infty+1/2)^2 - |\mu_\infty|^2}} \\ &\times \log\left\{\frac{\left[\sqrt{(\nu_\infty+1/2)^2-|\mu_\infty|^2}+1/2\right]^2}{\nu_\infty(\nu_\infty+1) - |\mu_\infty|^2}\right\}\end{aligned} \tag{134}$$

$$m = \frac{\mu_\infty}{\nu_\infty + 1/2}n\,, \tag{135}$$

where the asymptotic values ν_∞ and μ_∞ are easily obtained from (129) and (130). An interesting aspect of this stationary state is that it can show arbitrary quadrature squeezing. For example, the stationary variance of the quadrature X_ξ is given by

$$\left\langle X_\xi^2\right\rangle = \frac{1}{2}\left[\frac{1}{2} + \frac{\delta_a\left(\Gamma\delta_a + 2Fk_a\right)}{8k_a\left(k_a^2 - \Delta^2\right)}\right] \tag{136}$$

and one has squeezing when $\delta_a\delta_b < 0$ and $k_b/k_a < |\delta_b/\delta_a|$. It is easily seen that when $\delta_a = 0$ no squeezing is possible, while for $\delta_a \neq 0$ but $\delta_b = 0$ extra noise is added to the system. The possibility to obtain squeezing with this model is thus only due to the existence of detunings, which give a sort of implicit feedback.

12. Interference Fringes

Let us now focus on the detection of the interference fringes associated to a linear superposition of coherent states. These fringes can generally be seen from the marginal probability distribution of the quadrature X_ξ, $P(x_\xi) = \langle x_\xi|\rho(t)|x_\xi\rangle$, where $|x_\xi\rangle$ is the eigenstate of

X_ξ with eigenvalue x_ξ. As we have seen above, this probability distribution can be reconstructed from the homodyne measurement of the meter mode b and its general expression can be easily obtained from the characteristic function (126) [30, 34]

$$P(x_\xi,t)=\sum_{\alpha,\beta} N_{\alpha,\beta}\frac{\langle\beta|\alpha\rangle}{\sqrt{\pi\sigma_x^2(t)}}\exp\left\{-\frac{(x_\xi-\delta_{\alpha,\beta}(t))^2}{\sigma_x^2(t)}\right\}, \tag{137}$$

where

$$\sigma_x^2(t) = \frac{1}{2}+\nu(t)+\mathrm{Re}\left\{\mu(t)e^{2i\xi}\right\}, \tag{138}$$

$$\delta_{\alpha,\beta}(t) = \frac{A(t)e^{i\xi}+B^*(t)e^{-i\xi}}{2}. \tag{139}$$

As a special case we consider the initial superposition trated by Yurke and Stoler [22], produced by the unitary evolution of a coherent state in a Kerr medium

$$\rho(0)=\frac{1}{2}\left(e^{-i\pi/4}|\alpha\rangle+e^{i\pi/4}|-\alpha\rangle\right)\left(e^{i\pi/4}\langle\alpha|+e^{-i\pi/4}\langle-\alpha|\right). \tag{140}$$

With this choice (137) simplifies to and one has squeezin

$$P(x_\xi,t)=\frac{1}{2}\left\{p_+^2(x_\xi,t)+p_-^2(x_\xi,t)\right.$$
$$\left.+2p_+(x_\xi,t)p_-(x_\xi,t)\sin\left[\Omega(x_\xi,t)\right]|\langle\alpha|-\alpha\rangle|^{\eta(t)}\right\}. \tag{141}$$

The first two terms $p_\pm^2(x_\xi,t)$ describe the two Gaussian peaks corresponding to the two coherent states $|\pm\alpha\rangle$ of the initial superposition and they are explicitely given by

$$p_\pm^2(x_\xi,t)=\frac{1}{\sqrt{\pi\sigma_x^2(t)}}\exp\left\{-\frac{(x_\xi\mp\delta_{++}(t))^2}{\sigma_x^2(t)}\right\}, \tag{142}$$

where

$$\delta_{++}(t) = \mathrm{Re}\left\{\alpha e^{i\xi}r(t)\right\} \tag{143}$$

$$r(t) = e^{-k_a t}\left(\cosh\Delta t-i\frac{\delta_a}{\Delta}\sinh\Delta t\right). \tag{144}$$

The third term in (141) describes the quantum interference between the two coherent states, where the function

$$\Omega(x_\xi,t)=\frac{2x_\xi}{\sigma_x^2(t)}\mathrm{Im}\left\{\alpha e^{i\xi}r(t)\right\} \tag{145}$$

gives the probability oscillations associated with the interference fringes and the factor $|\langle\alpha|-\alpha\rangle|^{\eta(t)}=\exp\left\{-2|\alpha|^2\eta(t)\right\}$ describes the suppression of quantum coherence due to dissipation. It is clear that this suppression is practically immediate for macroscopically distinguishable states (i.e., large $|\alpha|$), unless $\eta(t)\simeq 0$. It is therefore important to analyze the behavior of this decoherence function $\eta(t)$, which is equal to

$$\eta(t)=1-\frac{|r(t)|^2}{2\sigma_x^2(t)}. \tag{146}$$

To be more specific, if we want to determine the conditions under which the detection of macroscopic quantum coherence is facilitated, we have to compare $\eta(t)$ with the corresponding decoherence function of a standard vacuum bath, which is given by [30]

$$\eta_{vac}(t) = 1 - e^{-2k_a t} . \qquad (147)$$

This function shows that in the standard case, after a time $t \simeq 1/(2k_a)$, it is $\eta_{vac}(t) \simeq 1$ and therefore the quantum interference is quickly washed out. On the contrary, in the present model it is possible that $\eta(t)$ assumes much smaller values, so to significantly slow down the destruction of the interference pattern.

Acknowledgments

This work has been partially supported by European Community under the Human Capital and Mobility (HCM) programme. Most of the work presented in this series of lectures was done in collaboration with my colleagues G. M. D'Ariano, M. Fortunato, S. Mancini and D. Vitali.

References

1. R. J. Glauber, Phys. Rev. Lett. **10**, 84 (1963).
2. E. C. G. Sudarshan, Phys. Rev. Lett. **10**, 177 (1963).
3. K. Husimi, Proc. Phys. Math. Soc. Japan **23**, 264 (1940).
4. E. Wigner, Phys. Rev. **40**, 749 (1932).
5. K. E. Cahill, and R. J. Glauber, Phys. Rev. **177**, 1882 (1969).
6. K. Vogel, and H. Risken, Phys. Rev. A **40**, 2847 (1989).
7. J.H. Shapiro and H.P.Yuen, IEEE Trans. Inf. Theory **IT25**, 179 (1979)
8. P.L. Kelley and W.H. Kleiner, Phys. Rev. **136**, 316 (1964).
9. J.H. Shapiro, G. Saplakoglu, S.-T. Ho, P. Kumar. B.E.A. Saleh and M.C. Teich, J.Opt. Soc. Am. B **4**, 1604 (1987).
10. H.M. Wiseman, and G.J. Milburn, Phys. Rev. A **47**, 642 (1993).
11. S. Mancini, V. I. Man'ko and P. Tombesi, to appear on Quantum Semiclass. Opt.
12. see for example W. H. Louisell, *Quantum Statistical Properties of Radiation* (Wiley, New York, 1973).
13. C. W. Gardiner, *Quantum Noise* (Springer, Heidelberg, 1991).
14. D. T. Smithey, M. Beck, M. G. Raymer, and A. Faridani, Phys. Rev. Lett. **70**, 1244 (1993).
15. F. Natterer, *The Mathematics of Computerized Tomography* (Wiley, New York, 1986).
16. G. M. D'Ariano, C. Macchiavello, and M. G. A. Paris, Phys. Rev. A **50**, 4298 (1994).
17. H. Kühn, D. G. Welsch and W. Vogel, J. of Mod. Opt. **41**, 1607 (1994).
18. G. M. D'Ariano, U. Leonhardt and H. Paul, Phys. Rev. A **52** (1995) (in press).
19. G.M. D'Ariano, S. Mancini, V.I. Man'ko and P. Tombesi, Quantum Semiclass. Opt. (submitted)
20. U. Leonhardt and H. Paul, Phys. Rev. A **48**, 4598 (1993).
21. G. M. D'Ariano, Int. J. Mod. Phys. B **6**, 1291 (1992).
22. B. Yurke and D. Stoler, Phys. Rev. Lett. **57**, 13 (1986);
23. S. Song, C.M. Caves, B. Yurke, Phys. Rev. A **41**, 5261 (1990); B. Yurke, W. Schleich, D.F. Walls, Phys. Rev. A **42**, 1703 (1990); M. Brune, S. Haroche, J.M. Raimond, L. Davidovich and N. Zagury, Phys. Rev. A **45**, 5193 (1992); L. Krippner, W.J. Munro and M.D. Reid, Phys. Rev. A **50**, 4330 (1994).
24.
25. A.M. Fox, J.J. Baumberg, M. Dabbicco, B. Huttner, and J.F. Ryan, Phys. Rev. Lett. **74**, 1728(1995).
26. S.T. Ho, C.E. Soccolich, M.N. Islam, W.S. Hobson, A.F.J. Levi, and R.E. Slusher, Appl. Phys. Lett. **59**, 2558(1991)
27. G.M. D'Ariano, M. Fortunato,and P. Tombesi, Quantum Semicl. Opt. (to appear)
28. P. Tombesi and D. Vitali, Phys. Rev. A **50**, 4253 (1994).

29. S. Mancini, and P. Tombesi, Phys. Rev. A (to appear)
30. T.A.B. Kennedy and D.F. Walls, Phys. Rev. A **37**, 152 (1988).
31. A. Mecozzi and P. Tombesi, Phys. Lett. A **121**, 101 (1987); J. Opt. Soc. Am. B **4**,1700 (1987).
32. P. Tombesi and D. Vitali, Phys. Rev. Lett. submitted
33. P. Alsing, G. J. Milburn and D. F. Walls, Phys. Rev. A **37**, 2970 (1988).
34. P. Tombesi and D. Vitali, Appl. Phys. B **60**, S69 (1995); Phys. Rev. A **51**, 4913 (1995).
35. H.M. Wiseman, and G.J. Milburn, Phys. Rev. A **47**, 642 (1993).
36. H. Risken, *The Fokker Planck Equation* 2nd ed. (Springer, Berlin 1989).

HARMONIC OSCILLATOR STATES IN FINITE DIMENSIONAL HILBERT SPACE

A. MIRANOWICZ
Nonlinear Optics Division, Institute of Physics, A. Mickiewicz University, 60-780 Poznań, Poland

T. OPATRNÝ
Department of Theor. Physics, Palacký University, 771 46 Olomouc, Czech Republic

AND

J. BAJER
Laboratory of Quantum Optics, Palacký University, 772 07 Olomouc, Czech Republic

Abstract. We consider various harmonic-oscillator states in a finite-dimensional Hilbert space. In greater detail we analyze finite-dimensional coherent states defined either by truncation of a number-state expansion of the standard coherent states or by the action of the generalized finite-dimensional displacement operator on vacuum. Within these two approaches other finite-dimensional states are analyzed, including displaced number states, and even and odd coherent states. We propose some new definitions of the states in a finite-dimensional Hilbert space and construct explicitly all these states in Fock representation. The number-phase Wigner function is computed for several states. Analytical results and numerically computed graphs are presented.

1. Introduction

Weyl's formulation of quantum mechanics [1], with his representation of the canonical commutation relations, has opened the possibility of studying quantum mechanics in a finite-dimensional space of operators which are bounded and have a discrete spectrum. Weyl's approach, extended by Schwinger [2], is based on the fact that the kinematical structure of a physical system can be expressed by an irreducible Abelian group of unitary representations of system space. For a given finite Abelian group there is a unique class of unitarily equivalent, irreducible representations in *a finite-dimensional Hilbert space* (FDHS). Hence, this formulation has provided the basis for studies of the behavior of the harmonic oscillator in a FDHS. In the 1970s, Santhanam and co-workers contributed to the above formulation in a series of papers (e.g., Ref. [3]). Recently, interest therein

T. Hakioğlu and A. S. Shumovsky (eds.), Quantum Optics and the Spectroscopy of Solids, 225–236.

has increased due to the Pegg-Barnett proposal of constructing the quantum phase for the linear harmonic oscillator [4]. Their idea proved very fruitful and numerous recent articles deal with the properties of various states of the finite-dimensional harmonic oscillator (see [5, 6] and references therein), including some interesting approaches to, e.g., generalized coherent states or finite-state Wigner functions.

This work was intended as an attempt to present two essentially different constructions of various harmonic-oscillator states in a FDHS. We propose some new definitions of the states and find their explicit forms in number-state representation. For the convenience of the Reader we also bring together several known finite-dimensional states, thus making our exposition more self-contained. We shall discuss coherent states, displaced number states, and Schrödinger cats. We shall show some intriguing properties of the states with the help of the generalized finite-state Wigner function.

It is not our aim to discuss methods of generation of the finite-dimensional harmonic oscillator states. We shall only mention that several methods for preparation of a number state have been developed recently [7, 8]. So, in principle, the finite-dimensional states discussed here can be generated and detected using various quantum-state engineering techniques [9]. Schemes for a direct generation of the finite-dimensional Glauber coherent states in quantum systems have recently been proposed by Leoński and Tanaś [8, 10]. The Leoński-Tanaś method can be generalized for preparation of various harmonic oscillator states.

2. Finite-Dimensional Hilbert Space

We shall discuss various states constructed in the finite-dimensional Hilbert space (FDHS) of a harmonic oscillator. This space, denoted by $\mathcal{H}^{(s)}$, is spanned by $(s+1)$ Fock states fulfilling the completeness and orthogonality relations

$$\hat{1} = \sum_{n=0}^{s} |n\rangle\langle n|\,, \qquad \langle n|m\rangle = \delta_{n,m}\,. \tag{1}$$

Equivalently, the space $\mathcal{H}^{(s)}$ is spanned by the phase states

$$|\theta_m\rangle = (s+1)^{-1/2}\sum_{n=0}^{s} \exp(in\theta_m)|n\rangle, \qquad \theta_m = \theta_0 + \frac{2\pi}{s+1}m, \tag{2}$$

which also form a complete and orthonormal basis

$$\hat{1} = \sum_{m=0}^{s} |\theta_m\rangle\langle\theta_m|\,, \qquad \langle\theta_m|\theta_n\rangle = \delta_{m,n}\,. \tag{3}$$

Here, θ_0 is the initial reference phase. The states (2) were applied by Pegg and Barnett [4] in their formalism of the Hermitian optical phase operator. On the other hand, the phase states can be used in the construction of a discrete Wigner function.

The generalized (finite-dimensional) annihilation and creation operators are defined in $\mathcal{H}^{(s)}$ by

$$\hat{a} \equiv \hat{a}^{(s)} = \sum_{n=1}^{s} \sqrt{n}|n-1\rangle\langle n|, \qquad \hat{a}^{\dagger} \equiv (\hat{a}^{(s)})^{\dagger} = \sum_{n=1}^{s} \sqrt{n}|n\rangle\langle n-1|. \tag{4}$$

As can readily be checked, the commutator $[\widehat{a}, \widehat{a}^{\dagger}]$ differs from $\widehat{1}$. So, the operators (4) are not related to the Weyl-Heisenberg algebra. Moreover, the commutators $[[\widehat{a}, \widehat{a}^{\dagger}], \widehat{a}]$ and $[[\widehat{a}, \widehat{a}^{\dagger}], \widehat{a}^{\dagger}]$ do not vanish, hence the Baker-Campbell-Hausdorff identity does not hold for the operators (4). These properties of the operators $\widehat{a}$ and $\widehat{a}^{\dagger}$ really complicate analytical approach to quantum mechanics in $\mathcal{H}^{(s)}$, including an explicit construction of various finite-dimensional states.

3. Discrete Number-Phase Wigner Function

Wigner function is widely used in non-relativistic quantum mechanics as an alternative to the density matrix of quantum systems [11]. Although original Wigner function applies only to systems with continuous degrees of freedom, it can be transcribed for finite-state systems as well. The generalized Wigner function for arbitrary finite-state systems was proposed by Wootters [12]. His definition takes the simplest form for prime-number-dimensional systems. A similar construction of a discrete Wigner function for odd-dimensional systems was suggested by a Cohendet et al. [13]. Another general approach (for odd dimensions equivalent to that of Wootters) was recently suggested by Leonhardt [14]. A number-phase discrete Wigner function, a special case of the Wootters definition, was analyzed in detail by Vaccaro and Pegg [15]. In this work we shall use odd-dimensional number-phase Wigner functions only.

The number-phase characteristic function in $\mathcal{H}^{(s)}$ can be defined in terms of the phase states (2) as [14]

$$C(\nu, \theta_{\mu}) = \sum_{m=0}^{s} \exp\left(-\frac{4\pi i}{s+1}\nu(m+\mu)\right) \langle\theta_m|\widehat{\rho}|\theta_{m+2\mu}\rangle\,, \tag{5}$$

A discrete Fourier transform applied to $C(\nu, \theta_{\mu})$ leads to the discrete (finite-dimensional) Wigner function for phase and number in the following form (for odd s)

$$W(n, \theta_m) = \frac{1}{(s+1)^2} \sum_{\nu=0}^{s} \sum_{\mu=0}^{s} \exp\left(\frac{4\pi i}{s+1}(n\mu + \nu m)\right) C(\nu, \theta_{\mu}) \tag{6}$$

or explicitly as [12]

$$W(n, \theta_m) = \frac{1}{s+1} \sum_{\mu=0}^{s} \exp\left(\frac{4\pi i}{s+1} n\mu\right) \langle\theta_{m+\mu}|\widehat{\rho}|\theta_{m-\mu}\rangle. \tag{7}$$

The Wigner function $W(n, \theta_m)$ is periodic both in n and θ_m,

$$W(n, \theta_m) = W(n \pm (s+1), \theta_m) = W(n, \theta_{m\pm(s+1)}) = W(n, \theta_m \pm 2\pi), \tag{8}$$

which leads to an interesting graphical representation on torus (see [16]).

For arbitrary pure states, with the following number-state expansion

$$|\psi\rangle_{(s)} = \sum_{n=0}^{s} b_n^{(s)} e^{i\varphi_n} |n\rangle, \tag{9}$$

the discrete Wigner function can be expressed in terms of the decomposition coefficients [15]:

$$W(n,\theta_m) = \frac{1}{(s+1)}\left\{ \sum_{k=M+1}^{s} b_k^{(s)} b_{M-k+s+1}^{(s)} \exp[i(2k-M-s-1)\theta_m + \varphi_{M-k+s+1} - \varphi_k] + \sum_{k=0}^{M} b_k^{(s)} b_{M-k}^{(s)} \exp[i(2k-M)\theta_m + \varphi_{M-k} - \varphi_k] \right\}, \quad (10)$$

where $M \equiv 2n \bmod(s+1)$. Several graphs of these functions of various states are presented in the next Section. The physical interpretation of the Wigner functions is based on the fact that the marginal sum of their values over a generalized line gives probability of the system to be in some special state [12, 14, 16]. Particular cases of these marginal distributions are the photon number and phase distributions.

4. Special States of Finite-Dimensional Hilbert Space

In quantum optics, the most common are Glauber coherent states (CS). The literature on coherent states and their generalizations is truly prodigious and has been summarized in a number of excellent monographs [17, 18] and reviews [19]. There exist several ways of generalizing these states comprising the finite-dimensional case. It is possible to define the CS using the concept of Lie group representations (see, e.g., Ref. [20]), or to postulate the validity of some properties of the infinite-dimensional Hilbert space CS for the finite-dimensional CS. In this paper we shall be interested in two definitions of the latter kind. First, coherent states in a FDHS are usually treated as the displaced vacuum, where the displacement operator is defined analogously to the usual displacement operator in the infinite space (the Glauber treatment of CS [21]). This idea was applied in the work of Bužek and co-workers [22] and further studied by Miranowicz et al. [23] and Opatrný et al. [16]. Here, we shall refer to such states as finite-dimensional Glauber coherent states or simply *coherent states* in a FDHS. Similarly, other states of the electromagnetic field are defined in FDHS such as displaced phase states and coherent phase states as discussed by Gangopadhyay [24], as well as displaced number states and even and odd coherent states in a FDHS as proposed and analyzed explicitly here. Another definition is based on the postulate that the Fock expansion of the finite-dimensional CS shall be equal to the truncated expansion of the usual infinite-dimensional CS. This approach was extensively developed by Kuang et al. [25] and Opatrný et al. [16]. Here we shall refer to states of this kind as *truncated coherent states* in a FDHS. The same construction was used in the analysis of other finite-dimensional states, namely even and odd coherent states by Zhu and Kuang [26] and phase coherent states by Kuang and Chen [27], whereas here we propose truncated displaced number states.

4.1. COHERENT STATES

The coherent states $|\alpha\rangle_{(s)}$ in the $(s+1)$-dimensional Hilbert space of a harmonic oscillator can be defined in a Glauber sense by the action of the analogue of the Glauber displacement operator

$$\hat{D}^{(s)}(\alpha) = \exp(\alpha\hat{a}^\dagger - \alpha^*\hat{a}) \quad (11)$$

on the vacuum state as was suggested by Bužek et al. [22]. The operator $\widehat{D}^{(s)}(\alpha)$ is given in terms of the finite-dimensional annihilation and creation operators (4). The coherent states $|\alpha\rangle_{(s)}$ are the close *analogues* of Glauber's (i.e., infinite-dimensional) coherent states $|\alpha\rangle$. They were introduced and discussed by Bužek et al. [22] and their number-state representation was found by Miranowicz et al. [23]. In this basis the coherent state $|\alpha\rangle_{(s)}$, with $\alpha = |\alpha|\exp(i\varphi)$, is

$$|\alpha\rangle_{(s)} = \widehat{D}^{(s)}(\alpha)|0\rangle = \sum_{n=0}^{s} \mathrm{e}^{in\varphi}\, b_n^{(s)}|n\rangle, \tag{12}$$

where

$$b_n^{(s)} = \frac{s!}{s+1}\frac{(-i)^n}{\sqrt{n!}}\sum_{k=0}^{s} \mathrm{e}^{ix_k|\alpha|}\,\mathrm{He}_n(x_k)\mathrm{He}_s^{-2}(x_k). \tag{13}$$

Here, $x_k \equiv x_k^{(s+1)}$ are the roots of the Hermite polynomial of order $(s+1)$, $\mathrm{He}_{s+1}(x_k) = 0$. To avoid confusion, we note that Hermite polynomials $\mathrm{He}(x)$ and $\mathrm{H}(x)$ are related by $\mathrm{He}_n(x) \equiv 2^{-n/2}\mathrm{H}_n(x/\sqrt{2})$. Numerical analysis of photon-number statistics of the CS (12) in comparison with the Poissonian photon statistics of the standard infinite-dimensional CS was given in Ref. [22]. Comparative study of the phase properties of the finite- and infinite-dimensional CS within the Pegg-Barnett phase formalism was presented in Ref. [23]. Thorough description of these states, together with their discrete number-phase Wigner functions was presented in Ref. [16]. The finite-dimensional coherent states (12) approach the standard infinite-dimensional coherent states for $|\alpha|^2 \ll s$ as was shown numerically in Ref. [23] and analytically in Ref. [16]. Wigner function of (12) for various values of α is depicted in Fig. 1. It is seen that the coherent state (12) for $|\alpha| = 8.8$ (Fig. 1(c)) is almost a vacuum state as a sign of a quasi-periodic behavior. This interesting property of the coherent states $|\alpha\rangle_{(s)}$, in an arbitrary-dimensional Hilbert space, was predicted and analytically explained by us in Ref. [16].

4.2. TRUNCATED COHERENT STATES

Kuang, Wang and Zhou [25] defined the normalized finite-dimensional CS truncating the Fock expansion of the Glauber infinite-dimensional CS or equivalently by the action of the operator $\exp(\overline{\alpha}\hat{a}^\dagger)$ on vacuum state. We shall pay some attention to the Kuang et al. approach because it is the most similar to the Vaccaro-Pegg treatment of the coherent states Wigner function. The states $|\overline{\alpha}\rangle_{(s)}$, where $\overline{\alpha} = |\overline{\alpha}|\exp(i\varphi)$, can be defined in the Fock expansion (9) as follows [25]

$$|\overline{\alpha}\rangle_{(s)} = \mathcal{N}^{(s)}\exp(\overline{\alpha}\hat{a}^\dagger)|0\rangle = \sum_{n=0}^{s} \mathrm{e}^{in\varphi}\, b_n^{(s)}|n\rangle, \tag{14}$$

where

$$b_n^{(s)} = \mathcal{N}^{(s)}|\overline{\alpha}|^n(n!)^{-1/2}. \tag{15}$$

Here, we rewrite the normalization constant in the form [16]

$$\mathcal{N}^{(s)} = \left\{(-1)^s\mathrm{L}_s^{-s-1}(|\overline{\alpha}|^2)\right\}^{-1/2} \tag{16}$$

in terms of Laguerre polynomials $\mathrm{L}_s^n(x)$. The superposition coefficients $b_n^{(s)}$ are just the Fock expansion coefficients of the standard (infinite-dimensional) CS truncated at a $(s+1)$th

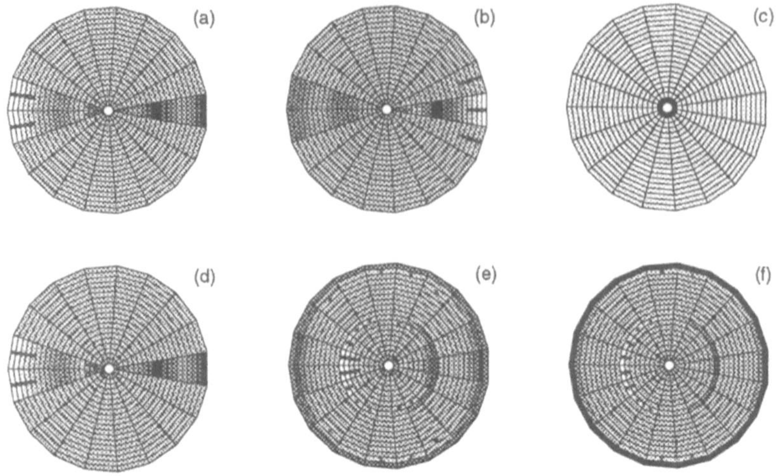

Figure 1. Discrete number-phase Wigner function for the coherent states $|\alpha\rangle_{(s)}$ (a,b,c) and $|\overline{\alpha}\rangle_{(s)}$ (d,e,f) for $s = 18$, and various values of real α and $\overline{\alpha}$: (a) $\alpha = (1/3)T$; (b) $\alpha = (2/3)T$; (c) $\alpha = T$; (d) $\overline{\alpha} = (1/3)T$; (e) $\overline{\alpha} = (2/3)T$; (f) $\overline{\alpha} = T$, where the quasi-period is $T = 8.8$. The polar coordinates are (n, θ_m). The darker is a region the higher is the value of the Wigner-function. Negative values are marked by crosses.

term and properly normalized. This explains our concept of *truncated coherent states* for the states $|\overline{\alpha}\rangle_{(s)}$ defined by (14). In Fig. 1 (d,e,f) we present the Wigner function (10) for the truncated coherent states $|\overline{\alpha}\rangle_{(s)}$ for the same values of $|\overline{\alpha}|$ as the Wigner function for $|\alpha\rangle_{(s)}$. The number-phase Wigner function for the states (12) and (14), calculated here from the general expression (10), was explicitly given by us in Ref. [16]. For better comparison of these two definitions, the squared inner products between the coherent states $|\alpha\rangle_{(s)}$ and $|\overline{\alpha}\rangle_{(s)}$ are depicted in Fig. 2. It is seen that these states coincide for $|\alpha|^2 = |\overline{\alpha}|^2 \ll s$, which corresponds to unity of their inner product and almost the same Wigner representations (see Figs. 1(a) and 1(d)). However, they exhibit qualitatively different behavior for $|\alpha|^2 = |\overline{\alpha}|^2$ of order s (Figs. 1(b) and 1(e)) or greater (Figs. 1(c) and 1(f)). The truncated coherent states (14) are aperiodic in $|\alpha|$ (see Fig. 1(f)) contrary to the quasi-periodical behavior of (12), represented by the almost-vacuum state in Fig. 1(c) and the oscillations of the inner product in Fig. 2.

4.3. DISPLACED NUMBER STATES

Let us propose a definition of the finite-dimensional displaced number states in a way similar to the definition (12) of the coherent states $|\alpha\rangle_{(s)}$, namely, by the action of the displacement operator $\widehat{D}^{(s)}(\alpha)$ on the number state $|n_d\rangle$ (not only on the vacuum state). Using the method developed in Ref. [23], we find the following explicit Fock representation of the displaced number states in the FDHS:

$$|\alpha, n_d\rangle_{(s)} = \widehat{D}^{(s)}(\alpha)|n_d\rangle = \sum_{n=0}^{s} e^{i(n-n_d)\varphi}\, b_n^{(s)}|n\rangle, \tag{17}$$

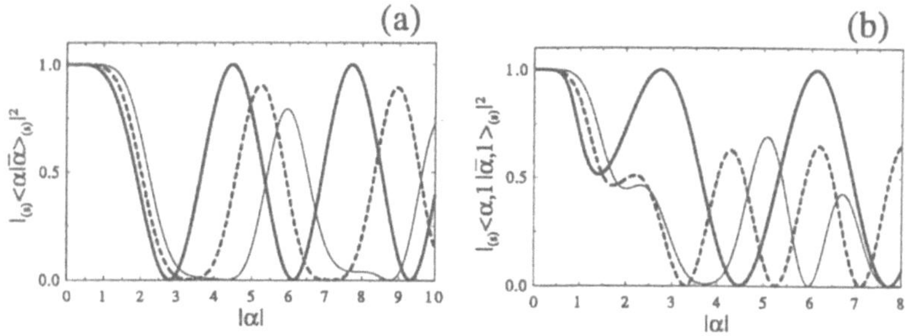

Figure 2. Squared inner products (a) between the coherent states $|\alpha\rangle_{(s)}$ and $|\overline{\alpha}\rangle_{(s)}$ and (b) between the displaced single-photon states $|\alpha,1\rangle_{(s)}$ and $|\overline{\alpha},1\rangle_{(s)}$ as a function of $|\alpha|=|\overline{\alpha}|$ for s=1 (thick solid lines), s=2 (dashed lines), and s=3 (thin solid lines).

where

$$b_n^{(s)} \equiv b_n^{(s)}(n_d) = \frac{s!}{s+1}\,\frac{i^{n_d}(-i)^n}{\sqrt{n!n_d!}}\sum_{k=0}^{s} \mathrm{e}^{ix_k|\alpha|}\,\frac{\mathrm{He}_n(x_k)\mathrm{He}_{n_d}(x_k)}{\mathrm{He}_s^2(x_k)}, \tag{18}$$

and $\alpha=|\alpha|\exp(i\varphi)$ as in §4.1. For the same reasons as $|\alpha\rangle_{(s)}$, the finite-dimensional states $|\alpha,n_d\rangle_{(s)}$ go over into the ordinary (infinite-dimensional) displaced number states in the dimension limit $s\to\infty$. Obviously, in the special case of $n_d=0$, the displaced number states $|\alpha,n_d\rangle_{(s)}$ reduce to the coherent states $|\alpha\rangle_{(s)}$, defined by Eq. (12). In Fig. 3 we present a few examples of the Wigner function for the states $|\alpha,n_d\rangle_{(s)}$ for (n_d=1, 2, 12). A typical shape (for small displacement parameter $|\alpha|$) has a simple interpretation: the circle of the ordinary number state splits into a pair of crescents; their mutual orientation depends on the phase, and their distance on the modulus of the displacement parameter. Simultaneously, another pair of crescents appears. Their radii are dependent on the value of n: for n smaller than $s/2$ they are greater than the double-crescent radius and vice versa. The occurrence of a double crescent is connected with the fact that on a slight displacement of the number state $|n_d\rangle$, the probability of finding a slightly smaller or greater number of photons increases. The crescent orientation refers to the preferred phase of such a state.

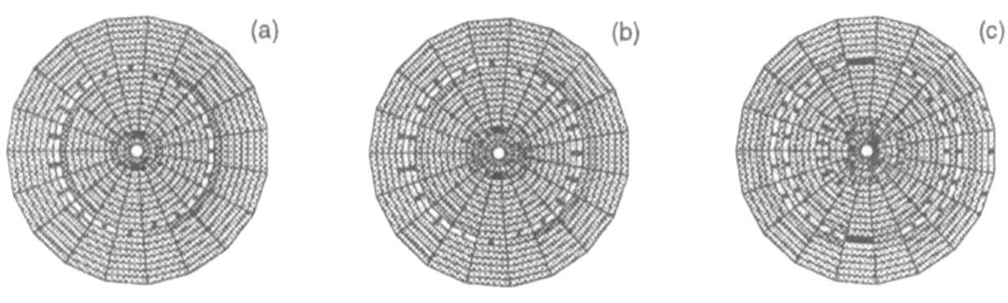

Figure 3. Wigner function of the displaced number states $|\alpha,n_d\rangle_{(s)}$ for $s=18$, $\alpha=0.5$, and (a) $n_d=1$; (b) $n_d=2$, and (c) $n_d=12$.

4.4. TRUNCATED NUMBER STATES

Here, we introduce another definition of finite-dimensional displaced number states which in a special case goes over into the truncated coherent states $|\overline{\alpha}\rangle_{(s)}$ of Kuang et al. [25] and into ordinary displaced number states [28] in the limit $s \to \infty$. We define the normalized truncated displaced number states $|\overline{\alpha}, n_d\rangle_{(s)}$ by the following number-state representation:

$$|\overline{\alpha}, n_d\rangle_{(s)} = \sum_{n=0}^{s} b_n^{(s)} e^{i(n-n_d)\varphi} |n\rangle, \tag{19}$$

$$b_n^{(s)} \equiv b_n^{(s)}(n_d) = \mathcal{N}_{n_d}^{(s)} \left(\frac{n_1!}{n_2!}\right)^{1/2} (-1)^{n_2-n} |\overline{\alpha}|^{n_2-n_1} L_{n_1}^{n_2-n_1}(|\overline{\alpha}|^2), \tag{20}$$

where the normalization constant $\mathcal{N}_{n_d}^{(s)}$ takes the form

$$\mathcal{N}_{n_d}^{(s)} = \left(\sum_{n=0}^{s} \frac{n_1!}{n_2!} |\overline{\alpha}|^{2(n_2-n_1)} \left[L_{n_1}^{n_2-n_1}(|\overline{\alpha}|^2)\right]^2\right)^{-1/2}. \tag{21}$$

For brevity, we have introduced the indices $n_1 \equiv \min(n, n_d)$ and $n_2 \equiv \max(n, n_d)$. The state (19) is, in fact, given by the Fock expansion of the standard (i.e., infinite-dimensional) displaced number states [28, 6] truncated at the $(s+1)$th term and properly normalized. This construction justifies our concept of the *truncated* displaced number states. Alternatively, the states (19) can be defined by the action of the finite-dimensional, formally factorized, displacement operator $\exp(\overline{\alpha}\,\widehat{a}^\dagger)\exp(-\overline{\alpha}^*\widehat{a})$ on a Fock state $|n_d\rangle$, i.e.,

$$|\overline{\alpha}, n_d\rangle_{(s)} = \mathcal{N}_{n_d}^{(s)} \exp(\overline{\alpha}\widehat{a}^\dagger)\exp(-\overline{\alpha}^*\widehat{a})|n_d\rangle. \tag{22}$$

Eq. (22) explicitly shows how the concept of the truncated coherent states (14) is generalized. The states $|\overline{\alpha}, n_d\rangle_{(s)}$ are different from the finite-dimensional displaced number states $|\alpha, n_d\rangle_{(s)}$ introduced in §4.3. The differences are particularly distinct for values $|\overline{\alpha}|^2 \equiv |\alpha|^2$ of the order s or greater. However, for $|\overline{\alpha}|^2 \equiv |\alpha|^2 \ll s$, the finite-dimensional displaced number states $|\alpha, n_d\rangle_{(s)}$ and $|\overline{\alpha}, n_d\rangle_{(s)}$ approach each other. This is represented, in Fig. 2(b), by unity of the squared inner products between the states (for brevity, we analyze only the case of $n_d = 1$). In Fig. 3 we have presented the Wigner representations for $|\alpha, n_d\rangle_{(s)}$. We have omitted corresponding figures for the truncated displaced number states. The Wigner functions for $|\alpha, n_d\rangle_{(s)}$ and $|\overline{\alpha}, n_d\rangle_{(s)}$ are almost indistinguishable for the displacement parameter $|\alpha| = |\overline{\alpha}|$ and n_d much less than s, which are the cases presented in Fig. 3(a) and 3(b). Slight differences in the Wigner functions are observed for $|\alpha, 12\rangle_{(s)}$ (Fig. 3(c)) and the corresponding $|\overline{\alpha}, 12\rangle_{(s)}$.

4.5. EVEN AND ODD COHERENT STATES

Superpositions of two coherent states attract much attention [18, 29] as a simple example of Schrödinger cats. Let us define the finite-dimensional even coherent states by

$$|\alpha_+\rangle_{(s)} = \mathcal{M}_+^{(s)} \left(|\alpha\rangle_{(s)} + |-\alpha\rangle_{(s)}\right), \tag{23}$$

and odd coherent states by

$$|\alpha_-\rangle_{(s)} = \mathcal{M}_-^{(s)} \left(|\alpha\rangle_{(s)} - |-\alpha\rangle_{(s)}\right), \tag{24}$$

where the normalization is guaranteed by $\mathcal{M}_{\pm}^{(s)}$. On inserting the finite-dimensional coherent states $|\alpha\rangle_{(s)}$, given by Eq. (12), into (23) and (24), we find expressions in Fock basis for $|\alpha_+\rangle_{(s)}$ and $|\alpha_-\rangle_{(s)}$, respectively:

$$|\alpha_+\rangle_{(s)} = \mathcal{N}_+^{(s)} \sum_{n=0}^{[[s/2]]} e^{i2n\varphi}\, b_{2n}^{(s)} |2n\rangle, \tag{25}$$

$$|\alpha_-\rangle_{(s)} = \mathcal{N}_-^{(s)} \sum_{n=0}^{[[s/2]]} e^{i(2n+1)\varphi}\, b_{2n+1}^{(s)} |2n+1\rangle, \tag{26}$$

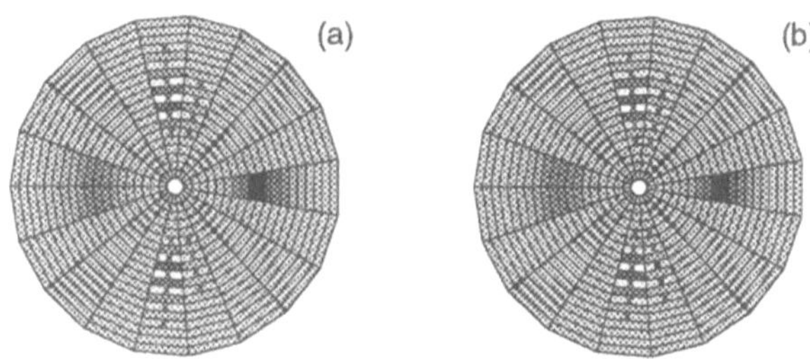

Figure 4. Wigner function of (a) the even CS $|\alpha_+\rangle_{(s)}$ and (b) the odd CS $|\alpha_-\rangle_{(s)}$ for $s = 18$, $\alpha = T/3 \approx 2.9$

where the coefficients b_n are given by Eq. (13), $[\![s/2]\!]$ is the largest integer not exceeding $s/2$, and the normalization is

$$\mathcal{N}_{\pm}^{(s)} = \left(\sum_{n=0}^{[[s/2]]} b_{2n+1/2\mp 1/2}^2 \right)^{-1/2}. \tag{27}$$

Analogously one can construct finite-dimensional superpositions of several coherent states (i.e., Schrödinger kitten states) which in a special case go over into infinite-dimensional ones [29, 30]. Two examples of the Wigner function for the even and odd states are given in Fig. 4. Their interpretation is quite clear. These are two-peak structures with many interference fringes. The fringes in Wigner function are typical of superposition states, and are not observed for a mixture of states. The main difference between the Wigner functions presented in Figs. 2(a) and 2(b) consists in a shift of the interference fringes. Let us note that the even CS for $|\alpha| = T/2$ is almost the CS (12).

4.6. TRUNCATED EVEN AND ODD COHERENT STATES

Finite-dimensional even, (23), and odd, (24), coherent states can be constructed in a slightly different way from those presented in §4.5. The truncated coherent states $|\pm\overline{\alpha}\rangle_{(s)}$ can be used in the definitions (23)–(24) instead of the coherent states $|\pm\alpha\rangle_{(s)}$. This approach was explored by Zhu and Kuang [26]. We rewrite briefly their explicit expressions for $|\overline{\alpha}_+\rangle_{(s)}$ and $|\overline{\alpha}_-\rangle_{(s)}$ in Fock representation:

$$|\overline{\alpha}_\pm\rangle_{(s)} = \mathcal{N}_{\pm}^{(s)} \sum_{n=0}^{[[s/2]]} \frac{\overline{\alpha}^{2n+\delta}}{\sqrt{(2n+\delta)!}} |2n+\delta\rangle, \tag{28}$$

where $\delta = 0$ for even cats and $\delta = 1$ for odd cats. The normalization is

$$\mathcal{N}_{\pm}^{(s)} = \left(\sum_{n=0}^{[[s/2]]} \frac{|\overline{\alpha}|^{2(2n+\delta)}}{(2n+\delta)!} \right)^{-1/2} \tag{29}$$

Eqs. (28) and (29) can be directly calculated from the general definitions (23)–(24) after insertion of the Fock expansion (14) for $|\pm\overline{\alpha}\rangle_{(s)}$. Therefore, we refer to the states (28) as the *truncated* even, $|\overline{\alpha}_{+}\rangle_{(s)}$, and odd, $|\overline{\alpha}_{-}\rangle_{(s)}$, coherent states. In Fig. 5(a) we compare the finite-dimensional even coherent states $|\alpha_{+}\rangle_{(s)}$ with $|\overline{\alpha}_{+}\rangle_{(s)}$, whereas in Fig. 5(b) we compare the finite-dimensional odd coherent states $|\alpha_{-}\rangle_{(s)}$ and $|\overline{\alpha}_{-}\rangle_{(s)}$ by virtue of the inner products in their dependence on the parameter $|\alpha| \equiv |\overline{\alpha}|$ for several dimensions $s+1$. In Fig. 5, we have not included the "pathological" (i.e., expressed by a single Fock state only) two- and three-dimensional odd coherent states $|\alpha_{-}\rangle_{(s)}$ and $|\overline{\alpha}_{-}\rangle_{(s)}$, and two-dimensional even coherent states $|\alpha_{+}\rangle_{(1)}$ and $|\overline{\alpha}_{+}\rangle_{(1)}$. Again, as in Fig. 2 for the finite-

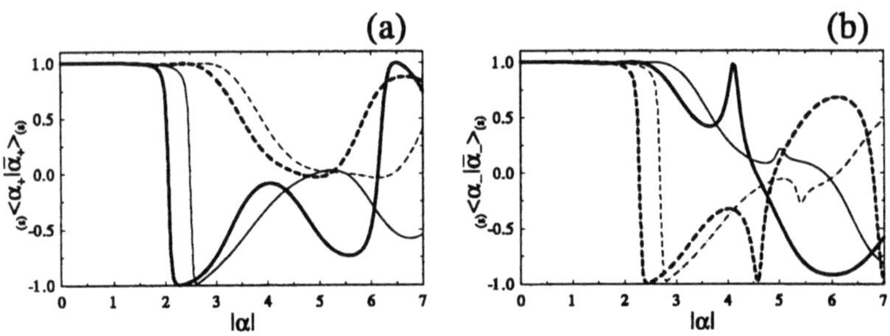

Figure 5. Inner products (a) between the even coherent states $|\alpha_{+}\rangle_{(s)}$ and $|\overline{\alpha}_{+}\rangle_{(s)}$, and (b) between the odd coherent states $|\alpha_{-}\rangle_{(s)}$ and $|\overline{\alpha}_{-}\rangle_{(s)}$ as a function of $|\alpha| = |\overline{\alpha}|$ ($\varphi = \overline{\varphi}$) for s=3 (thick solid lines), s=4 (thick dashed lines), s=5 (thin solid lines), and s =6 (thin dashed lines).

dimensional CS and displaced number-states, it is clearly seen that, for $|\alpha|^2 \equiv |\overline{\alpha}|^2 \ll s$, the states $|\alpha_{-}\rangle_{(s)}$ ($|\alpha_{+}\rangle_{(s)}$) and $|\overline{\alpha}_{-}\rangle_{(s)}$ ($|\overline{\alpha}_{+}\rangle_{(s)}$) approach each other. However, the even CS approach each other faster in the space $\mathcal{H}^{(2s)}$ (with arbitrary s) than in $\mathcal{H}^{(2s+1)}$. Whereas, the odd CS are more similar in $\mathcal{H}^{(2s-1)}$ than in $\mathcal{H}^{(2s)}$ (see Fig. 5). The Wigner representations of the truncated Schrödinger cats $|\overline{\alpha}_{+}\rangle_{(s)}$ and $|\overline{\alpha}_{-}\rangle_{(s)}$, for $|\alpha| = |\overline{\alpha}| = T/3 \approx 2.9$ and s=18, are very similar to the representations of the Schrödinger cats $|\alpha_{+}\rangle_{(s)}$ and $|\alpha_{-}\rangle_{(s)}$ depicted in Fig. 4, since the displacement parameter is indeed much less than s.

5. Concluding Remarks

We have analyzed various *finite-dimensional* states, viz. (1) Glauber coherent states, in the sense of displaced vacuum, $|\alpha\rangle_{(s)}$, (2) truncated coherent states $|\overline{\alpha}\rangle_{(s)}$, (3) displaced number states $|\alpha, n_d\rangle_{(s)}$, (4) truncated displaced number states $|\overline{\alpha}, n_d\rangle_{(s)}$, (5) even, $|\alpha_{+}\rangle_{(s)}$, and odd, $|\alpha_{-}\rangle_{(s)}$, coherent states; and (6) truncated even, $|\alpha_{+}\rangle_{(s)}$, and odd, $|\alpha_{-}\rangle_{(s)}$, coherent states. We have confronted two essentially different ways of defining states in a finite-dimensional Hilbert space $\mathcal{H}^{(s)}$. We have proposed definitions of the finite-dimensional displaced number states, and the finite-dimensional even and odd coherent states. We have constructed explicitly all of these states generated by various finite-dimensional displacement operators using the method developed in Ref. [23] for the coherent states $|\alpha\rangle_{(s)}$.

The expression of various quantum states by quasidistributions enables a very intuitive understanding of some of their properties. Here, we have also tried to propose graphical representations of number-phase Wigner function for several kinds of states, often discussed in recent works. There are numerous other interesting finite-dimensional states (like phase coherent states, Schrödinger kittens, squeezed states, etc.) which deserve deeper study of their properties, but here they were not at the focus of our interest.

Acknowledgements

The authors gratefully acknowledge the many helpful comments of J. Peřina, R. Tanaś, K. Piątek, W. Leoński, and S. M. Barnett, V. Bužek, P. L. Knight, G. Kurizki, A. Lukš, Yu. P. Malakyan and W. P. Schleich during the preparation of the paper. This work was carried out under the KBN Project 2 PO3B 188 08 and the Palacký University grant 31403003-IG. A. M. thanks the Foundation for Polish Science for the Fellowship. J. B. thanks the Czech Academy of Science for partial support.

References

1. H. Weyl, *Theory of Groups and Quantum Mechanics* (New York: Dover, 1931).
2. J. Schwinger, Proc. Nat. Acad. Sci. **46**, 570 (1960); reprinted in: J. Schwinger, *Quantum Kinematics and Dynamics* (Benjamin: New York, 1970), pp. 63-72.
3. T. S. Santhanam and A. R. Tekumalla, Found. Phys. **6**, 583 (1976); T. S. Santhanam, Phys. Lett. A **56**, 345 (1976); T. S. Santhanam and K. B. Sinha, Aust. J. Phys. **31**, 233 (1978).
4. D. T. Pegg and S. M. Barnett, Europhys. Lett. **6**, 483 (1988); Phys. Rev. A **41**, 3427 (1989).
5. W. P. Schleich and S. M. Barnett (eds), *Quantum Phase and Phase Dependent Measurements*, special issue of Physica Scripta, vol. **T48** (1993).
6. R. Tanaś, A. Miranowicz, and Ts. Gantsog, Progress in Opt. **35**, 355 (1996).
7. J. Krause, M. O. Scully, T. Walther, and H. Walther, Phys. Rev. Lett. **39**, 1915 (1989); M. Brune, S. Haroshe, V. Lefevre, J. M. Raimond, and N. Zagury, Phys. Rev. Lett. **65**, 976 (1990); M. J. Holland, D. F. Walls, P. Zoller, Phys. Rev. Lett. **67**, 1716 (1991).
8. W. Leoński and R. Tanaś, Phys. Rev. A **49**, R20 (1994).
9. K. Vogel, V. M. Akulin, and W. P. Schleich, Phys. Rev. Lett. **71**, 1816 (1993); B. M. Garraway, B. Sherman, H. Moya-Cessa, P. L. Knight, and G. Kurizki, Phys. Rev. A **49**, 535 (1994); J. Janszky, P. Domokos, S. Szabó, and P. Adam, Phys. Rev. A **51**, 4191 (1995).
10. W. Leoński, Acta Phys. Slov. **45**, 383 (1995).
11. M. Hillery, R. F. O'Connell, M. O. Scully, and E. P. Wigner, Phys. Rep. **106**, 121 (1984); V. I. Tatarskii, Sov. Phys. Usp. **26**, 311 (1983).
12. W. K. Wootters, Ann. Phys. **176**, 1 (1987).
13. O. Cohendet, Ph. Combe, M. Sirugue, and M. Sirugue-Collin, J. Phys. A**21**, 2875 (1988).
14. U. Leonhardt, Phys. Rev. Lett. **74**, 4101 (1995).
15. J. A. Vaccaro, and D. T. Pegg, Phys. Rev. A **41**, 5156 (1990).
16. T. Opatrný, A. Miranowicz, and J. Bajer, J. Mod. Opt. **43**, 417 (1996).
17. J. R. Klauder and B. S. Skagerstam (eds), *Coherent states: Applications in Physics and Mathematical Physics* (Singapore: World Scientific, 1985); A. M. Perelomov, *Generalized Coherent States and their Applications* (Berlin: Springer Verlag, 1986).
18. I. A. Malkin and V. I. Man'ko, *Dynamical Symmetries and Coherent States of Quantum Systems* (Moscow: Nauka, 1979).
19. C. Arecchi, E. Courtens, R. Gilmore and H. Thomas, Phys. Rev. A **6**, 2211 (1972); W. M. Zhang, D. H. Feng, and R. Gilmore, Rev. Mod. Phys. **62**, 867 (1990); T. S. Santhanam, *Generalized coherent states*, in: B. Gruber and S. Millmann (eds), *Symmetries in Science* (Plenum: New York, 1980), p. 337.
20. J. Peřina, Z. Hradil, and B. Jurčo, *Quantum Optics and Fundamentals of Physics*, vol. 63 of *Fundamental Theories in Physics* (Dordrecht: Kluwer Acad., 1994).

21. R. J. Glauber, Phys. Rev. **131**, 2766 (1963).
22. V. Bužek, A. D Wilson-Gordon, P. L. Knight, and W. K. Lai, 1992, Phys. Rev. A **45**, 8079 (1992).
23. A. Miranowicz, K. Piątek, and R. Tanaś, Phys. Rev. A **50**, 3423 (1994).
24. G. Gangopadhyay, J. Mod. Opt. **41**, 525 (1994).
25. L. M. Kuang, F. B. Wang and Y. G. Zhou, Phys. Lett. A **183**, 1 (1993); J. Mod. Opt. **41**, 1307 (1994).
26. J. Y. Zhu, and L. M. Kuang, Phys. Lett. A **193**, 227 (1994); Chinese Phys. Lett. A **11**, 424 (1994).
27. L. M. Kuang, and X. Chen, Phys. Rev. A **50**, 4228 (1994); Phys. Lett. A **186**, 8 (1994).
28. F. A. M. de Oliveira, M. S. Kim, P. L. Knight, and V. Bužek, Phys. Rev. A **41**, 2645 (1990).
29. V. Bužek, and P. L. Knight, Progress in Opt. **34**, 1 (1995).
30. A. Miranowicz, R. Tanaś, and S. Kielich, Quantum Opt. **2**, 253 (1990); M. Paprzycka and R. Tanaś, Quantum Opt. **4**, 331 (1992).

OPTICAL SPECTROSCOPY OF MICROCAVITIES

A.SERPENGÜZEL
Physics Department, Bilkent University, Ankara, 06533 Turkey
S.ARNOLD
Physics Department, Polytechnic University, Brooklyn, New York 11201 USA
G.GRIFFEL
Department of Electrical Engineering, Polytechnic University, Brooklyn, New York 11201 USA
AND
J.A.LOCK
Physics Department, Cleveland State University, Cleveland, Ohio 44115 USA

Abstract.
Microcavities, such as microspheres, possess morphology-dependent resonances (MDR's). In this chapter light coupling mechanisms to MDR's is examined. The novel mechanism of excitation of the microsphere with an optical fiber coupler (OFC) provides spatially and spectrally selective, as well as enhanced light coupling to the MDR's, when compared with the conventional plane wave excitation. The plane wave excitation is described by Lorenz-Mie Theory. However, Generalized Lorenz-Mie Theory and the Localization Principle are used for the OFC excitation.

1. Introduction

In recent years, the quantum confinement and localization effects in small scale systems have attracted substantial attention. Physical properties of these systems make them promising for applications in nonlinear optics and optoelectronic device physics. Available examples are micrometer-size droplets, polystyrene microspheres, doped and undoped optical fibers, colloids, glasses doped with semiconductor microcrystallites, and quantum well, wire, dot, and superlattice structures. These structures have well characterized electromagnetic properties, and due to their morphology they act as optical cavities at specific frequencies, thereby resulting in the localization of light, and affecting the dispersive properties of the medium.

Microspheres, specially, continue to enjoy the attention of the optical spectroscopy community, because of their unique optical properties. [1] In the terminology of Photonic Band Gaps (PBG's), [2, 3] these microspheres are the equivalent of "Photonic Atoms", [4] since they can be used as a building block for the "Photonic Crystals". [5]

T. Hakioğlu and A. S. Shumovsky (eds.), Quantum Optics and the Spectroscopy of Solids, 237–248.

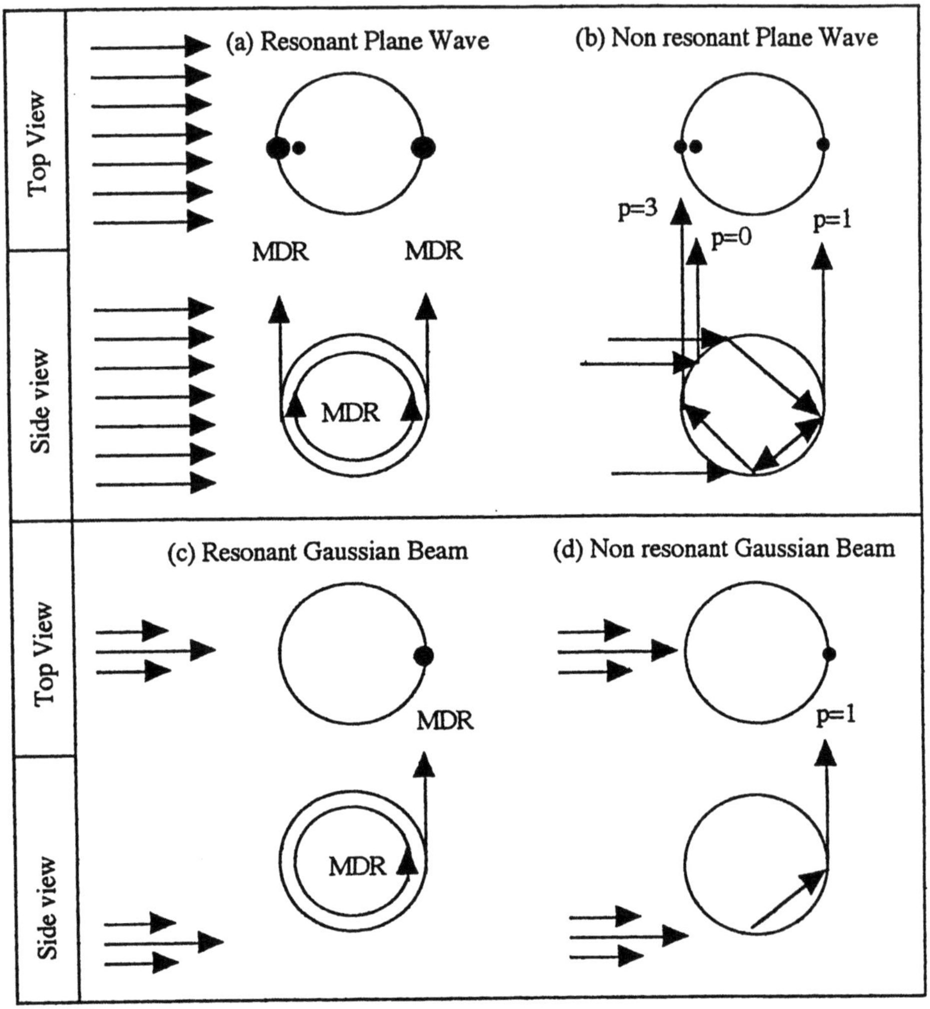

Figure 1. Schematic of the top and side views of the microsphere depicting the non- resonant (p=0,1,3) glare spots and MDR glare spots when excited by (a) a resonant plane wave, (b) a non-resonant plane wave, (c) a resonant off-axis Gaussian Beam, and (d) a non-resonant off-axis Gaussian Beam. Note that the glare spots are not totally correlated with the internal intensity distributions.

The curved interface of the microsphere is responsible for three electromagnetic and quantum-electrodynamic (QED) effects. First, for a plane wave illumination, the internal intensity is concentrated along the principal diameter near the front and the back surfaces of the microsphere (Fig. 1(b) and Fig. 2).

For a droplet with radius $a = 35\mu m$, and refractive index N=1.36, an intensity enhance-

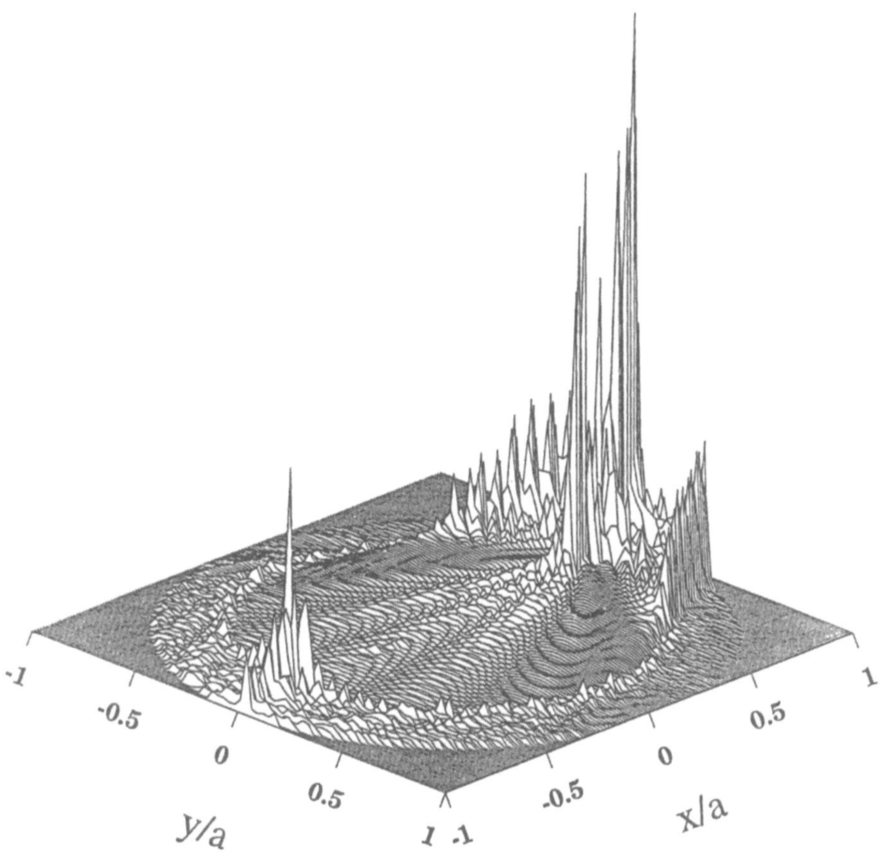

Figure 2. The internal input-laser intensity distribution in the equatorial plane of the droplet for a plane-wave input (off resonance) propagating in the x-direction with size parameter x =418 and N=1.36.

ment of approximately 100X (and 10X) is expected in the localized region just within the droplet shadow (and illuminated) face. These intensity enhancement factors increase with the droplet radius, because more of the incident radiation is intercepted by the droplet illuminated face or hemisphere. Second, the microsphere acts as an optical cavity for specific wavelengths, which satisfy the morphology-dependent resonance (MDR) condition. MDR's can be considered as standing waves, which may be decomposed into two coun-

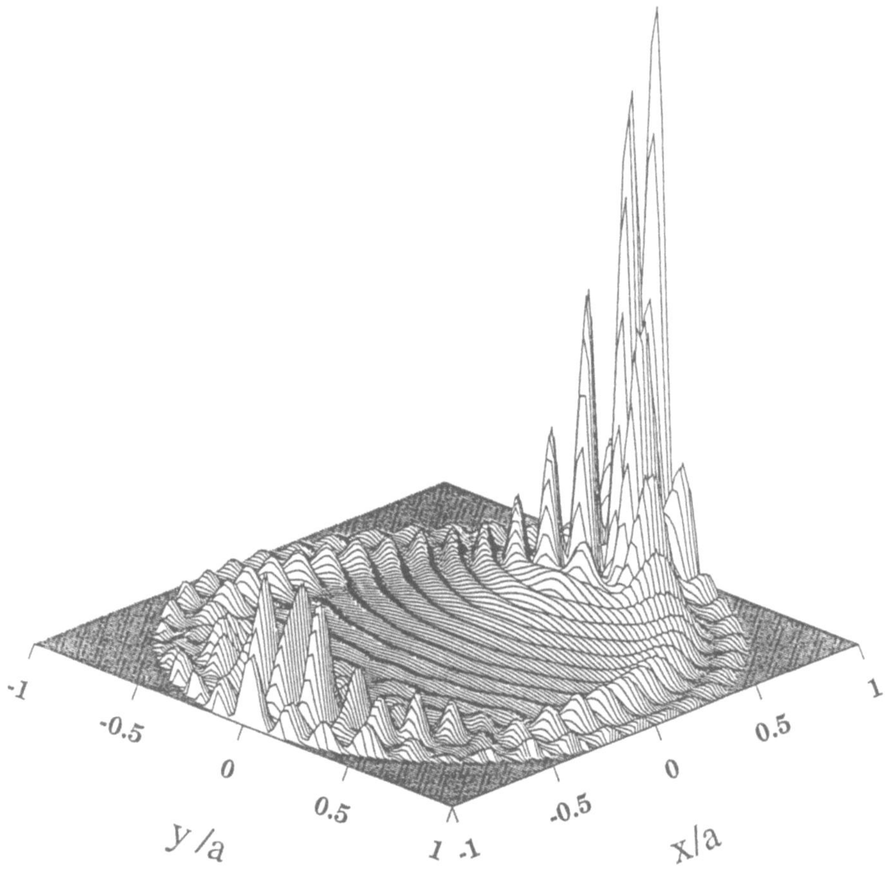

Figure 3. The internal intensity distribution in the equatorial plane of the droplet for a plane-wave input [on an input (TE) resonance] propagating in the x-direction with N=1.59, x=18.83. The maximum value of the internal intensity is 159.

terpropagating waves traveling around the microsphere rim (Fig. 1(a) and Figs. 3-4). The counterpropagating traveling waves must experience quasi-total-internal reflection repeatedly at the spherical droplet interface and return to the starting point with their initial phase. For wavelengths that are on MDR's and within the medium's gain profile, the generated output waves circulate around the droplet rim and experience gain at the two high internal intensity regions. Occasionally, when the monochromatic input-laser frequency is

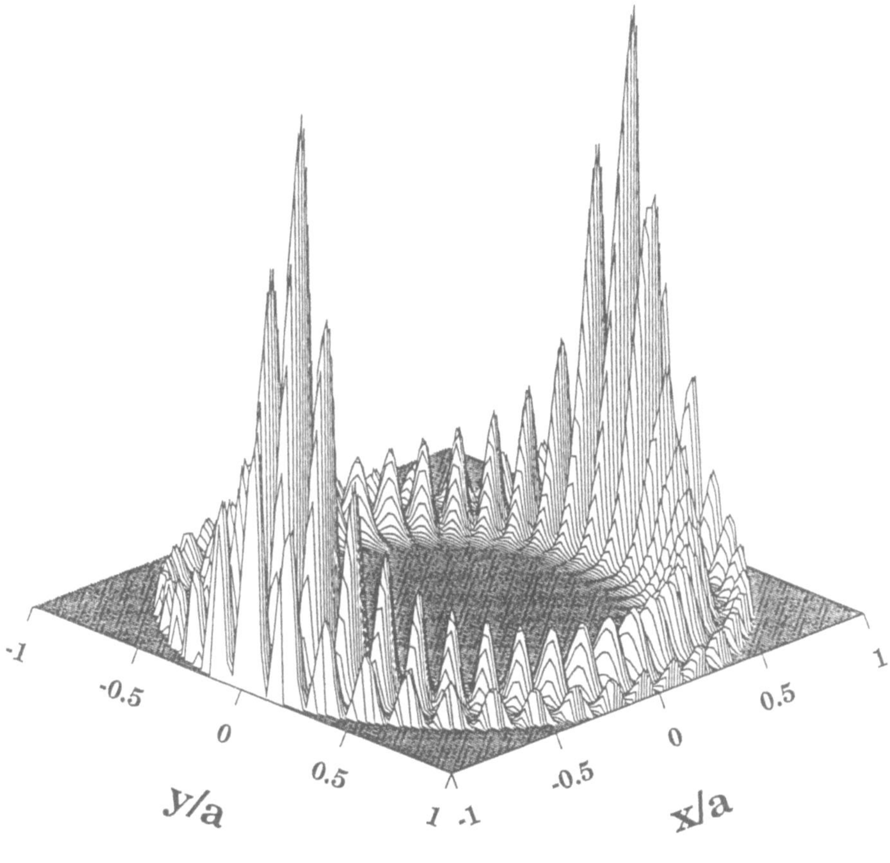

Figure 4. The internal intensity distribution in the equatorial plane of the droplet for a MDR [an output (TE) resonance] with N=1.59 and x =18.83. Notice that the internal intensity due to the focusing of the plane wave input of the Fig. 3 is missing, and the maximum value of the internal intensity is 49 as opposed to 159 of Fig. 3.

tuned to a MDR (an input resonance), there can be an even larger enhancement the input intensity (Fig. 4). Third, the optical transition cross-sections in the microspheres can be larger than bulk optical transition cross-sections, because of the modified density of final electromagnetic states. [6] In the microsphere, the final electromagnetic states correspond to the microsphere cavity resonances, which are described by MDR's. For a bulk sample, however, the final electromagnetic states are the continuum modes of an infinite system. [7, 8]

The SRS threshold for a water microdroplet is lower than that of water in an 11-cm optical cell.[9] In addition, these three effects are responsible for the occurrence of a number of other nonlinear optical effects such as stimulated anti-Stokes Raman scattering (SARS),[10] coherent anti-Stokes Raman scattering (CARS),[11] stimulated Brillouin scattering (SBS),[12] third-harmonic generation (THG),[13, 14] and lasing [15, 16, 17] in microdroplets.

If the incident plane wave is resonant with a MDR (i.e., on-resonance), there will, in addition, be a uniform intensity distribution within the rim of the microsphere in the volume determined by the MDR (Fig. 1(a)). [18] However, if an off-axis Gaussian beam is used at a resonant wavelength, the internal intensity is only distributed within the rim of the microsphere in the volume determined by the MDR, and is no longer concentrated near the front and the back surfaces of the microsphere (Fig. 1(c)). [19] Therefore, a resonant off-axis Gaussian beam excites the MDR's more uniformly and more efficiently than a plane wave. Even edge illumination with a focused beam excites MDR's more efficiently than plane waves (Fig. 1(d)). [20, 21] The off-axis Gaussian beam calculations have recently been realized using generalized Lorenz-Mie theory (GLMT). [22, 23]

2. Experiments

The first experimental realization of the off-axis Gaussian beam excitation geometry was performed by using an optical fiber coupler (OFC). [24] The frequency shift and linewidth broadening of the MDR's due to the OFC-microsphere interaction were also studied. [25] In these OFC- microsphere experiments the cladding of the fiber was not index matched to the media surrounding the microsphere, so that there was an optical interface at the OFC surface. In this chapter, we describe the excitation of the MDR's of a microsphere resting on the surface of an OFC, whose surface has been wetted with an index matching oil to eliminate the air- fiber interface. Thereby, the beam in the optical fiber effectively becomes the equivalent of a Gaussian Beam with an infinitely long skirt length.

Figure 5 is a schematic of our experimental setup. A polystyrene (PS) microsphere with an approximate radius $a = 12$ μm and refractive index $N = 1.59$ is placed on an OFC. The OFC is made from a single-mode optical fiber (SMOF) with a core diameter of 3.8 μm ($N = 1.462$), and a cladding diameter of 125 μm ($N = 1.457$). The cladding below the microsphere is shaved down to 0.7 μm. The SMOF mode has approximately a Gaussian intensity profile and is doubly degenerate with both horizontal and vertical polarizations. The OFC surface and the microsphere were wetted by a few millimeters of index matching oil ($N = 1.456$, same as the cladding). The excitation is provided by a tunable and linearly polarized CW dye laser with a linewidth of 0.025 nm. The dye laser is coupled to the SMOF with a microscope objective. Although the output of the dye laser was linearly polarized, the output from the SMOF was observed to be elliptically polarized due to the fiber birefringence. Therefore, the OFC provides both linear polarizations to the microsphere. The scattering from the microsphere was collected at 90° with a microscope objective followed by a polarizer, and detected with a photomultiplier tube.

If plane wave illumination were used, the image of the microsphere (either on resonance or off resonance) would show three principal glare spots (Fig. 1(a-b)). [26] However, in our

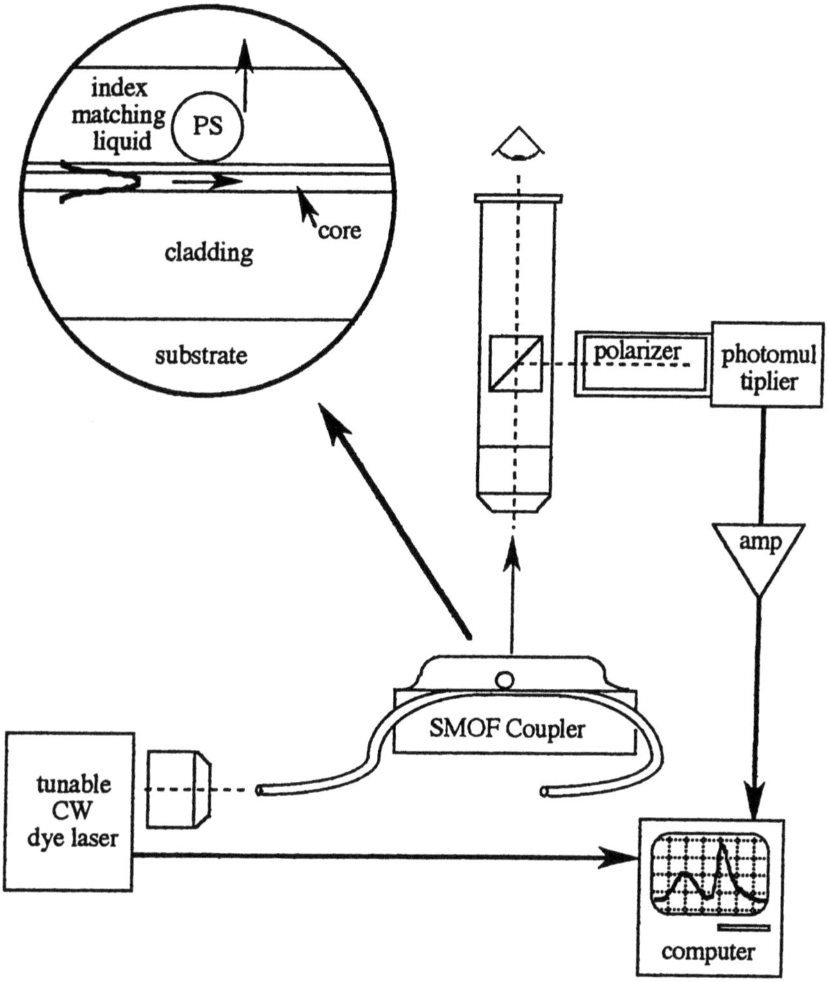

Figure 5. Schematic of the experimental setup with inset to depict the microsphere on the wetted surface of the SMOF coupler.

case of coupling an external beam from the SMOF, we observe only one glare spot on the far side of the microsphere (Fig. 1(c)). In contrast to the non-index matched case, this far side glare spot is observed for all laser wavelengths, even when the incident wavelength is not on a MDR (i.e., off-resonance). However, when the incident light is on-resonance, the far side glare spot intensity is enhanced by a factor of about 2. Apparently, the standing wave pattern which was set up by a plane wave excitation of a MDR, with its counterpropagating traveling waves (Fig. 1(a)), is replaced with a single counter-clockwise traveling wave in the Gaussian beam excitation (Fig. 1(c)). Also, in the Gaussian beam excitation (Fig. 1(c-d)), the off-resonance glare spots are due to refraction only (i.e., p=1 rays), while for a plane wave illumination the off-resonance glare spots are due to both refraction (i.e., p=1,3 rays) and specular reflection (i.e., p=0 rays).

3. Theory of off-axis Gaussian beam excitation

Figure 6(a) displays the experimentally observed transverse electric (TE) scattering spectrum from the microsphere excited with the OFC. Figure 6(b) depicts the theoretically calculated TE scattering spectrum for an off-axis Gaussian beam placed at the location of the OFC. Comparing the spectra in Fig. 6 to the scattering spectra obtained with a plane wave, one notices two features: (1) there is a large background intensity, which cannot be explained by the light scattered due to OFC surface imperfections; and (2) the MDR's have nearly Lorentzian lineshapes as opposed to the Fano lineshapes [27] of the plane wave scattering spectra. Most of the prominent features of these scattering spectra are described by the interaction of the microsphere with an external beam having a Gaussian intensity profile and propagating at an impact parameter (b), which is slightly greater than the microsphere radius (a). Since the excitation by such a beam occurs beyond the edge of the microsphere, the light scattering can be calculated by (i) removing the partial waves with angular momentum quantum numbers (n) less than the size parameter (x, which is the ratio of the perimeter of the microsphere to the incident light wavelength) from the conventional Lorenz-Mie (plane wave excitation) infinite series, and (ii) applying generalized Lorenz-Mie theory (GLMT) to parametrize the incident beam profile. This removal of partial waves is justified by the localization principle, [28] which associates a light ray having an impact parameter (b) with a partial wave with mode number (n).

Plane wave Lorenz-Mie theory restricts the angular momentum quantum number (n) range of the light rays, passing by the microsphere surface but yet interacting with it, to have their mode number (n) between x and Nx, where N is the relative refractive index of the microsphere with respect to the outside medium. This condition, together with the localization principle, restricts the impact parameter (b) range to be between a and Na. Therefore, only the light rays within this impact parameter range can couple to the MDR's of the microsphere.

To check the validity of this approach and to simulate our experimental results of Fig. 6(a), we have used the GLMT computation algorithm, [29] which can be applied to on- and off-axis focused Gaussian beam excitation geometries. In GLMT, the plane wave expansion coefficients (a_n for TM or b_n for TE resonances) are replaced by the partial wave expansion coefficients (a_{nm} for TM coefficients or b_{nm} for TE coefficients). For example for TE contributions, $b_{nm} = b_n \, B_{nm}$, where B_{nm} describes the angular overlap of the excitation field and the spherical harmonics.

For our external off-axis Gaussian beam excitation geometry, B_{nm} become significant as n exceeds x. As n is increased further, B_{nm} goes through a series of resonances (i.e., MDR's), with none seen beyond n=Nx. The background in the spectra of Fig. 6(b) is due to the refraction for non-resonant partial waves with n¡x. For a plane wave excitation, B_{nm} would have been significant for all n. For the calculations of Fig. 6(b), we used a Gaussian beam (with an infinite skirt-length and a beamwaist with a half-width w_o = 2.176 μm) propagating at an impact parameter of 14.94 μm from a microsphere (with a = 12.34 μm and a relative refractive index of 1.09). The calculation of the scattered

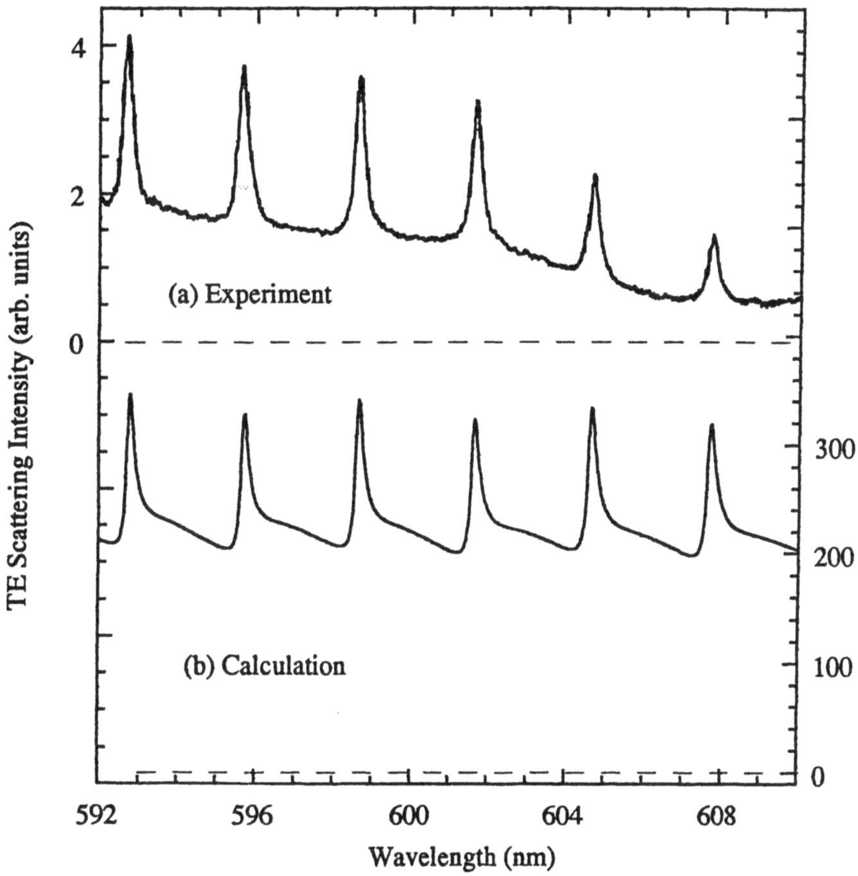

Figure 6. Scattering intensity of (a) experimental, and (b) calculated TE polarized spectrum for a PS microsphere in index matching liquid (N=1.456).

intensity is averaged over the scattering angle around 90°. The results of this calculation are compelling. The theoretical MDR's for an off-axis incidence bear a good relationship to the experimental data [Figs. 6(a)], and appear to correspond to first order MDR's with theoretical quality factors (Q's) of approximately 2000. The mode numbers (n) for the MDR's are within the range of n = 194 -198.

By using the localization principle, we can estimate, where the beam should have been

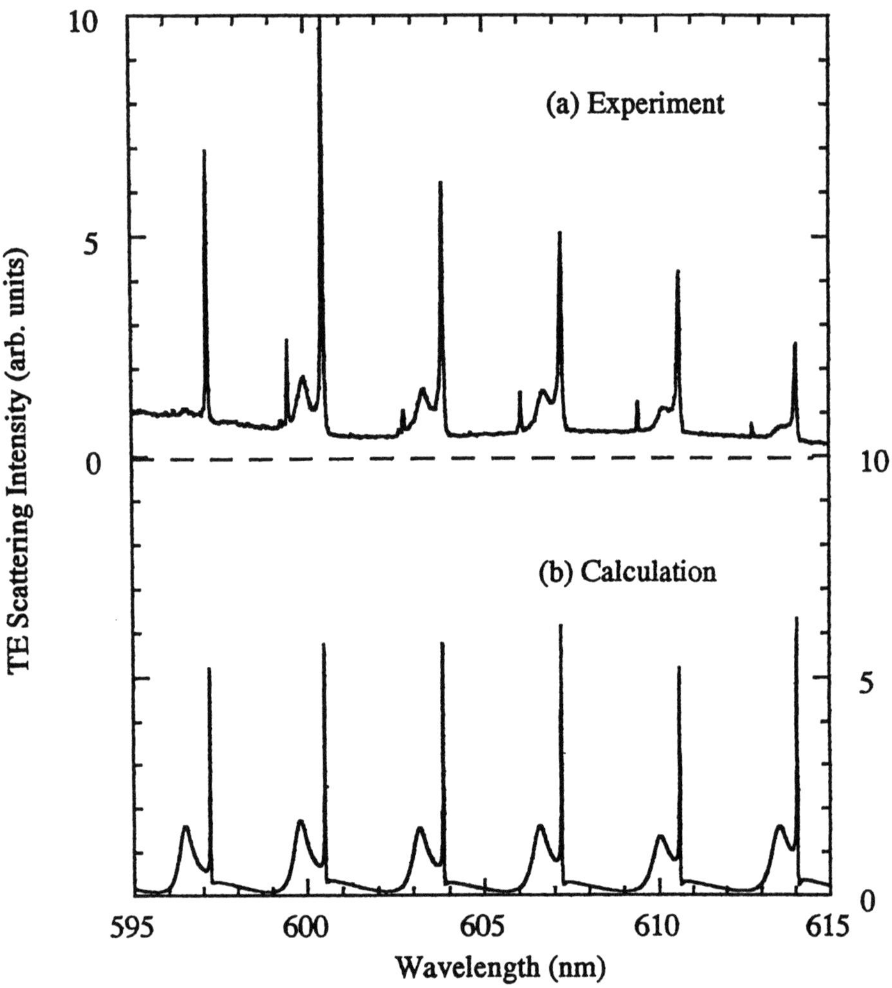

Figure 7. Scattering intensity of (a) experimental, and (b) calculated TE polarized spectrum for a PS microsphere in water (N=1.18).

placed to maximally excite these MDR's with n = 194 -198. We obtain b = n/k = 12.9 μm, which is smaller than our experimental value of b = 14.94 μm. Therefore, we are not coupling to these MDR's with n = 194 -198 with maximal efficiency. Different mode numbers are excited at different impact parameters. For our modes with n = 194-198, in accordance with the localization principle, the maximum of the scattering efficiency occurs at an impact parameter of 12.9 μm. At our impact parameter of 14.94 μm, the scattering efficiency is less than 0.01. In order to couple most efficiently to our modes with n = 194-198, we would have had to excite the microsphere at an impact parameter of 12.9 μm.

However, efficient coupling is not sufficient for maximum energy storage in the microsphere. The total energy that can be stored in the mode is proportional to the lifetime of that mode, which can further be enhanced with the use of a higher refractive index contrast between the PS microsphere and its surrounding medium, e.g., when N = 1.09 (index matching liquid) is replaced by N = 1.18 (water). Although the water results are more difficult to simulate due to the refractive index step at the cladding-water interface, and the excitation electric field being evanescent rather than being Gaussian, it is possible to have modes with higher measured Q's of 24000. Figure 7(a) shows the experimental elastic scattering intensity of the TE polarized spectrum for a PS microsphere in water (N = 1.18), while Figure 7(b) displays the corresponding calculated spectrum. A noteworthy feature of the spectra in Fig. 7(a) is that, MDR's with linewidths as narrow as 0.04 nm were observed. Since our dye laser has a linewidth of 0.025 nm, the measured MDR linewidths are clearly limited by the convolution of our dye laser linewidth.

4. Conclusion

In conclusion, the microsphere-optical fiber system with the index matching geometry has proved to be very useful in the verification of the generalized Lorenz-Mie theory (GLMT) and the localization principle. The microsphere-optical fiber system also shows promise as a possible building block for photonic memories, [30] and can be used as an external cavity feedback system to line narrow a broader light source such as a diode laser. [31, 32]

Acknowledgements

We are grateful for the partial support of this research from the United States Air Force Office of Scientific Research (Grant Number F49620-94- 0195).

References

1. P. W. Barber and R. K. Chang, Eds., *Optical Effects Associated with Small Particles* (World Scientific 1988) Singapore.
2. E. Yablonovitch, Phys. Rev. Lett. **58**, 2059 (1987).
3. C. M. Soukoulis, *Photonic Band Gaps and Localization* (Plenum Press 1993) New York.
4. S. Arnold, J. Comunale, W.B. Whitten, J.M. Ramsey, and K.A. Fuller, J. Opt. Soc. Am. B **9** 819 (1992).
5. W. Hu, H. Li, B. Cheng, J. Yang, Z. Li, and D. Zhang, Opt. Lett. **20** 964 (1995).
6. E. M. Purcell, Phys. Rev. **69** 681 (1946).
7. S. C. Ching, H. M. Lai, and K. Young, J. Opt. Soc. Am. B **4** 1995 (1987).
8. S. C. Ching, H. M. Lai, and K. Young, J. Opt. Soc. Am. B **4** 2004 (1987).
9. J.B.Snow, S.-X.Qian, and R.K.Chang, Opt. Lett. **10** 37 (1985).
10. D.H. Leach, R.K.Chang, and W.P. Acker, Opt. Lett. **17** 387 (1992).
11. S.-X.Qian, J.B.Snow, and R.K.Chang, Opt. Lett. **10** 499 (1985).
12. J.-Z. Zhang and R.K.Chang, J. Opt. Soc. Am. B **6** 151 (1989).
13. W.P. Acker, D.H. Leach, and R.K.Chang, Opt. Lett. **14** 402 (1989).
14. D.H. Leach, W.P. Acker, and R.K.Chang, Opt. Lett. **15** 894 (1990).
15. H.-M. Tzeng, K.F. Wall, M.B. Long, and R.K.Chang, Opt. Lett. **9** 499 (1984).
16. S.-X.Qian, J.B.Snow, H.-M. TzEng, and R.K.Chang, Science **231** 486 (1986).
17. H.-B.Lin, A.L.Huston, B.L.Justus, and A.J.Campillo, Opt. Lett. **11** 614 (1986).
18. D.S. Benincasa, P.W. Barber, J.-Z. Zhang, W.-F. Hsieh, and R.K. Chang, Appl. Opt. **26** 1348 (1987).
19. E. E. M. Khaled, S. C. Hill, and P. W. Barber, Appl. Opt. **33** 524 (1994).
20. J. -Z. Zhang, D. H. Leach, and R. K. Chang, Opt. Lett. **13** 270 (1988).
21. J. P. Barton, D. R. Alexander, and S. A. Schaub, J. Appl. Phys. **64** 1632 (1988).
22. J. A. Lock and G. Gouesbet, J. Opt. Soc. Am. A **11** 2503 (1994).
23. G. Gouesbet and J. A. Lock, J. Opt. Soc. Am. A **11** 2516 (1994).
24. A. Serpengüzel, S. Arnold, and G. Griffel, Opt. Lett. **20** 654 (1995).
25. N. Dubreuil, J.C. Knight, D.K. Leventhal, V. Sandoghar, J. Hare, and V. Lefevre, Opt. Lett. 813 **20** (1995).
26. S. Arnold, S. Holler, J. H. Li, A. Serpengüzel, W. F. Auffermann, and S. C. Hill, Opt. Lett. **20** 773 (1995).
27. P. W. Barber and R. K. Chang, Eds., *Optical Effects Associated with Small Particles* (World Scientific 1988) Singapore, p.20.
28. H. C. van de Hulst, *Light Scattering by Small Particles* (Dover, 1981) New York, p. 208.
29. J. A. Lock, Appl. Opt. **34** 559 (1995).
30. S. Arnold, C. T. Liu, W. B. Whitten and J. M. Ramsey, Opt. Lett. **16** 420 (1991).
31. G. Griffel, A. Serpengüzel, and S. Arnold, "Quenching of semiconductor lasers linewidth by detuned loading using spherical cavities morphology dependent resonances," *Proceedings of the IEEE: Frequency Control Conf.*, San Francisco, CA, USA (1995).
32. G. Griffel, S. Arnold, D. Taskent, A. Serpengüzel, J. Connolly, and N. Morris, Opt. Lett. **21** 695 (1996).

Fundamental Theories of Physics

Series Editor: Alwyn van der Merwe, *University of Denver, USA*

1. M. Sachs: *General Relativity and Matter.* A Spinor Field Theory from Fermis to Light-Years. With a Foreword by C. Kilmister. 1982 ISBN 90-277-1381-2
2. G.H. Duffey: *A Development of Quantum Mechanics.* Based on Symmetry Considerations. 1985 ISBN 90-277-1587-4
3. S. Diner, D. Fargue, G. Lochak and F. Selleri (eds.): *The Wave-Particle Dualism.* A Tribute to Louis de Broglie on his 90th Birthday. 1984 ISBN 90-277-1664-1
4. E. Prugovečki: *Stochastic Quantum Mechanics and Quantum Spacetime.* A Consistent Unification of Relativity and Quantum Theory based on Stochastic Spaces. 1984; 2nd printing 1986 ISBN 90-277-1617-X
5. D. Hestenes and G. Sobczyk: *Clifford Algebra to Geometric Calculus.* A Unified Language for Mathematics and Physics. 1984 ISBN 90-277-1673-0; Pb (1987) 90-277-2561-6
6. P. Exner: *Open Quantum Systems and Feynman Integrals.* 1985 ISBN 90-277-1678-1
7. L. Mayants: *The Enigma of Probability and Physics.* 1984 ISBN 90-277-1674-9
8. E. Tocaci: *Relativistic Mechanics, Time and Inertia.* Translated from Romanian. Edited and with a Foreword by C.W. Kilmister. 1985 ISBN 90-277-1769-9
9. B. Bertotti, F. de Felice and A. Pascolini (eds.): *General Relativity and Gravitation.* Proceedings of the 10th International Conference (Padova, Italy, 1983). 1984 ISBN 90-277-1819-9
10. G. Tarozzi and A. van der Merwe (eds.): *Open Questions in Quantum Physics.* 1985 ISBN 90-277-1853-9
11. J.V. Narlikar and T. Padmanabhan: *Gravity, Gauge Theories and Quantum Cosmology.* 1986 ISBN 90-277-1948-9
12. G.S. Asanov: *Finsler Geometry, Relativity and Gauge Theories.* 1985 ISBN 90-277-1960-8
13. K. Namsrai: *Nonlocal Quantum Field Theory and Stochastic Quantum Mechanics.* 1986 ISBN 90-277-2001-0
14. C. Ray Smith and W.T. Grandy, Jr. (eds.): *Maximum-Entropy and Bayesian Methods in Inverse Problems.* Proceedings of the 1st and 2nd International Workshop (Laramie, Wyoming, USA). 1985 ISBN 90-277-2074-6
15. D. Hestenes: *New Foundations for Classical Mechanics.* 1986 ISBN 90-277-2090-8; Pb (1987) 90-277-2526-8
16. S.J. Prokhovnik: *Light in Einstein's Universe.* The Role of Energy in Cosmology and Relativity. 1985 ISBN 90-277-2093-2
17. Y.S. Kim and M.E. Noz: *Theory and Applications of the Poincaré Group.* 1986 ISBN 90-277-2141-6
18. M. Sachs: *Quantum Mechanics from General Relativity.* An Approximation for a Theory of Inertia. 1986 ISBN 90-277-2247-1
19. W.T. Grandy, Jr.: *Foundations of Statistical Mechanics.* Vol. I: *Equilibrium Theory.* 1987 ISBN 90-277-2489-X
20. H.-H von Borzeszkowski and H.-J. Treder: *The Meaning of Quantum Gravity.* 1988 ISBN 90-277-2518-7
21. C. Ray Smith and G.J. Erickson (eds.): *Maximum-Entropy and Bayesian Spectral Analysis and Estimation Problems.* Proceedings of the 3rd International Workshop (Laramie, Wyoming, USA, 1983). 1987 ISBN 90-277-2579-9

Fundamental Theories of Physics

22. A.O. Barut and A. van der Merwe (eds.): *Selected Scientific Papers of Alfred Landé. [1888-1975]*. 1988 ISBN 90-277-2594-2
23. W.T. Grandy, Jr.: *Foundations of Statistical Mechanics.* Vol. II: *Nonequilibrium Phenomena.* 1988 ISBN 90-277-2649-3
24. E.I. Bitsakis and C.A. Nicolaides (eds.): *The Concept of Probability.* Proceedings of the Delphi Conference (Delphi, Greece, 1987). 1989 ISBN 90-277-2679-5
25. A. van der Merwe, F. Selleri and G. Tarozzi (eds.): *Microphysical Reality and Quantum Formalism, Vol. 1.* Proceedings of the International Conference (Urbino, Italy, 1985). 1988 ISBN 90-277-2683-3
26. A. van der Merwe, F. Selleri and G. Tarozzi (eds.): *Microphysical Reality and Quantum Formalism, Vol. 2.* Proceedings of the International Conference (Urbino, Italy, 1985). 1988 ISBN 90-277-2684-1
27. I.D. Novikov and V.P. Frolov: *Physics of Black Holes.* 1989 ISBN 90-277-2685-X
28. G. Tarozzi and A. van der Merwe (eds.): *The Nature of Quantum Paradoxes.* Italian Studies in the Foundations and Philosophy of Modern Physics. 1988 ISBN 90-277-2703-1
29. B.R. Iyer, N. Mukunda and C.V. Vishveshwara (eds.): *Gravitation, Gauge Theories and the Early Universe.* 1989 ISBN 90-277-2710-4
30. H. Mark and L. Wood (eds.): *Energy in Physics, War and Peace.* A Festschrift celebrating Edward Teller's 80th Birthday. 1988 ISBN 90-277-2775-9
31. G.J. Erickson and C.R. Smith (eds.): *Maximum-Entropy and Bayesian Methods in Science and Engineering.* Vol. I: *Foundations.* 1988 ISBN 90-277-2793-7
32. G.J. Erickson and C.R. Smith (eds.): *Maximum-Entropy and Bayesian Methods in Science and Engineering.* Vol. II: *Applications.* 1988 ISBN 90-277-2794-5
33. M.E. Noz and Y.S. Kim (eds.): *Special Relativity and Quantum Theory.* A Collection of Papers on the Poincaré Group. 1988 ISBN 90-277-2799-6
34. I.Yu. Kobzarev and Yu.I. Manin: *Elementary Particles. Mathematics, Physics and Philosophy.* 1989 ISBN 0-7923-0098-X
35. F. Selleri: *Quantum Paradoxes and Physical Reality.* 1990 ISBN 0-7923-0253-2
36. J. Skilling (ed.): *Maximum-Entropy and Bayesian Methods.* Proceedings of the 8th International Workshop (Cambridge, UK, 1988). 1989 ISBN 0-7923-0224-9
37. M. Kafatos (ed.): *Bell's Theorem, Quantum Theory and Conceptions of the Universe.* 1989 ISBN 0-7923-0496-9
38. Yu.A. Izyumov and V.N. Syromyatnikov: *Phase Transitions and Crystal Symmetry.* 1990 ISBN 0-7923-0542-6
39. P.F. Fougère (ed.): *Maximum-Entropy and Bayesian Methods.* Proceedings of the 9th International Workshop (Dartmouth, Massachusetts, USA, 1989). 1990 ISBN 0-7923-0928-6
40. L. de Broglie: *Heisenberg's Uncertainties and the Probabilistic Interpretation of Wave Mechanics.* With Critical Notes of the Author. 1990 ISBN 0-7923-0929-4
41. W.T. Grandy, Jr.: *Relativistic Quantum Mechanics of Leptons and Fields.* 1991 ISBN 0-7923-1049-7
42. Yu.L. Klimontovich: *Turbulent Motion and the Structure of Chaos.* A New Approach to the Statistical Theory of Open Systems. 1991 ISBN 0-7923-1114-0

Fundamental Theories of Physics

43. W.T. Grandy, Jr. and L.H. Schick (eds.): *Maximum-Entropy and Bayesian Methods*. Proceedings of the 10th International Workshop (Laramie, Wyoming, USA, 1990). 1991 ISBN 0-7923-1140-X
44. P.Pták and S. Pulmannová: *Orthomodular Structures as Quantum Logics*. Intrinsic Properties, State Space and Probabilistic Topics. 1991 ISBN 0-7923-1207-4
45. D. Hestenes and A. Weingartshofer (eds.): *The Electron*. New Theory and Experiment. 1991 ISBN 0-7923-1356-9
46. P.P.J.M. Schram: *Kinetic Theory of Gases and Plasmas*. 1991 ISBN 0-7923-1392-5
47. A. Micali, R. Boudet and J. Helmstetter (eds.): *Clifford Algebras and their Applications in Mathematical Physics*. 1992 ISBN 0-7923-1623-1
48. E. Prugovečki: *Quantum Geometry*. A Framework for Quantum General Relativity. 1992 ISBN 0-7923-1640-1
49. M.H. Mac Gregor: *The Enigmatic Electron*. 1992 ISBN 0-7923-1982-6
50. C.R. Smith, G.J. Erickson and P.O. Neudorfer (eds.): *Maximum Entropy and Bayesian Methods*. Proceedings of the 11th International Workshop (Seattle, 1991). 1993 ISBN 0-7923-2031-X
51. D.J. Hoekzema: *The Quantum Labyrinth*. 1993 ISBN 0-7923-2066-2
52. Z. Oziewicz, B. Jancewicz and A. Borowiec (eds.): *Spinors, Twistors, Clifford Algebras and Quantum Deformations*. Proceedings of the Second Max Born Symposium (Wrocław, Poland, 1992). 1993 ISBN 0-7923-2251-7
53. A. Mohammad-Djafari and G. Demoment (eds.): *Maximum Entropy and Bayesian Methods*. Proceedings of the 12th International Workshop (Paris, France, 1992). 1993 ISBN 0-7923-2280-0
54. M. Riesz: *Clifford Numbers and Spinors* with Riesz' Private Lectures to E. Folke Bolinder and a Historical Review by Pertti Lounesto. E.F. Bolinder and P. Lounesto (eds.). 1993 ISBN 0-7923-2299-1
55. F. Brackx, R. Delanghe and H. Serras (eds.): *Clifford Algebras and their Applications in Mathematical Physics*. Proceedings of the Third Conference (Deinze, 1993) 1993 ISBN 0-7923-2347-5
56. J.R. Fanchi: *Parametrized Relativistic Quantum Theory*. 1993 ISBN 0-7923-2376-9
57. A. Peres: *Quantum Theory: Concepts and Methods*. 1993 ISBN 0-7923-2549-4
58. P.L. Antonelli, R.S. Ingarden and M. Matsumoto: *The Theory of Sprays and Finsler Spaces with Applications in Physics and Biology*. 1993 ISBN 0-7923-2577-X
59. R. Miron and M. Anastasiei: *The Geometry of Lagrange Spaces: Theory and Applications*. 1994 ISBN 0-7923-2591-5
60. G. Adomian: *Solving Frontier Problems of Physics: The Decomposition Method*. 1994 ISBN 0-7923-2644-X
61 B.S. Kerner and V.V. Osipov: *Autosolitons*. A New Approach to Problems of Self-Organization and Turbulence. 1994 ISBN 0-7923-2816-7
62. G.R. Heidbreder (ed.): *Maximum Entropy and Bayesian Methods*. Proceedings of the 13th International Workshop (Santa Barbara, USA, 1993) 1996 ISBN 0-7923-2851-5
63. J. Peřina, Z. Hradil and B. Jurčo: *Quantum Optics and Fundamentals of Physics*. 1994 ISBN 0-7923-3000-5

Fundamental Theories of Physics

64. M. Evans and J.-P. Vigier: *The Enigmatic Photon.* Volume 1: The Field $\boldsymbol{B}^{(3)}$. 1994
ISBN 0-7923-3049-8
65. C.K. Raju: *Time: Towards a Constistent Theory.* 1994 ISBN 0-7923-3103-6
66. A.K.T. Assis: *Weber's Electrodynamics.* 1994 ISBN 0-7923-3137-0
67. Yu. L. Klimontovich: *Statistical Theory of Open Systems.* Volume 1: A Unified Approach to Kinetic Description of Processes in Active Systems. 1995
ISBN 0-7923-3199-0; Pb: ISBN 0-7923-3242-3
68. M. Evans and J.-P. Vigier: *The Enigmatic Photon.* Volume 2: Non-Abelian Electrodynamics. 1995 ISBN 0-7923-3288-1
69. G. Esposito: *Complex General Relativity.* 1995 ISBN 0-7923-3340-3
70. J. Skilling and S. Sibisi (eds.): *Maximum Entropy and Bayesian Methods.* Proceedings of the Fourteenth International Workshop on Maximum Entropy and Bayesian Methods. 1996 ISBN 0-7923-3452-3
71. C. Garola and A. Rossi (eds.): *The Foundations of Quantum Mechanics – Historical Analysis and Open Questions.* 1995 ISBN 0-7923-3480-9
72. A. Peres: *Quantum Theory: Concepts and Methods.* 1995 (see for hardback edition, Vol. 57) ISBN Pb 0-7923-3632-1
73. M. Ferrero and A. van der Merwe (eds.): *Fundamental Problems in Quantum Physics.* 1995 ISBN 0-7923-3670-4
74. F.E. Schroeck, Jr.: *Quantum Mechanics on Phase Space.* 1996 ISBN 0-7923-3794-8
75. L. de la Peña and A.M. Cetto: *The Quantum Dice.* An Introduction to Stochastic Electrodynamics. 1996 ISBN 0-7923-3818-9
76. P.L. Antonelli and R. Miron (eds.): *Lagrange and Finsler Geometry.* Applications to Physics and Biology. 1996 ISBN 0-7923-3873-1
77. M.W. Evans, J.-P. Vigier, S. Roy and S. Jeffers: *The Enigmatic Photon.* Volume 3: Theory and Practice of the $\boldsymbol{B}^{(3)}$ Field. 1996 ISBN 0-7923-4044-2
78. W.G.V. Rosser: *Interpretation of Classical Electromagnetism.* 1996
ISBN 0-7923-4187-2
79. K.M. Hanson and R.N. Silver (eds.): *Maximum Entropy and Bayesian Methods.* 1996
ISBN 0-7923-4311-5
80. S. Jeffers, S. Roy, J.-P. Vigier and G. Hunter (eds.): *The Present Status of the Quantum Theory of Light.* Proceedings of a Symposium in Honour of Jean-Pierre Vigier. 1997
ISBN 0-7923-4337-9
81. *Still to be published*
82. R. Miron: *The Geometry of Higher-Order Lagrange Spaces.* Applications to Mechanics and Physics. 1997 ISBN 0-7923-4393-X
83. T. Hakioğlu and A.S. Shumovsky (eds.): *Quantum Optics and the Spectroscopy of Solids.* Concepts and Advances. 1997 ISBN 0-7923-4414-6

KLUWER ACADEMIC PUBLISHERS – DORDRECHT / BOSTON / LONDON

MIX
Papier aus verantwortungsvollen Quellen
Paper from responsible sources
FSC® C105338

If you have any concerns about our products,
you can contact us on
ProductSafety@springernature.com

In case Publisher is established outside the EU,
the EU authorized representative is:
Springer Nature Customer Service Center GmbH
Europaplatz 3, 69115 Heidelberg, Germany

Printed by Libri Plureos GmbH
in Hamburg, Germany